T0271279

Rarefied Gas Dynamics

The aim of this book is to present the concepts, methods, and applications of kinetic theory to rarefied gas dynamics. After introducing the basic tools, problems in plane geometry are treated using approximation techniques (perturbation and numerical methods). These same techniques are later used to deal with two- and three-dimensional problems. The models include not only monatomic but also polyatomic gases, mixtures, and chemical reactions. A special chapter is devoted to evaporation and condensation phenomena.

Each section is accompanied by problems that are mainly intended to demonstrate the use of the material in the text and to outline additional subjects, results, and equations. This will help ensure that the book can be used for a range of graduate courses in aerospace engineering or applied mathematics.

Cambridge Texts in Applied Mathematics

Maximum and Minimum Principles
M. J. SEWELL

Solitons
P. G. DRAZIN AND R. S. JOHNSON

The Kinematics of Mixing
J. M. OTTINO

Introduction to Numerical Linear Algebra and Optimisation
PHILIPPE G. CIARLET

Integral Equations
DAVID PORTER AND DAVID S. G. STIRLING

Perturbation Methods
E. J. HINCH

The Thermomechanics of Plasticity and Fracture
GERARD A. MAUGIN

Boundary Integral and Singularity Methods for Linearized Viscous Flow
C. POZRIKIDIS

Nonlinear Wave Processes in Acoustics
K. NAUGOLNYKH AND L. OSTROVSKY

Nonlinear Systems
P. G. DRAZIN

Stability, Instability and Chaos
PAUL GLENDINNING

Applied Analysis of the Navier–Stokes Equations
C. R. DOERING AND J. D. GIBBON

Viscous Flow
H. OCKENDON AND J. R. OCKENDON

Scaling, Self-Similarity and Intermediate Asymptotics
G. I. BARENBLATT

A First Course in the Numerical Analysis of Differential Equations
A. ISERLES

Complex Variables: Introduction and Applications
M. J. ABLOWITZ AND A. S. FOKAS

Mathematical Models in the Applied Sciences
A. FOWLER

Thinking About Ordinary Differential Equations
R. O'MALLEY

A Modern Introduction to the Mathematical Theory of Water Waves
R. S. JOHNSON

Rarefied Gas Dynamics
CARLO CERCIGNANI

Rarefied Gas Dynamics
From Basic Concepts to Actual Calculations

CARLO CERCIGNANI
Department of Mathematics
Politechnic University of Milan

CAMBRIDGE
UNIVERSITY PRESS

32 Avenue of the Americas, New York NY 10013-2473, USA

Cambridge University Press is part of the University of Cambridge.

It furthers the University's mission by disseminating knowledge in the pursuit of education, learning and research at the highest international levels of excellence.

www.cambridge.org
Information on this title: www.cambridge.org/9780521650083

First published 2000

A catalogue record for this publication is available from the British Library

Library of Congress Cataloguing in Publication data
Cercignani, Carlo.
Rarefied gas dynamics : from basic concepts to actual calculations / Carlo Cercignani.
p. cm. – (Cambridge texts in applied mathematics)
Includes bibliographical references.
ISBN 0-521-65008-9 (hardback). – ISBN 0-521-65992-2 (pbk.)
1. Gas dynamics. I. Title. II. Series.
QA930.C34 2000
533′.2 – dc21 99-15475
CIP

ISBN 978-0-521-65008-3 Hardback
ISBN 978-0-521-65992-5 Paperback

Contents

Preface

The present volume is intended to cover the present status of the theoretical tools of rarefied gas dynamics. The meaning and usefulness of the subject, and the extent to which it is covered in the book, are discussed in some detail in the introduction. In short, I tried to present the basic concepts and the techniques used in probing difficult, and partly unsolved, questions and in attacking difficult problems posed by aerospace research, environmental sciences, aerosol reactors, micromachines, and vacuum technology. For the book to be up-to-date without being excessively large, it was necessary to omit some topics, which are treated elsewhere, as indicated in the introduction. Their omission does not alter the aim of the book: to provide a working understanding of the essentials of rarefied gas dynamics and to form the foundation and give the background for a study of the original literature.

The tables and figures should not be considered as a complete collection of the information required to attack practical problems or as a substitute for the calculations that must be carried out in each specific instance. They rather illustrate the kind of results to be expected when using a particular method. General principles and essentials can be given in a book such as the present one, but the coverage of different applications must be left to the scientist or engineer dealing with that application.

The problems that are given at the end of the various sections are mainly intended to demonstrate the use of the material in the text and to outline additional subjects, results, and equations.

The general prerequisites of the book are such that it can also be used in a graduate course in engineering or applied mathematics.

Although I have tried to provide a rather complete bibliography, the choice of the topics and of the references certainly reflects a personal bias and I apologize in advance for any omission.

During the work on the manuscript, I have benefited from personal and electronic contacts with many colleagues. Since any list would be incomplete, I restrict my acknowledgment here to those who have helped me with the figures: K. Aoki, M. S. Ivanov, and K. Koura, who gave me permission to reproduce pictures from some of their papers in Chapter 7 and provided the relevant files, and L. Valdettaro, who helped me in the preparation of the software required to produce the remaining pictures.

Introduction

The subject of rarefied gas dynamics can be conveniently defined as the study of gas flows in which the average value of the distance between two subsequent collisions of a molecule (the so-called mean free path) is not negligible in comparison with a length typical of the structure of the flow being considered (e.g., the radius of a pipe or the radius of curvature of the nose of a space shuttle). The field is thus seen to be one that intrinsically requires the use of statistical ideas typical of the kinetic theory of gases as embodied in the integro-differential equation proposed by Boltzmann in 1872 and bearing his name.

Limiting cases of the Boltzmann equation yield the continuum description, commonly based on the Euler or Navier–Stokes equations, at the extreme of small mean free paths, and the collisionless (or free-molecular) flow, at the extreme of large mean free paths. Although most problems in rarefied gas dynamics involve the central region between these two limiting behaviors, commonly called the transition regime, there are interesting problems at the two extremes as well.

In addition to the description based on the Boltzmann equation, the study of rarefied flows requires an additional piece of information concerning the interaction of the gas molecules with the solid (or, possibly, liquid) surfaces that bound the gas expanse. It is to this interaction that one can trace the origin of the drag and lift exerted by the gas on the body and the heat transfer between the gas and the solid boundary.

The study of gas–surface interaction may be regarded as a bridge between the kinetic theory of gases and solid state physics and is an area of research by itself. The difficulties of a theoretical investigation are due, mainly, to our lack of knowledge of the structure of surface layers of solid bodies and hence of the effective interaction potential of the gas molecules with the wall. When a molecule impinges upon a surface, it is adsorbed and may form chemical bonds, dissociate, become ionized, or displace surface molecules. Its interaction with the solid surface depends on the surface finish, the cleanliness of the surface, its temperature, etc. It may also vary with time because of outgassing from the surface. Preliminary heating of a surface also promotes purification of the surface through emission of adsorbed molecules. In general, adsorbed layers may be present; in this case, the interaction of a given molecule with the surface may also depend on the distribution of molecules impinging on a surface element.

This physical aspect has a mathematical counterpart: The Boltzmann equation must be accompanied by boundary conditions, which describe the aforementioned interaction of the gas molecules with the solid walls.

If we add the fact that polyatomic molecules (such as oxygen and nitrogen in the atmosphere) are far from being the hard spheres or centers of forces at distance commonly considered in elementary kinetic theory, it becomes clear that rarefied gas dynamics is a rather complex subject. Fortunately, there is a simplifying aspect, which has slowly been discovered and confirmed during forty years of research: It is possible to adopt simplified models that embody certain basic physical features and forget about other complexities, in order to obtain essentially accurate results for the purpose of engineering applications. Whereas modeling is more an art than a science, a long practice has repeatedly shown what should be embodied in a model and what can be forgotten: Basic physics, embodied in a certain number of general ideas and principles, should be respected; detailed information about interactions can be replaced by a qualitatively correct description that contains a few parameters to be adjusted so as to represent the behavior of real molecules within the accuracy required by applications.

Rarefied gas dynamics has existed, in principle, since the nineteenth century but came in the foreground with space exploration. One can even give a birthdate, July 1958, when the first international symposium on rarefied gas dynamics was held in Nice, France. Since then, these symposia have been held regularly every other year. The corresponding proceedings constitute a set of useful references as a whole and the list of them is given at the end of this introduction.

When glancing through the aforementioned volumes, one should not be surprised to find a shift of topics. The first few volumes contain a considerable amount of experimental papers, and the theoretical papers contain very general surveys on the Boltzmann equations but very few papers dealing with explicit solutions of some elementary problems. Also, the gas speeds in experiments were rather low, usually slightly supersonic. The first numerical solutions of some interest appear in 1962, but even in the late 1960s these were few in number and not so accurate. Then one witnesses the reduction of experimental work and the increasing importance of numerical simulation. In the most recent issues, experiments regrettably occupy just a few pages of the proceedings. This is compensated for by the fact that numerical simulations have spread through all the subfields, indicating the maturity reached by our theoretical understanding of the subject. Increasingly complicated phenomena, such as reacting flows or evaporation and condensation, are the object of widespread interests.

Thus it is clear that the Boltzmann equation became a practical tool for the aerospace engineers, when they started to remark that flight in the upper atmosphere must face the problem of a decrease in the ambient density with increasing height. This density reduction would alleviate the aerodynamic forces and heat fluxes that a flying vehicle would have to withstand. However, for virtually all missions, the increase of altitude is accompanied by an increase in speed; thus it is not uncommon for a spacecraft to experience its peak heating at considerable altitudes (such as, e.g., 70 km).

In the area of environmental problems, rarefied gas dynamics is also required. Understanding and controlling the formation, motion, reactions, and evolution of particles of varying composition and shapes, ranging from a diameter of the order of .001 μm to 50 μm, as well as their space–time distribution under gradients of concentration, pressure, temperature, and the action of radiation, has grown in importance. This is because of the increasing awareness of the local and global problems related to the emission of particles from electric power plants, chemical plants, and vehicles as well as of the role played by small particles in the formation of fog and clouds, in the release of radioactivity from nuclear reactor accidents, and in the problems arising from the exhaust streams of aerosol reactors, such as those used to produce optical fibers, catalysts, ceramics, silicon chips, and carbon whiskers.

One cubic centimeter of atmospheric air at ground level contains approximately 2.5×10^{19} molecules. About a thousand of them may be charged (ions). A typical molecular diameter is 3×10^{-10} m (3×10^{-4} μm) and the average distance between the molecules is about ten times as much. The mean free path is of the order of 10^{-8} m, or 10^{-2} μm. In addition to molecules and ions one cubic centimeter of air also contains a significant number of particles varying in size, as indicated above. In relatively clean air, these particles, which include pollen, bacteria, dust, and industrial emissions, can number 10^5 or more. They can be both beneficial and detrimental, and they arise from a number of natural sources as well as from the activities of all living organisms, especially humans. The particles can have complex chemical compositions and shapes and may even be toxic or radioactive.

A suspension of particles in a gas is known as an aerosol. Atmospheric aerosols are of global interest and have important impact on our lives. Aerosols are also of great interest in numerous scientific and engineering applications.

A third area of application of rarefied gas dynamics has emerged in the past few years. Small machines, called micromachines, are being designed and built. Their typical sizes range from a few microns to a few millimeters. Rarefied flow phenomena that are more or less laboratory curiosities in machines of more usual size can form the basis of important systems in the micromechanical domain.

One should not forget the design and simulation of aerosol reactors, used to produce optical fibers, catalysts, ceramics, silicon chips, and carbon whiskers, which have been mentioned above as sources of air pollution.

A further area of interest occurs in the vacuum industry. Although this area existed for a long time, the expense of the early computations with kinetic theory precluded applications of numerical methods. The latter could develop only in the context of the aerospace industry, because until recently the big budgets required were available only there.

The present volume is an attempt to cover the theoretical aspects of rarefied gas dynamics. The most theoretical and mathematical aspects have been left out, and the aim is the understanding of the basic concepts and the computation techniques. Readers interested to the difficult but interesting rigorous mathematics concerning the Boltzmann equation are advised to consult the following book:

C. Cercignani, R. Illner, M. Pulvirenti, *The Mathematical Theory of Dilute Gases*, Springer-Verlag, New York (1994).

The analytical techniques used to obtain approximate solutions are dealt with in some detail, but only insofar as they are useful to introduce or illustrate important concepts or show properties of the solutions of the Boltzmann equation. For a more detailed study of these techniques as well as for a discussion of general properties of the Boltzmann equation and its models, two other books should be consulted:

C. Cercignani, *Mathematical Methods in Kinetic Theory*, Plenum Press, New York (1969; revised edition 1990),

C. Cercignani, *The Boltzmann Equation and Its Applications*, Springer-Verlag, New York (1988).

The former is a rather simple introduction to the theory of the Boltzmann equation; the latter is a rather complete monograph, written when a unified presentation of the mathematical and practical aspects of the theory of the Boltzmann equation looked feasible. Nowadays, with the great explosion of the subject in different directions, it seems appropriate to have a book aimed at applications, leaving out the mathematical theory. In particular, computational methods such as the Direct Simulation Monte Carlo and important phenomena such as those related to polyatomic gases, chemical reactions, evaporation, and condensation were barely mentioned in the last reference and are treated here in detail.

Having defined more or less the general subject of the book and the main topics dealt with, it remains to say what is *not* treated here. Ionization phenomena, although discussed in Chapter 6, are only briefly touched upon and viewed as a neighboring topic. The great and important field of rarefied plasmas is left completely untouched. The same can be said for the interaction between gases and thermal radiation.

Finally, one should mention that the basic idea of a kinetic equation, having the Boltzmann equation as prototype, has spread to many other fields, not only to the two aforementioned areas, ionized gases and radiative transfer, and to the well-known area of neutron transport, but also to the behavior of electrons and holes in semiconductors, of dense gases, and of more sizable particles, such as those met in the theory of granular materials and car traffic. These topics are of great interest but are, of course, not dealt with here. Yet, the author hopes that people working in other areas of the growing field of kinetic equations can find suggestions and inspiration for problems in the aforementioned areas by consulting the present book.

List of the proceedings of the Symposia on Rarefied Gas Dynamics

1. *Rarefied Gas Dynamics*, M. Devienne, ed., Pergamon Press, New York (1960).
2. *Rarefied Gas Dynamics*, L. Talbot, ed., Academic Press, New York (1961).
3. *Rarefied Gas Dynamics*, J. A. Laurman, ed., Vols. I and II, Academic Press, New York (1963).
4. *Rarefied Gas Dynamics*, J. H. de Leeuw, ed., Vols. I and II, Academic Press, New York (1965).
5. *Rarefied Gas Dynamics*, C. L. Brundin, ed., Vols. I and II, Academic Press, New York (1967).
6. *Rarefied Gas Dynamics*, L. Trilling and H. Y. Wachman, eds., Vols. I and II, Academic Press, New York (1969).
7. *Rarefied Gas Dynamics*, Dini et al., eds., Vols. I and II, Editrice Tecnico-Scientifica, Pisa (1974).
8. *Rarefied Gas Dynamics*, K. Karamcheti, ed., Academic Press, New York (1974).

9. *Rarefied Gas Dynamics*, M. Becker and M. Fiebig, eds., Vols. I and II, DFLVR Press, Porz-Wahn (1974).
10. *Rarefied Gas Dynamics*, J. L. Potter, ed., Parts I and II, AIAA, New York (1977).
11. *Rarefied Gas Dynamics*, R. Campargue, ed., Vols. I and II, CEA, Paris (1979).
12. *Rarefied Gas Dynamics*, S. S. Fisher, ed., Parts I and II, AIAA, New York (1981).
13. *Rarefied Gas Dynamics*, M. Belotserkovski, M. N. Kogan, S. S. Kutateladze, and A. K. Rebrov, eds., Vols. I and II, Plenum Press, New York (1985).
14. *Rarefied Gas Dynamics*, H. Oguchi, ed., Vols. I and II, University of Tokyo Press, Tokyo (1984).
15. *Rarefied Gas Dynamics*, V. Boffi and C. Cercignani, eds., Vols. I and II, B. G. Teubner, Stuttgart (1986).
16. *Rarefied Gas Dynamics: Space Related Studies*, E. P. Muntz, D. P. Weaver, and D. H. Campbell, eds., AIAA, Washington (1989). *Rarefied Gas Dynamics: Theoretical and Computational Techniques*, E. P. Muntz, D. P. Weaver, and D. H. Campbell, eds., AIAA, Washington (1989). *Rarefied Gas Dynamics: Physical Phenomena*, E. P. Muntz, D. P. Weaver, and D. H. Campbell, eds., AIAA, Washington (1989).
17. *Rarefied Gas Dynamics*, A. E. Beylich, ed., VCH, Weinheim (1991).
18. *Rarefied Gas Dynamics: Experimental Techniques and Physical Systems*, B. D. Shizgal and D. P. Weaver, eds., AIAA, Washington (1994). *Rarefied Gas Dynamics: Theory and Simulations*, B. D. Shizgal and D. P. Weaver, eds., AIAA, Washington (1994). *Rarefied Gas Dynamics: Space Science and Engineering*, B. D. Shizgal and D. P. Weaver, eds., AIAA, Washington (1994).
19. *Rarefied Gas Dynamics*, J. Harvey and G. Lord, eds., Vols. 1 and 2, Oxford University Press, Oxford (1995).
20. *Rarefied Gas Dynamics Symposium 20*, Ching Shen, ed., Peking University Press, Beijing (1997).
21. *Rarefied Gas Dynamics*, Proceedings of the 21st Symposium, to appear (1999).

1

Boltzmann Equation and Gas–Surface Interaction

1.1. Introduction

According to kinetic theory, a gas in normal conditions (no chemical reactions, no ionization phenomena, etc.) is formed of elastic molecules rushing hither and thither at high speed, colliding and rebounding according to the laws of elementary mechanics. In this and the next section, the molecules of a gas will be assumed to be hard, elastic, and perfectly smooth spheres. Later we shall consider molecules as centers of forces that move according to the laws of classical mechanics and, starting with Chapter 6, more complex models describing polyatomic molecules.

The rules generating the dynamics of many spheres are easy to describe: Thus, for example, if no external forces, such as gravity, are assumed to act on the molecules, each of them will move in a straight line unless it happens to strike another molecule or a solid wall. The phenomena associated with this dynamics are not so simple, especially when the number of spheres is large. It turns out that this complication is always present when dealing with a gas, because the number of molecules usually considered is extremely large: There are about $2.7 \cdot 10^{19}$ in a cubic centimeter of a gas at atmospheric pressure and a temperature of 0° C.

Given the vast number of particles to be considered, it would of course be a hopeless task to attempt to describe the state of the gas by specifying the so-called microscopic state (i.e., the position and velocity of every individual sphere); we must have recourse to statistics. A description of this kind is made possible because in practice all that our typical observations can detect are changes in the macroscopic state of the gas, described by quantities such as density, bulk velocity, temperature, stresses, and heat flow, and these are related to some suitable averages of quantities depending on the microscopic state.

1.2. The Boltzmann Equation

The exact dynamics of N particles is a useful conceptual tool, but it cannot in any way be used in practical calculations because it requires a huge number of real variables (of the order of 10^{20}). This was realized by Maxwell and Boltzmann when they started to work with the one-particle probability density, or distribution function $P^{(1)}(\mathbf{x}, \boldsymbol{\xi}, t)$. The latter is a function of seven variables: the components of the two vectors $\mathbf{x}$ and $\boldsymbol{\xi}$ and time t. In particular, Boltzmann wrote an evolution equation for $P^{(1)}$ by means of a heuristic argument, which we shall try to present in such a way as to show where extra assumptions are introduced.

Let us first consider the meaning of $P^{(1)}(\mathbf{x}, \boldsymbol{\xi}, t)$; it gives the probability density of finding one fixed particle (say, the one labeled by 1) at a certain point $(\mathbf{x}, \boldsymbol{\xi})$ of the six-dimensional reduced phase space associated with the position and velocity of that molecule. To simplify the treatment, we shall for the moment assume that the molecules are hard spheres, whose center has position $\mathbf{x}$. When the molecules collide, momentum and kinetic energy must be conserved; thus (Problem 1.2.2) the velocities after the impact, $\boldsymbol{\xi}_1'$ and $\boldsymbol{\xi}_2'$, are related to those before the impact, $\boldsymbol{\xi}_1$ and $\boldsymbol{\xi}_2$, by

$$\begin{aligned} \boldsymbol{\xi}_1' &= \boldsymbol{\xi}_1 - \mathbf{n}[\mathbf{n} \cdot (\boldsymbol{\xi}_1 - \boldsymbol{\xi}_2)], \\ \boldsymbol{\xi}_2' &= \boldsymbol{\xi}_2 + \mathbf{n}[\mathbf{n} \cdot (\boldsymbol{\xi}_1 - \boldsymbol{\xi}_2)], \end{aligned} \tag{1.2.1}$$

where $\mathbf{n}$ is the unit vector along $\boldsymbol{\xi}_1 - \boldsymbol{\xi}_1'$. Note that the relative velocity

$$\mathbf{V} = \boldsymbol{\xi}_1 - \boldsymbol{\xi}_2 \tag{1.2.2}$$

satisfies

$$\mathbf{V}' = \mathbf{V} - 2\mathbf{n}(\mathbf{n} \cdot \mathbf{V}), \tag{1.2.3}$$

that is, it undergoes a specular reflection at the impact. This means that if we split $\mathbf{V}$ at the point of impact into a normal component $\mathbf{V}_n$, directed along $\mathbf{n}$ and a tangential component $\mathbf{V}_t$ (in the plane normal to $\mathbf{n}$), then $\mathbf{V}_n$ changes sign and $\mathbf{V}_t$ remains unchanged in a collision (Problem 1.2.4). We can also say that $\mathbf{n}$ bisects the directions of $\mathbf{V}$ and $-\mathbf{V}' = -(\boldsymbol{\xi}_1' - \boldsymbol{\xi}_2')$ (see Fig. 1.1).

Let us remark that, in the absence of collisions, $P^{(1)}$ would remain unchanged along the trajectory of a particle (see Problem 1.2.1). Accordingly we must evaluate the effects of collisions on the time evolution of $P^{(1)}$. Note that the probability of occurrence of a collision is related to the probability of finding another molecule with a center at exactly one diameter from the center of the first one, whose distribution function is $P^{(1)}$. Thus, generally speaking, in order

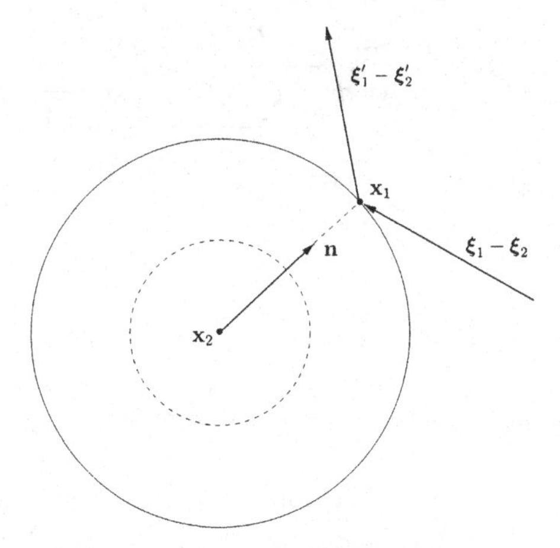

Figure 1.1. The directions of the relative velocities before and after the impact are bisected by the unit vector **n**.

to write the evolution equation for $P^{(1)}$ we shall need another function, $P^{(2)}$, that gives the probability density of finding, at time t, the first molecule at $\mathbf{x}_1$ with velocity $\boldsymbol{\xi}_1$ and the second at $\mathbf{x}_2$ with velocity $\boldsymbol{\xi}_2$; obviously $P^{(2)} = P^{(2)}(\mathbf{x}_1, \mathbf{x}_2, \boldsymbol{\xi}_1, \boldsymbol{\xi}_2, t)$. Hence $P^{(1)}$ satisfies an equation of the following form:

$$\frac{\partial P^{(1)}}{\partial t} + \boldsymbol{\xi}_1 \cdot \frac{\partial P^{(1)}}{\partial \mathbf{x}_1} = G - L. \tag{1.2.4}$$

Here $L d\mathbf{x}_1 d\boldsymbol{\xi}_1 dt$ gives the expected number of particles with position between $\mathbf{x}_1$ and $\mathbf{x}_1 + d\mathbf{x}_1$ and velocity between $\boldsymbol{\xi}_1$ and $\boldsymbol{\xi}_1 + d\boldsymbol{\xi}_1$ that disappear from these ranges of values because of a collision in the time interval between t and $t + dt$, and $G d\mathbf{x}_1 d\boldsymbol{\xi}_1 dt$ gives the analogous number of particles entering the same range in the same time interval. Counting these numbers is easy, provided we use the trick of imagining particle 1 as a sphere at rest and endowed with twice the actual diameter σ and the other particles to be point masses with velocity $(\boldsymbol{\xi}_i - \boldsymbol{\xi}_1) = \mathbf{V}_i$. In fact, each collision will send particle 1 out of the above range and the number of the collisions of particle 1 will be the number of expected collisions of any other particle with that sphere. Since there are exactly $(N - 1)$ identical point masses and multiple collisions are disregarded, $G = (N - 1)g$ and $L = (N - 1)l$, where the lowercase letters indicate the contribution of a fixed particle, say particle 2. We shall then compute the effect of the collisions of particle 2 with particle 1. Let $\mathbf{x}_2$ be a point of the sphere such that the vector joining the center of the sphere with $\mathbf{x}_2$ is $\sigma\mathbf{n}$, where $\mathbf{n}$ is a unit vector. A cylinder with height $|\mathbf{V} \cdot \mathbf{n}| dt$ (where we write just $\mathbf{V}$ for $\mathbf{V}_2$) and base area $dS = \sigma^2 d\mathbf{n}$ (where $d\mathbf{n}$ is the area of a surface

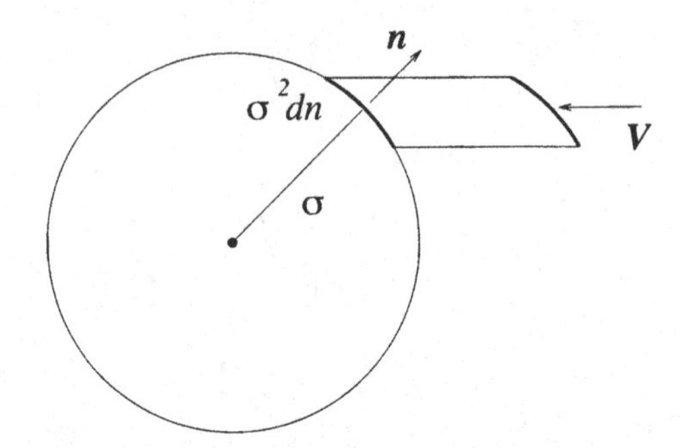

Figure 1.2. Calculation of the number of collisions between two molecules.

element of the unit sphere about $\mathbf{n}$) will contain the particles with velocity $\boldsymbol{\xi}_2$ hitting the base dS in the time interval $(t, t+dt)$ (see Fig. 1.2); its volume is $\sigma^2 d\mathbf{n}|\mathbf{V}\cdot\mathbf{n}|dt$. Thus the number of collisions of particle 2 with particle 1 in the ranges $(\mathbf{x}_1, \mathbf{x}_1 + d\mathbf{x}_1)$, $(\boldsymbol{\xi}_1, \boldsymbol{\xi}_1 + d\boldsymbol{\xi}_1)$, $(\mathbf{x}_2, \mathbf{x}_2 + d\mathbf{x}_2)$, $(\boldsymbol{\xi}_2, \boldsymbol{\xi}_2 + d\boldsymbol{\xi}_2)$, $(t, t+dt)$ occuring at points of dS is $P^{(2)}(\mathbf{x}_1, \mathbf{x}_2, \boldsymbol{\xi}_1, \boldsymbol{\xi}_2, t)\, d\mathbf{x}_1 d\boldsymbol{\xi}_1 d\boldsymbol{\xi}_2 \sigma^2 d\mathbf{n}|\mathbf{V}_2\cdot\mathbf{n}|dt$. If we want the number of collisions of particle 1 with 2, when the range of the former is fixed but the latter may have any velocity $\boldsymbol{\xi}_2$ and any position $\mathbf{x}_2$ on the sphere (i.e., any $\mathbf{n}$), we integrate over the sphere and all the possible velocities of particle 2 to obtain

$$l d\mathbf{x}_1 d\boldsymbol{\xi}_1 dt = d\mathbf{x}_1 d\boldsymbol{\xi}_1 dt \int_{R^3}\int_{\mathcal{B}^-} P^{(2)}(\mathbf{x}_1, \mathbf{x}_1 + \sigma\mathbf{n}, \boldsymbol{\xi}_1, \boldsymbol{\xi}_2, t)|\mathbf{V}\cdot\mathbf{n}|\sigma^2 d\mathbf{n} d\boldsymbol{\xi}_2, \tag{1.2.5}$$

where $\mathcal{B}^-$ is the hemisphere corresponding to $\mathbf{V}\cdot\mathbf{n} < 0$ (the particles are moving toward each other before the collision). Thus we have the following result:

$$L = (N-1)\sigma^2 \int_{R^3}\int_{\mathcal{B}^-} P^{(2)}(\mathbf{x}_1, \mathbf{x}_1 + \sigma\mathbf{n}, \boldsymbol{\xi}_1, \boldsymbol{\xi}_2, t)|(\boldsymbol{\xi}_2 - \boldsymbol{\xi}_1)\cdot\mathbf{n}| d\boldsymbol{\xi}_2 d\mathbf{n}. \tag{1.2.6}$$

The calculation of the gain term G is exactly the same as the one for L, except for the fact that we have to integrate over the hemisphere B^+, defined by $\mathbf{V}\cdot\mathbf{n} > 0$ (the particles are moving away from each other after the collision). Thus we have

$$G = (N-1)\sigma^2 \int_{R^3}\int_{\mathcal{B}^+} P^{(2)}(\mathbf{x}_1, \mathbf{x}_1 + \sigma\mathbf{n}, \boldsymbol{\xi}_1, \boldsymbol{\xi}_2, t)|(\boldsymbol{\xi}_2 - \boldsymbol{\xi}_1)\cdot\mathbf{n}| d\boldsymbol{\xi}_2 d\mathbf{n}. \tag{1.2.7}$$

We thus could write the right-hand side of Eq. (1.2.4) as a single expression:

$$G - L = (N-1)\sigma^2 \int_{R^3}\int_{\mathcal{B}} P^{(2)}(\mathbf{x}_1, \mathbf{x}_1 + \sigma\mathbf{n}, \boldsymbol{\xi}_1, \boldsymbol{\xi}_2, t)(\boldsymbol{\xi}_2 - \boldsymbol{\xi}_1)\cdot\mathbf{n} d\boldsymbol{\xi}_2 d\mathbf{n}, \tag{1.2.8}$$

where now $\mathcal{B}$ is the entire unit sphere and we have abolished the bars of absolute value in the right-hand side.

Equation (1.2.8), although absolutely correct, is not so useful. It turns out to be much more convenient to keep the gain and loss terms separated. Only in this way, in fact, can we insert in Eq. (1.2.4) the information that the probability density $P^{(2)}$ is continuous at a collision; in other words, although the velocities of the particles undergo the discontinuous change described by Eqs. (1.2.1), we can write

$$P^{(2)}(\mathbf{x}_1, \boldsymbol{\xi}_1, \mathbf{x}_2, \boldsymbol{\xi}_2, t) = P^{(2)}(\mathbf{x}_1, \boldsymbol{\xi}_1 - \mathbf{n}(\mathbf{n}\cdot\mathbf{V}), \mathbf{x}_2, \boldsymbol{\xi}_2 + \mathbf{n}(\mathbf{n}\cdot\mathbf{V}), t) \quad \text{if } |\mathbf{x}_1 - \mathbf{x}_2| = \sigma. \tag{1.2.9}$$

For brevity, we write (in agreement with Eq. (1.2.1)

$$\boldsymbol{\xi}'_1 = \boldsymbol{\xi}_1 - \mathbf{n}(\mathbf{n}\cdot\mathbf{V}), \qquad \boldsymbol{\xi}'_2 = \boldsymbol{\xi}_2 + \mathbf{n}(\mathbf{n}\cdot\mathbf{V}). \tag{1.2.10}$$

Inserting Eq. (1.2.8) in Eq. (1.2.5) we thus obtain

$$G = (N-1)\sigma^2 \int_{R^3}\int_{\mathcal{B}^+} P^{(2)}(\mathbf{x}_1, \mathbf{x}_1 + \sigma\mathbf{n}, \boldsymbol{\xi}'_1, \boldsymbol{\xi}'_2, t)|(\boldsymbol{\xi}_2 - \boldsymbol{\xi}_1)\cdot\mathbf{n}|d\boldsymbol{\xi}_2 d\mathbf{n}, \tag{1.2.11}$$

which is a frequently used form. Sometimes $\mathbf{n}$ is changed into $-\mathbf{n}$ to have the same integration range as in L; the only change (in addition to the change in the range) is in the second argument of $P^{(2)}$, which becomes $\mathbf{x}_1 - \sigma\mathbf{n}$.

At this point we are ready to understand Boltzmann's argument. N is a very large number and σ (expressed in common units, such as, e.g., centimeters) is very small; to fix the ideas, let us consider a box whose volume is 1 cm^3 at room temperature and atmospheric pressure. Then $N \cong 10^{20}$ and $\sigma \cong 10^{-8}$ cm. Then $(N-1)\sigma^2 \cong N\sigma^2 \cong 10^4 \text{ cm}^2 = 1 \text{ m}^2$ is a sizable quantity, while we can neglect the difference between $\mathbf{x}_1$ and $\mathbf{x}_1 + \sigma\mathbf{n}$. This means that the equation to be written can be rigorously valid only in the so-called *Boltzmann–Grad limit*, when $N \to \infty$, $\sigma \to 0$ with $N\sigma^2$ finite.

In addition, the collisions between two preselected particles are rather rare events. Thus two spheres that happen to collide can be thought to be two randomly chosen particles and it makes sense to assume that the probability density of finding the first molecule at $\mathbf{x}_1$ with velocity $\boldsymbol{\xi}_1$ and the second at $\mathbf{x}_2$ with velocity $\boldsymbol{\xi}_2$ is the product of the probability density of finding the first molecule at $\mathbf{x}_1$ with velocity $\boldsymbol{\xi}_1$ times the probability density of finding the second molecule at $\mathbf{x}_2$ with velocity $\boldsymbol{\xi}_2$. If we accept this we can write (assumption of *molecular chaos*)

$$P^{(2)}(\mathbf{x}_1, \boldsymbol{\xi}_1, \mathbf{x}_2, \boldsymbol{\xi}_2, t) = P^{(1)}(\mathbf{x}_1, \boldsymbol{\xi}_1, t)P^{(1)}(\mathbf{x}_2, \boldsymbol{\xi}_2, t) \tag{1.2.12}$$

for two particles that are about to collide, or

$$P^{(2)}(\mathbf{x}_1, \boldsymbol{\xi}_1, \mathbf{x}_1 + \sigma\mathbf{n}, \boldsymbol{\xi}_2, t) = P^{(1)}(\mathbf{x}_1, \boldsymbol{\xi}_1, t)P^{(1)}(\mathbf{x}_1, \boldsymbol{\xi}_2, t) \quad \text{for } (\boldsymbol{\xi}_2 - \boldsymbol{\xi}_1)\cdot\mathbf{n} < 0. \tag{1.2.13}$$

Thus we can apply this recipe to the loss term (1.2.4) but not to the gain term in the form (1.2.5). It is possible, however, to apply Eq. (1.2.13) (with $\boldsymbol{\xi}_1'$, $\boldsymbol{\xi}_2'$ in place of $\boldsymbol{\xi}_1$, $\boldsymbol{\xi}_2$) to the form (1.2.9) of the gain term, because the transformation (1.2.10) maps the hemisphere $\mathcal{B}^+$ onto the hemisphere $\mathcal{B}^-$.

If we accept all the simplifying assumptions made by Boltzmann, we obtain the following form for the gain and loss terms:

$$G = N\sigma^2 \int_{R^3}\int_{\mathcal{B}^-} P^{(1)}(\mathbf{x}_1, \boldsymbol{\xi}_1', t)P^{(1)}(\mathbf{x}_1, \boldsymbol{\xi}_2', t)|(\boldsymbol{\xi}_2 - \boldsymbol{\xi}_1)\cdot\mathbf{n}|d\boldsymbol{\xi}_2 d\mathbf{n}, \tag{1.2.14}$$

$$L = N\sigma^2 \int_{R^3}\int_{\mathcal{B}^-} P^{(1)}(\mathbf{x}_1, \boldsymbol{\xi}_1, t)P^{(1)}(\mathbf{x}_1, \boldsymbol{\xi}_2, t)|(\boldsymbol{\xi}_2 - \boldsymbol{\xi}_1)\cdot\mathbf{n}|d\boldsymbol{\xi}_2 d\mathbf{n}. \tag{1.2.15}$$

By inserting these expressions in Eq. (1.2.6) we can write the *Boltzmann equation* in the following form:

$$\frac{\partial P^{(1)}}{\partial t} + \boldsymbol{\xi}_1 \cdot \frac{\partial P^{(1)}}{\partial \mathbf{x}_1} = N\sigma^2 \int_{R^3}\int_{\mathcal{B}^-} \left[P^{(1)}(\mathbf{x}_1, \boldsymbol{\xi}_1', t)P^{(1)}(\mathbf{x}_1, \boldsymbol{\xi}_2', t) - P^{(1)}(\mathbf{x}_1, \boldsymbol{\xi}_1, t)P^{(1)}(\mathbf{x}_1, \boldsymbol{\xi}_2, t)\right] |(\boldsymbol{\xi}_2 - \boldsymbol{\xi}_1)\cdot\mathbf{n}|d\boldsymbol{\xi}_2 d\mathbf{n}. \tag{1.2.16}$$

We remark that the expressions for $\boldsymbol{\xi}_1'$ and $\boldsymbol{\xi}_2'$ given in Eq. (1.2.1) are by no means the only possible ones. In fact we might use a different unit vector $\boldsymbol{\omega}$, directed as $\mathbf{V}'$, instead of $\mathbf{n}$. Then Eq. (1.2.1) is replaced by

$$\boldsymbol{\xi}_1' = \overline{\boldsymbol{\xi}} + \frac{1}{2}|\boldsymbol{\xi}_1 - \boldsymbol{\xi}_2|\boldsymbol{\omega},$$
$$\boldsymbol{\xi}_2' = \overline{\boldsymbol{\xi}} - \frac{1}{2}|\boldsymbol{\xi}_1 - \boldsymbol{\xi}_2|\boldsymbol{\omega}, \tag{1.2.17}$$

where $\overline{\boldsymbol{\xi}} = \frac{1}{2}(\boldsymbol{\xi}_1 + \boldsymbol{\xi}_2)$ is the velocity of the center of mass. The relative velocity $\mathbf{V}$ satisfies

$$\mathbf{V}' = \boldsymbol{\omega}|\mathbf{V}|. \tag{1.2.18}$$

The Boltzmann equation is an evolution equation for $P^{(1)}$, without any reference to $P^{(2)}$. This is its main advantage. However, it has been obtained at the price of several assumptions; the chaos assumption present in Eqs. (1.2.12) and (1.2.13) is particularly strong and requires discussion.

The molecular chaos assumption is clearly a property of randomness. Intuitively, one feels that collisions exert a randomizing influence, but it would be completely wrong to argue that the statistical independence described by Eq. (1.2.12) is a consequence of the dynamics. It is quite clear that we cannot expect every choice of the initial distribution of positions and velocities of the molecules to give a $P^{(1)}$ that agrees with the solution of the Boltzmann equation in the Boltzmann–Grad limit. In other words molecular chaos must be present initially and we can only ask whether it is preserved by the time evolution of the system of hard spheres.

It is evident that the chaos property (1.2.12), if initially present, is almost immediately destroyed, if we insist that it should be valid everywhere. In fact, if it were strictly valid everywhere, the gain and loss terms, in the Boltzmann–Grad limit, would be exactly equal. As a consequence, there would be no effect of the collisions on the time evolution of $P^{(1)}$. The essential point is that we need the chaos property only for molecules that are about to collide, that is, those in the precise form stated in Eq. (1.2.13). It is clear then that even if $P^{(1)}$ as predicted by the exact dynamics converges nicely to a solution of the Boltzmann equation, $P^{(2)}$ may converge to a product, as stated in Eq. (1.2.11), only in a way that is in a certain sense very singular. In fact, it is not enough to show that the convergence is almost everywhere, because we need to use the chaos property in a zero measure set. However, we cannot try to show that convergence holds everywhere, because this would be false; in fact, we have just remarked that Eq. (1.2.11) is, generally speaking, simply not true for molecules that have just collided.

How can we approach the question of justifying the Boltzmann equation without invoking the molecular chaos assumption as an a priori hypothesis? Clearly, since $P^{(2)}$ appears in the evolution equation for $P^{(1)}$, we must investigate the time evolution for $P^{(2)}$; now, as is clear, the evolution equation for $P^{(2)}$ contains another function, $P^{(3)}$, which depends on time and the coordinates and velocities of three molecules and gives the probability density of finding, at time t, the first molecule at $\mathbf{x}_1$ with velocity $\boldsymbol{\xi}_1$, the second at $\mathbf{x}_2$ with velocity $\boldsymbol{\xi}_2$, and the third at $\mathbf{x}_3$ with velocity $\boldsymbol{\xi}_3$. In general if we introduce a function $P^{(s)} = P^{(s)}(\mathbf{x}_1, \mathbf{x}_2, \ldots, \mathbf{x}_s, \boldsymbol{\xi}_1, \boldsymbol{\xi}_2, \ldots, \boldsymbol{\xi}_s, t)$, the so-called *s-particle distribution function*, which gives the probability density of finding, at time t, the first molecule at $\mathbf{x}_1$ with velocity $\boldsymbol{\xi}_1$, the second at $\mathbf{x}_2$ with velocity $\boldsymbol{\xi}_2, \ldots$ and the sth at $\mathbf{x}_s$ with velocity $\boldsymbol{\xi}_s$, we find the evolution equation of $P^{(s)}$ contains the next function $P^{(s+1)}$, till we reach $s = N$; in fact $P^{(N)}$ satisfies a partial differential equation called the Liouville equation. Clearly we cannot proceed unless we handle all the $P^{(s)}$ at the same time and attempt to prove a generalized form of molecular chaos, that is,

$$P^{(s)}(\mathbf{x}_1, \mathbf{x}_2, \ldots, \mathbf{x}_s, \boldsymbol{\xi}_1, \boldsymbol{\xi}_2, \ldots, \boldsymbol{\xi}_s, t) = \prod_{j=1}^{s} P^{(1)}(\mathbf{x}_j, \boldsymbol{\xi}_j, t). \tag{1.2.19}$$

The task then becomes to show that, if true at $t = 0$, this property remains preserved (for any fixed s) in the Boltzmann–Grad limit. The discussion of this point is outside the scope of this book. The interested reader may consult Refs. 1–7.

There remains the problem of justifying the *initial chaos assumption*, according to which Eq. (1.2.19) is satisfied at $t = 0$. One can give two justifications, one of them being physical in nature and the second mathematical; essentially, they say the same thing, that is, it is hard to prepare an initial state for which Eq. (1.2.19) does not hold. The physical reason for this is that, in general, we cannot handle every single molecule, but rather we act on the gas as a whole, if we act at a macroscopic level, usually starting from an equilibrium state (for which Eq. (1.2.19) holds). The mathematical argument indicates that if we choose the initial data for the molecules at random, there is an overwhelming probability that Eq. (1.2.19) is satisfied for $t = 0$.[1,3]

A word should be said about boundary conditions. When proving that chaos is preserved in the limit, it is absolutely necessary to have a boundary condition compatible (at least in the limit) with Eq. (1.2.19). If the boundary conditions are those of periodicity or specular reflection, no problems arise. In general, it is sufficient that the particles are scattered without adsorption from the boundary in a way that does not depend on the state of the other molecules of the gas.[1,3]

Problems

1.2.1 Show that if there are no collisions (and no body forces), then $P^{(1)}$ satisfies

$$\frac{\partial P^{(1)}}{\partial t} + \boldsymbol{\xi}_1 \cdot \frac{\partial P^{(1)}}{\partial \mathbf{x}_1} = 0.$$

1.2.2 Show that Eqs. (1.2.1) hold. (Remark: Momentum and energy conservation imply $\boldsymbol{\xi}_1 + \boldsymbol{\xi}_2 = \boldsymbol{\xi}_1' + \boldsymbol{\xi}_2'$ and $|\boldsymbol{\xi}_1|^2 + |\boldsymbol{\xi}_2|^2 = |\boldsymbol{\xi}_1'|^2 + |\boldsymbol{\xi}_2'|^2$ and, by definition, we have $\boldsymbol{\xi}_1' = \boldsymbol{\xi}_1 - \mathbf{n}C$, where C is a scalar to be determined...).

1.2.3 Check that if we split the relative velocity $\mathbf{V}$ at the point of impact into a normal component $\mathbf{V}_n$, directed along $\mathbf{n}$, and a tangential component

$\mathbf{V}_t$ (in the plane normal to $\mathbf{n}$), then $\mathbf{V}_n$ changes sign and $\mathbf{V}_t$ remains unchanged in a collision.

1.2.4 Show that if we transform from the variables $\boldsymbol{\xi}_1, \boldsymbol{\xi}_2$ to the variables $\mathbf{V}$ (the relative velocity) and $\overline{\boldsymbol{\xi}} = \frac{1}{2}(\boldsymbol{\xi}_1 + \boldsymbol{\xi}_2)$ (the velocity of the center of mass), the transformation has unit Jacobian.

1.2.5 Check, by a direct calculation, that the Jacobian of the transformation (1.2.1) is unity, if the collision occurs in a plane (i.e., $\boldsymbol{\xi}_1, \boldsymbol{\xi}_2, \boldsymbol{\xi}'_1$, and $\boldsymbol{\xi}'_2$ have just two components, while the components of $\mathbf{n}$ can be written $(\cos\theta, \sin\theta)$, where θ is a suitable angle).

1.2.6 Check that the transformation (1.2.1) actually maps the hemisphere $\mathcal{B}^+$ onto $\mathcal{B}^-$.

1.2.7 Find a relation between the angles formed by $\mathbf{n}$ and $\boldsymbol{\omega}$ with $\mathbf{V}$.

1.2.8 Give a reasonable definition of probability for the initial data in terms of $P^{(N)}$ and show that it attains a constrained maximum (the constraint being that $P^{(1)}$ is assigned) when $P^{(N)}$ is chaotic, that is, satisfies Eq. (1.2.19) (with $s = N$ and $t = 0$). (See Ref. 3.)

1.3. Molecules Different from Hard Spheres

In the previous section we discussed the Boltzmann equation when the molecules are assumed to be identical hard spheres. There are several possible generalizations of this molecular model; the most obvious is the case of molecules that are identical point masses interacting with a central force – a good general model for monatomic gases. If the range of the force extends to infinity, there is a complication due to the fact that two molecules are always interacting and the analysis in terms of "collisions" is no longer possible. If, however, the gas is sufficiently dilute, we can take into account that the molecular interaction is negligible for distances larger than a certain σ (the "molecular diameter") and assume that when two molecules are at a distance smaller than σ, then no other molecule is interacting with them and the binary collision analysis considered in the previous section can be applied. The only difference arises in the factor $\sigma^2|(\boldsymbol{\xi}_2 - \boldsymbol{\xi}_1)\cdot\mathbf{n}|$, which turns out to be replaced by a function of $V = |\boldsymbol{\xi}_2 - \boldsymbol{\xi}_1|$ and the angle θ between $\mathbf{n}$ and $\mathbf{V}$ (Refs. 1, 6, and 7). Thus the Boltzmann equation for monatomic molecules takes on the following form:

$$\frac{\partial P^{(1)}}{\partial t} + \boldsymbol{\xi}_1 \cdot \frac{\partial P^{(1)}}{\partial \mathbf{x}_1} = N \int_{R^3}\int_{\mathcal{B}_-} \left[P^{(1)}(\mathbf{x}_1, \boldsymbol{\xi}'_1, t)P^{(1)}(\mathbf{x}_1, \boldsymbol{\xi}'_2, t)\right.$$
$$\left. - P^{(1)}(\mathbf{x}_1, \boldsymbol{\xi}_1, t)P^{(1)}(\mathbf{x}_1, \boldsymbol{\xi}_2, t)\right] B(\theta, |\boldsymbol{\xi}_2 - \boldsymbol{\xi}_1|)d\boldsymbol{\xi}_2 d\theta d\epsilon, \qquad (1.3.1)$$

where ϵ is the other angle which, together with θ, identifies the unit vector $\mathbf{n}$.

The function $B(\theta, V)$ depends, of course, on the specific law of interaction between the molecules. In the case of hard spheres, of course,

$$B(\theta, |\boldsymbol{\xi}_2 - \boldsymbol{\xi}_1|) = \cos\theta \sin\theta |\boldsymbol{\xi}_2 - \boldsymbol{\xi}_1|. \tag{1.3.2}$$

In spite of the fact that the force is cut at a finite range σ when writing the Boltzmann equation, infinite range forces are frequently used. This has the disadvantage of making the integral in Eq. (1.3.1) rather hard to handle; in fact, one cannot split it into the difference of two terms (the loss and the gain), because each of them would be a divergent integral. This disadvantage is compensated in the case of power-law forces, because one can separate the dependence on θ from the dependence upon V. In fact, one can show[1,6] that, if the intermolecular force varies as the nth inverse power of the distance, then (Problem 1.3.1)

$$B(\theta, |\boldsymbol{\xi}_2 - \boldsymbol{\xi}_1|) = \beta(\theta) |\boldsymbol{\xi}_2 - \boldsymbol{\xi}_1|^{\frac{n-5}{n-1}}, \tag{1.3.3}$$

where $\beta(\theta)$ is a nonelementary function of θ (in the simplest cases it can be expressed by elliptic functions). In particular, for $n = 5$ one has the so-called Maxwell molecules, for which the dependence on V disappears.

Sometimes the artifice of cutting the grazing collisions corresponding to small values of $|\theta - \pi/2|$ is used (angle cutoff). In this case one has both the advantage of being able to split the collision term and of preserving a relation of the form (1.3.3) for power-law potentials.

Since solving the Boltzmann equation with actual cross sections is complicated, in many numerical simulations use is made of the so-called variable hard sphere model in which the diameter of the spheres is an inverse power law function of the relative speed V (see Chapter 7).

Another important case is that of a mixture rather than a single gas. In this case we have n unknowns, if n is the number of the species, and n Boltzmann equations; in each of them there are n collision terms to describe the collision of a molecule with other molecules of all the possible species.[3]

If the gas is polyatomic, then the gas molecules have other degrees of freedom in addition to the translation ones. This in principle requires using quantum mechanics, but one can devise useful and accurate models in the classical framework as well. Frequently the internal energy E_i is the only additional variable that is needed, in which case one can think of the gas as of a mixture of species,[3] each differing from the other because of the value of E_i. If the latter variable is discrete we obtain a strict analogy with a mixture; otherwise we have a continuum of species. We remark that in both cases, kinetic energy is not preserved by collisions, because internal energy also enters into the balance; this means that a molecule changes its "species" when colliding. This is the simplest example of a "reacting collision," which may be generalized to actual chemical species when chemical reactions occur. The subject of mixtures and polyatomic gases will be taken up again in Chapter 6.

Problem

1.3.1 Show that Eq. (1.3.3) holds (see Refs. 3 and 6).

1.4. Collision Invariants

Before embarking on a discussion of the properties of the solutions of the Boltzmann equation we remark that the unknown of the latter is not always chosen to be a probability density as we have done so far; it may be multiplied by a suitable factor and transformed into an (expected) number density or an (expected) mass density (in phase space, of course). The only thing that changes is the factor in front of Eq. (1.3.1), which is no longer N. To avoid any commitment to a special choice of that factor we replace $NB(\theta, V)$ by $\mathcal{B}(\theta, V)$ and the unknown P by another letter, f (which is also the most commonly used letter to denote the one-particle distribution function, no matter what its normalization is). In addition, we replace the current velocity variable $\boldsymbol{\xi}_1$ simply by $\boldsymbol{\xi}$ and $\boldsymbol{\xi}_2$ by $\boldsymbol{\xi}_*$. Thus we rewrite Eq. (1.3.1) in the following form:

$$\frac{\partial f}{\partial t} + \boldsymbol{\xi} \cdot \frac{\partial f}{\partial \mathbf{x}} = \int_{R^3} \int_{\mathcal{B}^-} (f' f'_* - f f_*) \mathcal{B}(\theta, V) d\boldsymbol{\xi}_* d\theta d\epsilon, \tag{1.4.1}$$

where $V = |\boldsymbol{\xi} - \boldsymbol{\xi}_*|$. The velocity arguments $\boldsymbol{\xi}'$ and $\boldsymbol{\xi}'_*$ in f' and f'_* are of course given by Eqs. (1.2.1) (or (1.2.16)) with the suitable modification.

The right-hand side of Eq. (1.4.1) contains a quadratic expression $Q(f, f)$, given by

$$Q(f, f) = \int_{R^3} \int_{\mathcal{B}^-} (f' f'_* - f f_*) \mathcal{B}(\theta, V) d\boldsymbol{\xi}_* d\theta d\epsilon. \tag{1.4.2}$$

This expression is called the collision integral or, simply, the collision term; the quadratic operator Q goes under the name of collision operator. In this section we study some elementary properties of Q. It actually turns out to be more convenient to study the slightly more general bilinear expression associated with $Q(f, f)$, that is,

$$Q(f, g) = \frac{1}{2} \int_{R^3} \int_{\mathcal{B}^-} (f' g'_* + g' f'_* - f g_* - g f_*) \mathcal{B}(\theta, V) d\boldsymbol{\xi}_* d\theta d\epsilon. \tag{1.4.3}$$

It is clear that when $g = f$, Eq. (1.4.3) reduces to Eq. (1.4.2) and

$$Q(f, g) = Q(g, f). \tag{1.4.4}$$

Our first aim is to study the eightfold integral:

$$\int_{R^3} Q(f, g)\phi(\boldsymbol{\xi})d\boldsymbol{\xi} = \frac{1}{2}\int_{R^3}\int_{R^3}\int_{\mathcal{B}_-} (f'g'_* + g'f'_* - fg_* - gf_*)\phi(\boldsymbol{\xi})\mathcal{B}(\theta, V)d\boldsymbol{\xi}_* d\boldsymbol{\xi} d\theta d\epsilon, \tag{1.4.5}$$

where f, g, and ϕ are functions such that the indicated integrals exist and the order of integration does not matter. A simple interchange of the starred and unstarred variables (with a glance at Eqs. (1.2.1)) shows that

$$\int_{R^3} Q(f, g)\phi(\boldsymbol{\xi})d\boldsymbol{\xi} = \frac{1}{2}\int_{R^3}\int_{R^3}\int_{\mathcal{B}_-} (f'g'_* + g'f'_* - fg_* - gf_*)\phi(\boldsymbol{\xi}_*)\mathcal{B}(\theta, V)d\boldsymbol{\xi}_* d\boldsymbol{\xi} d\theta d\epsilon. \tag{1.4.6}$$

Next, we consider another transformation of variables, the exchange of primed and unprimed variables (which is possible because the transformation in Eq. (1.2.1) is linear and its own inverse, for any fixed $\mathbf{n}$). This gives

$$\int_{R^3} Q(f, g)\phi(\boldsymbol{\xi})d\boldsymbol{\xi} = \frac{1}{2}\int_{R^3}\int_{R^3}\int_{\mathcal{B}-} (fg_* + gf_* - f'g'_* - g'f_*)\phi(\boldsymbol{\xi}')\mathcal{B}(\theta, V)d\boldsymbol{\xi}'_* d\boldsymbol{\xi}' d\theta d\epsilon. \tag{1.4.7}$$

(Actually since $\mathbf{V}' \cdot \mathbf{n} = -\mathbf{V} \cdot \mathbf{n}$ we should write $\mathcal{B}^-$ in place of B^+; changing $\mathbf{n}$ into $-\mathbf{n}$, however, gives exactly the expression written here.)

The absolute value of the Jacobian from $\boldsymbol{\xi}, \boldsymbol{\xi}_*$ to $\boldsymbol{\xi}', \boldsymbol{\xi}'_*$ is unity (see Problems 1.2.4 and 1.2.5); thus we can write $d\boldsymbol{\xi}\, d\boldsymbol{\xi}_*$ in place of $d\boldsymbol{\xi}'\, d\boldsymbol{\xi}'_*$ and Eq. (1.4.7) becomes

$$\int_{R^3} Q(f, g)\phi(\boldsymbol{\xi})d\boldsymbol{\xi} = \frac{1}{2}\int_{R^3}\int_{R^3}\int_{\mathcal{B}-} (fg_* + gf_* - f'g'_* - g'f'_*)\phi(\boldsymbol{\xi}')\mathcal{B}(\theta, V)d\boldsymbol{\xi}_* d\boldsymbol{\xi} d\theta d\epsilon. \tag{1.4.8}$$

Finally, we can interchange the starred and unstarred variables in Eq. (1.4.8) to find

$$\int_{R^3} Q(f, g)\phi(\xi)d\xi$$
$$= \frac{1}{2}\int_{R^3}\int_{R^3}\int_{\mathcal{B}-}(fg_* + gf_* - f'g'_* - g'f'_*)\phi(\xi'_*)\mathcal{B}(\theta, V)d\xi_* d\xi d\theta d\epsilon. \tag{1.4.9}$$

Equations (1.4.6), (1.4.8), and (1.4.9) differ from Eq. (1.4.5) because the factor $\phi(\xi)$ is replaced by $\phi(\xi_*)$, $-\phi(\xi')$, and $-\phi(\xi_*)$ respectively. We can now obtain more expressions for the integral in the left-hand side by taking linear combinations of the four different expressions available. Among them, the most interesting one is the symmetric expression obtained by taking the sum of Eqs. (1.4.5), (1.4.6), (1.4.8), and (1.4.9) and dividing by four. The result is

$$\int_{R^3} Q(f, g)\phi(\xi)d\xi = \frac{1}{8}\int_{R^3}\int_{R^3}\int_{\mathcal{B}-}(f'g'_* + g'f'_* - fg_* - gf_*)$$
$$\times(\phi + \phi_* - \phi' - \phi'_*)\mathcal{B}(\theta, V)d\xi_* d\xi d\theta d\epsilon. \tag{1.4.10}$$

This relation expresses a basic property of the collision term, which is frequently used. In particular, when $g = f$, Eq. (1.4.10) reads

$$\int_{R^3} Q(f, f)\phi(\xi)d\xi$$
$$= \frac{1}{4}\int_{R^3}\int_{R^3}\int_{\mathcal{B}-}(f'f'_* - ff_*)(\phi + \phi_* - \phi' - \phi'_*)\mathcal{B}(\theta, V)|d\xi_* d\xi d\theta d\epsilon. \tag{1.4.11}$$

We remark that the following form also holds:

$$\int_{R^3} Q(f,f)\phi(\xi)d\xi = \frac{1}{2}\int_{R^3}\int_{R^3}\int_{\mathcal{B}_-} ff_*(\phi' + \phi'_*\phi - \phi_*)\mathcal{B}(\theta, V)d\xi_* d\xi d\theta d\epsilon. \tag{1.4.12}$$

In fact, the integral in Eq. (1.4.11) can be split into the difference of two integrals (one containing $f'f'_*$; the other ff_*); the two integrals are just the opposite of each other, as an exchange between primed and unprimed variables shows, and Eq. (1.4.12) holds.

We now observe that the integral in Eq. (1.4.10) is zero independent of the particular functions f and g, if

$$\phi + \phi_* = \phi' + \phi'_* \tag{1.4.13}$$

is valid almost everywhere in velocity space. Because the integral appearing in the left-hand side of Eq. (1.4.11) is the rate of change of the average value of the function ϕ due to collisions, the functions satisfying Eq. (1.4.13) are called "collision invariants." It can be shown (see, e.g., Ref. 3 and Problems 1.4.1–1.4.6) that a continuous function ϕ has the property expressed by Eq. (1.4.13) if and only if

$$\phi(\boldsymbol{\xi}) = a + \mathbf{b}\cdot\xi + c|\boldsymbol{\xi}|^2, \tag{1.4.14}$$

where a and c are constant scalars and $\mathbf{b}$ a constant vector. The assumption of continuity can be considerably relaxed.[8–10] The functions $\psi_0 = 1$, $(\psi_1, \psi_2, \psi_3) = \boldsymbol{\xi}$, $\psi_4 = |\boldsymbol{\xi}|^2$ are usually called the elementary collision invariants; they span the five-dimensional subspace of the collision invariants.

Thus, in summary, a collision invariant is a function ϕ such that

$$\int_{R^3} \phi(\boldsymbol{\xi}) Q(f, g) d\boldsymbol{\xi} = 0, \tag{1.4.15}$$

and the most general expression of a collision invariant is given by Eq. (1.4.14)

Problems

1.4.1 Show that if $\mathbf{x}$ is a vector in an n-dimensional vector space E_n and $f(\mathbf{x})$ a function continuous in at least one point and satisfying $f(\mathbf{x}) + f(\mathbf{y}) = f(\mathbf{x}+\mathbf{y})$ for any $\mathbf{x}, \mathbf{y} \in E_n$, then $f(\mathbf{x}) = \mathbf{A}\cdot\mathbf{x}$, where $\mathbf{A}$ is a constant vector. (Hint: Show that f is actually continuous everywhere and satisfies $f(r\mathbf{x}) = rf(\mathbf{x})$ for any integer r; extend this property to any rational and then to any real r; then use a basis in E_n; see Refs. 3 and 6.)

1.4.2 Show that the even part of a function ϕ satisfying Eq. (1.4.13) is a function of $|\boldsymbol{\xi}|^2$ alone (Hint: $\phi + \phi_*$ is constant if and only if $\boldsymbol{\xi} + \boldsymbol{\xi}_*$ and $|\boldsymbol{\xi}|^2 + |\boldsymbol{\xi}_*|^2$ are constant and $\boldsymbol{\xi} + \boldsymbol{\xi}_*$ vanishes for $\boldsymbol{\xi}_* = -\boldsymbol{\xi}$).

1.4.3 Show that the even part of a continuous function satisfying Eq. (1.4.13) has the form $a + c|\boldsymbol{\xi}|^2$, where a and c are constants. (Hint: Let $a = \phi(0)$ and use the results of the two previous problems.)

1.4.4 Show that if $\boldsymbol{\xi}$ and $\boldsymbol{\xi}_*$ are orthogonal then the odd part of a collision invariant ϕ satisfies $\phi(\boldsymbol{\xi}) + \phi(\boldsymbol{\xi}_*) = \phi(\boldsymbol{\xi} + \boldsymbol{\xi}_*)$.

1.4.5 Extend the result of the previous problem to a pair of vectors $\boldsymbol{\xi}$ and $\boldsymbol{\xi}_*$, not necessarily orthogonal. (Hint: Consider another vector $\boldsymbol{\xi}_o$ orthogonal to both of them with magnitude $|\boldsymbol{\xi}_*\cdot\boldsymbol{\xi}|^{1/2}$ and consider the vectors $\boldsymbol{\xi} + \boldsymbol{\xi}_o$, $\boldsymbol{\xi}_* \pm \boldsymbol{\xi}_o$, to which the result of the previous problem applies.)

1.4.6 Apply the results of Problems 1.4.1 and 1.4.5 to show that the odd part of a collision invariant, if continuous in $\boldsymbol{\xi}$, must have the form $\mathbf{b}\cdot\boldsymbol{\xi}$ where

b is a constant vector, so that, because of the result of Problem 1.4.3 a collision invariant must have the form shown in Eq. (1.4.14).

1.5. The Boltzmann Inequality and the Maxwell Distributions

In this section we investigate the existence of positive functions f that give a vanishing collision integral:

$$Q(f, f) = \int_{R^3} \int_{\mathcal{B}^-} (f' f'_* - f f_*) \mathcal{B}(\theta, V) d\boldsymbol{\xi}_* d\theta d\epsilon = 0. \tag{1.5.1}$$

To solve this equation, we prove a preliminary result, which plays an important role in the theory of the Boltzmann equation: If f is a nonnegative function such that $\log f Q(f, f)$ is integrable and the manipulations of the previous section hold when $\phi = \log f$, then the *Boltzmann inequality*

$$\int_{R^3} \log f Q(f, f) d\boldsymbol{\xi} \leq 0 \tag{1.5.2}$$

holds; further, the equality sign applies if, and only if, $\log f$ is a collision invariant, or, equivalently,

$$f = \exp(a + \mathbf{b} \cdot \boldsymbol{\xi} + c|\boldsymbol{\xi}|^2). \tag{1.5.3}$$

To prove Eq. (1.5.2) it is enough to use Eq. (1.4.11) with $\phi = \log f$:

$$\int_{R^3} \log f Q(f, f) d\boldsymbol{\xi} = \frac{1}{4} \int_{R^3} \int_{\mathcal{B}^-} \log(f f_* / f' f'_*)(f' f'_* - f f_*) \mathcal{B}(\theta, V) d\boldsymbol{\xi} d\boldsymbol{\xi}_* d\epsilon, \tag{1.5.4}$$

and Eq. (1.5.2) follows thanks to the elementary inequality

$$(z - y) \log(y/z) \leq 0 \quad (y, z \in R^+). \tag{1.5.5}$$

Equation (1.5.5) becomes an equality if and only if $y = z$; thus the equality sign holds in Eq. (1.5.2) if and only if

$$f' f'_* = f f_* \tag{1.5.6}$$

applies almost everywhere. But, taking the logarithms of both sides of Eq. (1.5.6), we find that $\phi = \log f$ satisfies Eq. (1.4.13) and is thus given by Eq. (1.4.14). The function $f = \exp(\phi)$ is then given by Eq. (1.5.3).

We remark that in the latter equation c must be negative, since f must be integrable. If we let $c = -\beta$ and $\mathbf{b} = 2\beta\mathbf{v}$ (where $\mathbf{v}$ is another constant vector)

Eq. (1.5.3) can be rewritten as follows:

$$f = A\exp(-\beta|\boldsymbol{\xi} - \mathbf{v}|^2), \tag{1.5.7}$$

where A is a positive constant related to a, c, and $|\mathbf{b}|^2$ (β, $\mathbf{v}$, and A constitute a new set of constants). The function appearing in Eq. (1.2.7) is the *Maxwell distribution* or *Maxwellian.* Frequently one considers Maxwellians with $\mathbf{v} = 0$ (nondrifting Maxwellians), which can be obtained from drifting Mawellians by a change of the origin in velocity space.

Let us return now to the problem of solving Eq. (1.5.1). Multiplying both sides by $\log f$ and integrating gives Eq. (1.5.2) with the equality sign. This implies that f is a Maxwellian, by the result that has just been proved. Suppose now that f is a Maxwellian; then $f = \exp(\phi)$, where ϕ is a collision invariant and Eq. (1.5.6) holds; Eq. (1.5.1) then also holds. Thus there are functions that satisfy Eq. (1.5.1) and they are all Maxwellians, Eq. (1.5.7).

Problem

1.5.1 Prove (1.5.5).

1.6. The Macroscopic Balance Equations

In this section we compare the microscopic description supplied by kinetic theory with the macroscopic description supplied by continuum gas dynamics. For definiteness, in this section f will be assumed to be an expected mass density in phase space. To obtain a density, $\rho = \rho(x, t)$, in ordinary space, we must integrate f with respect to $\boldsymbol{\xi}$:

$$\rho = \int_{R^3} f\, d\boldsymbol{\xi}. \tag{1.6.1}$$

The bulk velocity $\mathbf{v}$ of the gas (e.g., the velocity of a wind) is the average of the molecular velocities $\boldsymbol{\xi}$ at a certain point $\mathbf{x}$ and time instant t; since f is proportional to the probability for a molecule to have a given velocity, $\mathbf{v}$ is given by

$$\mathbf{v} = \frac{\int_{R^3} \boldsymbol{\xi} f\, d\boldsymbol{\xi}}{\int_{R^3} f\, d\boldsymbol{\xi}} \tag{1.6.2}$$

(the denominator is required even if f is taken to be a probability density in phase space, because we are considering a conditional probability, referring to the position $\mathbf{x}$). Equation (1.6.2) can also be written as follows:

$$\rho \mathbf{v} = \int_{R^3} \boldsymbol{\xi} f \, d\boldsymbol{\xi}, \tag{1.6.3}$$

or, using components,

$$\rho v_i = \int_{R^3} \xi_i f \, d\boldsymbol{\xi} \qquad (i = 1, 2, 3). \tag{1.6.4}$$

The bulk velocity $\mathbf{v}$ is what we can directly perceive of the molecular motion by means of macroscopic observations; it is zero for a gas in equilibrium in a box at rest. Each molecule has its own velocity $\boldsymbol{\xi}$, which can be decomposed into the sum of $\mathbf{v}$ and another velocity

$$\mathbf{c} = \boldsymbol{\xi} - \mathbf{v} \tag{1.6.5}$$

called the random or peculiar velocity; $\mathbf{c}$ is clearly due to the deviations of $\boldsymbol{\xi}$ from $\mathbf{v}$. It is also clear that the average of $\mathbf{c}$ is zero (Problem 1.6.1).

The quantity ρv_i that appears in Eq. (1.6.4) is the ith component of the mass flow or, alternatively, of the momentum density of the gas. Other quantities of similar nature are: the momentum flow

$$m_{ij} = \int_{R^3} \xi_i \xi_j f \, d\boldsymbol{\xi} \qquad (i, j = 1, 2, 3); \tag{1.6.6}$$

the energy density per unit volume:

$$w = \frac{1}{2} \int_{R^3} |\boldsymbol{\xi}|^2 f \, d\boldsymbol{\xi}; \tag{1.6.7}$$

and the energy flow:

$$r_i = \frac{1}{2} \int_{R^3} \xi_i |\boldsymbol{\xi}|^2 f \, d\boldsymbol{\xi} \qquad (i, j = 1, 2, 3). \tag{1.6.8}$$

Equation (1.6.8) shows that the momentum flow is described by the components of a symmetric tensor of second order, because we must describe the flow in the ith direction of the jth component of momentum. It is to be expected that in a macroscopic description only a part of this tensor will be identified as a bulk momentum flow, because, in general, m_{ij} will be different from zero even in the absence of a macroscopic motion ($\mathbf{v} = 0$). It is thus convenient to reexpress the integral in m_{ij} in terms of $\mathbf{c}$ and $\mathbf{v}$. Then we have (Problem 1.6.2)

$$m_{ij} = \rho v_i v_j + p_{ij}, \tag{1.6.9}$$

where

$$p_{ij} = \int_{R^3} c_i c_j f d\boldsymbol{\xi} \qquad (i, j = 1, 2, 3) \tag{1.6.10}$$

plays the role of the stress tensor (because the microscopic momentum flow associated with it is equivalent to forces distributed on the boundary of any region of gas, according to the macroscopic description).

Similarly one has

$$w = \frac{1}{2}\rho\,|\mathbf{v}|^2 + \rho e, \tag{1.6.11}$$

where e is the internal energy per unit mass (associated with random motions) defined by

$$\rho e = \frac{1}{2}\int_{R^3} |\mathbf{c}|^2 f d\boldsymbol{\xi}, \tag{1.6.12}$$

and (Problem 1.6.3)

$$r_i = \rho v_i \left(\frac{1}{2}|\mathbf{v}|^2 + e\right) + \sum_{j=1}^{3} v_j p_{ij} + q_i \qquad (i = 1, 2, 3)\,, \tag{1.6.13}$$

where q_i are the components of the so-called heat flow vector:

$$q_i = \frac{1}{2}\int_3 c_i |\mathbf{c}|^2 f d\boldsymbol{\xi}. \tag{1.6.14}$$

The decomposition in Eq. (1.6.13) shows that the microscopic energy flow is a sum of a macroscopic flow of energy (both kinetic and internal), of the work (per unit area und unit time) done by stresses, and of the heat flow.

To complete the connection, as a simple mathematical consequence of the Boltzmann equation, one can derive five differential relations satisfied by the macroscopic quantities introduced above; these relations describe the balance of mass, momentum, and energy and have the same form as in continuum mechanics. To this end let us consider the Boltzmann equation

$$\frac{\partial f}{\partial t} + \boldsymbol{\xi}\cdot\frac{\partial f}{\partial \mathbf{x}} = Q(f, f). \tag{1.6.15}$$

If we multiply both sides by one of the elementary collision invariants ψ_α ($\alpha = 0, 1, 2, 3, 4$), defined in Section 1.4, and integrate with respect to $\boldsymbol{\xi}$, we have, thanks to Eq. (1.1.15) with $g = f$ and $\phi = \psi_\alpha$:

$$\int_{R^3} \psi_\alpha(\boldsymbol{\xi}) Q(f, f) d\boldsymbol{\xi} = 0, \tag{1.6.16}$$

and hence, if it is permitted to change the order by which we differentiate with respect to t and integrate with respect to $\boldsymbol{\xi}$:

$$\frac{\partial}{\partial t}\int \psi_\alpha f d\boldsymbol{\xi} + \sum_{i=1}^{3} \frac{\partial}{\partial x_i}\int \xi_i \psi_\alpha f d\boldsymbol{\xi} = 0 \qquad (\alpha = 1, 2, 3, 4). \tag{1.6.17}$$

If we take successively $\alpha = 0, 1, 2, 3, 4$ and use the definitions introduced above, we obtain

$$\frac{\partial \rho}{\partial t} + \sum_{i=1}^{3} \frac{\partial}{\partial x_i}(\rho v_i) = 0, \tag{1.6.18}$$

$$\frac{\partial}{\partial t}(\rho v_j) + \sum_{i=1}^{3} \frac{\partial}{\partial x_i}(\rho v_i v_j + p_{ij}) = 0 \qquad (j = 1, 2, 3), \tag{1.6.19}$$

$$\frac{\partial}{\partial t}\left(\frac{1}{2}\rho|\mathbf{v}|^2 + \rho e\right) + \sum_{i=1}^{3} \frac{\partial}{\partial x_i}\left[\rho v_i \left(\frac{1}{2}|\mathbf{v}|^2 + e\right) + \sum_{j=1}^{3} v_j p_{ij} + q_i\right] = 0. \tag{1.6.20}$$

These equations have the so-called conservation form because they express the circumstance that a certain quantity (whose density appears differentiated with respect to time) is created or destroyed in a certain region Ω because something is flowing through the boundary $\partial\Omega$. In fact, when integrating both sides of the equations with respect to $\mathbf{x}$ over Ω, the terms differentiated with respect to the space coordinates can be replaced by surface integrals over $\partial\Omega$, thanks to the divergence theorem. If these surface integrals turn out to be zero then we obtain that the total mass

$$M = \int_\Omega \rho d\mathbf{x}, \tag{1.6.21}$$

the total momentum

$$\mathbf{Q} = \int_\Omega \rho \mathbf{v} d\mathbf{x}, \tag{1.6.22}$$

and the total energy

$$E = \int_\Omega \left(\frac{1}{2}\rho|\mathbf{v}|^2 + \rho e\right) d\mathbf{x} \tag{1.6.23}$$

are conserved in Ω. Tipical cases when this occurs are:

a) Ω is R^3 and suitable conditions at infinity ensure that the fluxes of the mass, momentum, and energy flow vectors through a large sphere vanish when the radius of the sphere tends to infinity;
b) Ω is a box with periodicity conditions, because essentially there are no boundaries.

When Ω is a compact domain with the condition of specular reflection on Ω then the boundary terms on $\partial\Omega$ disappear in the mass and energy equations but not in the momentum equation; thus only M and E are conserved.

We also remark that in the so-called space-homogeneous case, the various quantities do not depend on $\mathbf{x}$; all the space derivatives then disappear from Eqs. (1.6.18–1.6.20) and the densities ρ, $\rho\mathbf{v}$, and $\frac{1}{2}\rho \mid \mathbf{v} \mid^2 + \rho e$ are conserved, that is, thay do not change with time.

The considerations of this section apply to all the solutions of the Boltzmann equation. The definitions, however, can be applied to any positive function for which they make sense. In particular if we take f to be a Maxwellian in the form (1.5.7), we find that the constant vector $\mathbf{v}$ appearing there is actually the bulk velocity as defined in Eq. (1.6.2) while β and A are related to the internal energy e and the density ρ in the following way:

$$\beta = 3/(4e), \qquad A = \rho(4\pi e/3)^{-3/2}. \tag{1.6.24}$$

Furthermore, the stress tensor turns out to be diagonal ($p_{ij} = (\frac{2}{3}\rho e)\delta_{ij}$, where δ_{ij} is the *Kronecker delta* ($=1$ if $i = j$; $= 0$ if $i \neq j$)), while the heat flow vector is zero.

We end this section with the definition of pressure p in terms of f; p is nothing else than 1/3 of the spur or trace (i.e., the sum of the three diagonal terms) of p_{ij} and is thus given by

$$p = \frac{1}{3}\int_{R^3} |\mathbf{c}|^2 f d\boldsymbol{\xi}. \tag{1.6.25}$$

If we compare this with the definition of the specific internal energy e, given in Eq. (1.3.12), we obtain the relation

$$p = \frac{2}{3}\rho e. \tag{1.6.26}$$

This relation also suggests the definition of temperature, according to kinetic theory, $T = (\frac{2}{3}e)/R$, where R is the gas constant equal to the universal Boltzmann constant k divided by the molecular mass m. Thus

$$T = \frac{1}{3\rho R}\int_{R^3} |\mathbf{c}|^2 f d\boldsymbol{\xi}. \tag{1.6.27}$$

This definition is appropriate, because if we mix two different gases and let them achieve a state of equilibrium, T turns out to be the same for the two gases (see Ref. 3). Thus, if we take into account Eq. (1.6.24), a Maxwellian distribution (see Eq. (1.5.7)) can be written as follows:

$$f = \rho(2\pi RT)^{-3/2} \exp[-|\boldsymbol{\xi} - \mathbf{v}|^2/(2RT)]. \tag{1.6.28}$$

Problems

1.6.1 Prove that $\int_{R^3} \mathbf{c} f \, d\xi = \mathbf{0}$, where $\mathbf{c}$ is the random velocity given by Eq. (1.6.5).

1.6.2 Prove Eq. (1.6.9).

1.6.3 Prove Eq. (1.6.11).

1.6.4 Prove Eq. (1.6.12).

1.6.5 Check Eqs. (1.6.18)–(1.6.20).

1.6.6 Check that the flows of mass and energy vanish at a boundary where the molecules are specularly reflected.

1.6.7 Prove that Eqs. (1.6.24) hold for a Maxwellian.

1.6.8 Prove that the heat flow vector vanishes and the stress is diagonal, if f is a Maxwellian.

1.7. The *H*-Theorem

Let us consider a further application of the properties of the collision term $Q(f, f)$ of the Boltzmann equation:

$$\frac{\partial f}{\partial t} + \boldsymbol{\xi} \cdot \frac{\partial f}{\partial \mathbf{x}} = Q(f, f). \tag{1.7.1}$$

If we multiply both sides of this equation by $\log f$ and integrate with respect to $\boldsymbol{\xi}$, we obtain

$$\frac{\partial \mathcal{H}}{\partial t} + \frac{\partial}{\partial \mathbf{x}} \cdot \mathbf{J} = \mathcal{S}, \tag{1.7.2}$$

where

$$\mathcal{H} = \int_{R^3} f \log f \, d\boldsymbol{\xi}, \tag{1.7.3}$$

$$\mathbf{J} = \int_{R^3} \boldsymbol{\xi} f \log f \, d\boldsymbol{\xi}, \tag{1.7.4}$$

and

$$\mathcal{S} = \int_{R^3} \log f Q(f, f) d\boldsymbol{\xi}. \tag{1.7.5}$$

Equation (1.7.2) differs from the balance equations considered in the previous section because the right side, generally speaking, does not vanish. We know, however, that the Boltzmann inequality, Eq. (1.5.2), implies

$$\mathcal{S} \leq 0 \quad \text{and} \quad \mathcal{S} = 0 \quad \text{if and only if} \quad f \text{ is a Maxwellian.} \tag{1.7.6}$$

Because of this inequality, Eq. (1.7.2) plays an important role in the theory of the Boltzmann equation. We illustrate the role of Eq. (1.7.6) in the case of space-homogeneous solutions. In this case the various quantities do not depend on $\mathbf{x}$ and Eq. (1.7.2) reduces to

$$\frac{\partial \mathcal{H}}{\partial t} = \mathcal{S} \leq 0. \tag{1.7.7}$$

This implies the so-called H-theorem (for the space-homogeneous case): $\mathcal{H}$ is a decreasing quantity, unless f is a Maxwellian (in which case the time derivative of $\mathcal{H}$ is zero). Remember now that in this case the densities ρ, $\rho\mathbf{v}$, and ρe are constant in time; we can thus build a Maxwellian M that has, at any time, the same ρ, $\mathbf{v}$, and e as any solution f corresponding to given initial data. Since $\mathcal{H}$ decreases unless f is a Maxwellian (i.e., $f = M$), it is tempting to conclude that f tends to M when $t \to \infty$. The temptation is strengthened when we realize that $\mathcal{H}$ is bounded from below by $\mathcal{H}_M$ (Problem 1.7.1), the value taken by the functional $\mathcal{H}$ when $f = M$. In fact $\mathcal{H}$ is decreasing; its derivative is nonpositive unless it takes the value $\mathcal{H}_M$; one feels that $\mathcal{H}$ tends to $\mathcal{H}_M$! This conclusion is, however, unwarranted from a purely mathematical viewpoint, without a more detailed consideration of the source term $\mathcal{S}$ in Eq. (1.7.7), for which Ref. 8 should be consulted. If the state of the gas is not space homogeneous, the situation becomes more complicated. In this case it is convenient to introduce the quantity

$$H = \int_{\Omega} \mathcal{H} d\mathbf{x}, \tag{1.7.8}$$

where Ω is the space domain occupied by the gas (assumed here to be time independent). Then Eq. (1.7.2) implies

$$\frac{dH}{dt} \leq \int_{\partial\Omega} \mathbf{J} \cdot \mathbf{n} d\sigma, \tag{1.7.9}$$

where $\mathbf{n}$ is the inward normal and $d\sigma$ the surface element on $\partial\Omega$. Clearly, several situations may arise. Among the most typical ones, we quote:

1. Ω is a box with periodicity boundary conditions (flat torus). Then there is no boundary, $dH/dt \leq 0$, and one can repeat about H what was said about $\mathcal{H}$ in the space-homogeneous case. In particular, there is a natural (space-homogeneous) Maxwellian associated with the total mass, momentum, and energy (which are conserved as was remarked in the previous section).
2. Ω is a bounded box with specular reflection. In this case the boundary term also disappears because the integrand of $\mathbf{J} \cdot \mathbf{n}$ is odd on $\partial\Omega$ and the situation is similar to that in case 1. There might seem to be a difficulty for the choice of the natural Maxwellian because momentum is not conserved, but a simple argument shows that the total momentum must vanish when $t \to \infty$. Thus M is a nondrifting Maxwellian.
3. Ω is the entire space. Then the asymptotic behavior of the initial values at ∞ is of paramount importance. If the gas is initially more concentrated at finite distances from the origin, one physically expects that the gas escapes through infinity and the asymptotic state is a vacuum.
4. Ω is a compact domain but the boundary conditions on $\partial\Omega$ are different from specular reflection. Then the asymptotic state may be completely different from a Maxwellian because the gas may be forced to remain in a nonequilibrium state.

Boltzmann's H-theorem is of basic importance because it shows that his equation has a basic feature of irreversibility: The quantities $\mathcal{H}$ (in the space-homogeneous case) and H (in other cases where the gas does not exchange mass and energy with a solid boundary) always decrease in time. This result seems to be in conflict with the fact that the molecules constituting the gas follow the laws of classical mechanics, which are time reversible, but there is a way out of this paradox (see Refs. 1–7). It should be clear that H has the properties of entropy (except for the sign); this identification is strengthened when we evaluate H in an equilibrium state (see Problem 1.7.2) because it turns out to coincide with the expression of a perfect gas according to equilibrium thermodynamics, apart from a factor $-R$. A further check of this identification is given by an inequality satisfied by the right-hand side of Eq. (1.7.9) when the gas is able to exchange energy with a solid wall bounding Ω (see Section 1.11).

Problems

1.7.1 Show that if $\mathcal{H}(f)$ is the functional defined in Eq. (1.7.3) and M is the Maxwellian with the same density, velocity, and internal energy as f, then $\mathcal{H}(f) \geq \mathcal{H}(M)$. (Hint: Use the inequality $z \log z - z \log y + y - z \geq 0$, valid for nonnegative y and z, and the fact that $\log M$ is a collision invariant so that $\int_{R^3} \log M(f - M) d\xi = 0$.)

1.7.2 Evaluate H when f is a Maxwellian distribution and compare the result with the entropy η of the gas according to equilibrium thermodynamics, to conclude that $H = -\eta/R$, where R is the gas constant.

1.8. Equilibrium States and Maxwellian Distributions

The trend toward a Maxwellian distribution expressed by the H-theorem indicates that this particular distribution is a good candidate to describe a gas in a (statistical) equilibrium state. In order to prove that Maxwellians describe the equilibrium states of a gas, however, we must give a definition of equilibrium. Intuitively, a gas is in equilibrium if, in a situation where it does not exchange mass and energy with other bodies, its state does not change with time. Thus for the moment we define an equilibrium state to be one of a gas in a steady situation in a box with periodicity or specular reflection boundary conditions. It is then clear that the distribution function must be a Maxwellian; in fact, Eq. (1.7.2) implies (when $\mathcal{H}$ does not depend on time)

$$-\int_{\partial\Omega} \mathbf{J} \cdot \mathbf{n}\, d\sigma = \int_{\Omega} \mathcal{S} d\mathbf{x} \leq 0, \tag{1.8.1}$$

where $\mathbf{n}$ is the inward normal and equality holds if and only if f is Maxwellian. But $\mathbf{J} \cdot \mathbf{n}$ is zero for the situation under consideration and the only possibility is that f be a Maxwellian. We impose now the condition that this Maxwellian must be a steady solution of the Boltzmann equation, that is, it must satisfy

$$\boldsymbol{\xi} \cdot \frac{\partial f}{\partial \mathbf{x}} = Q(f, f). \tag{1.8.2}$$

This readily implies that both the right- and the left-hand sides of the Boltzmann equation must vanish; as a consequence, the parameters A, β, and $\mathbf{v}$ appearing in the Maxwellian

$$f = A \exp(-\beta|\boldsymbol{\xi} - \mathbf{v}|^2) \tag{1.8.3}$$

must be of the form $\mathbf{v} = \mathbf{v}_0 + \boldsymbol{\omega} \wedge \mathbf{x}$, $A = A_0 \exp[|\boldsymbol{\omega}|^2\, |\mathbf{x}|^2 - (\boldsymbol{\omega} \cdot \mathbf{x})^2]$, and $\beta =$ constant (where $\mathbf{v}_0$ and $\boldsymbol{\omega}$ are constant vectors, A_0 is a constant scalar, and $\wedge$ denotes the vector product; see Problem 1.8.1). If the periodicity condition in a box or the specular reflection boundary condition (in a general domain not rotationally invariant about the direction of $\boldsymbol{\omega}$) is imposed, it turns out that also A and $\mathbf{v}$ must be constant (and not space dependent). Thus, in general, a Maxwellian with constant parameters is the most general equilibrium solution of the Boltzmann equation.

The question immediately arises of whether there are solutions of the Boltzmann equation that are Maxwellians with parameters depending on $\mathbf{x}$ and t. Since the right-hand side of the Boltzmann equation vanishes identically if f is a Maxwellian, it turns out that a Maxwellian (i.e., a function of the form specified in Eq. (1.8.3)) can be a solution of the Boltzmann equation if, and only if, A, β, and $\mathbf{v}$ depend on t and $\mathbf{x}$ in such a way that f also satisfies

$$\frac{\partial f}{\partial t} + \boldsymbol{\xi} \cdot \frac{\partial f}{\partial \mathbf{x}} = 0. \tag{1.8.4}$$

Because the general solution of this equation has the form $f = f(\mathbf{x} - \boldsymbol{\xi}t, \mathbf{x} \wedge \boldsymbol{\xi}, \boldsymbol{\xi})$, there are several solutions of this form; they were investigated by Boltzmann[11] in 1876. Among them we quote the case met above in which $\mathbf{v} = \mathbf{v}_0 + \boldsymbol{\omega} \wedge \mathbf{x}$, $A = A_0 \exp[|\boldsymbol{\omega}|^2 |\mathbf{x}|^2 - (\boldsymbol{\omega} \cdot \mathbf{x})^2]$, and $\beta =$ constant (with $\mathbf{v}_0$, A_0, and $\boldsymbol{\omega}$ constants) and the case in which $A =$ constant, $\beta = \beta_0(1 + t/t_0)^2$, and $\mathbf{v} = (t + t_0)^{-1}\mathbf{x}$. The latter solution describes a compression if t_0 is negative (but the solution ceases to exist for $t > |t_0|$) and an expansion into a vacuum if $t_0 > 0$ (in which case the solution exists for any positive time). These Maxwellians are particular cases of the homoenergetic affine solutions discussed in the next chapter (Section 2.2).

Problems

1.8.1 Show that if a Maxwellian M with parameters depending upon position satisfies $\boldsymbol{\xi} \cdot \partial_{\mathbf{x}} M = 0$, then it must be given by (1.8.3) with $\mathbf{v} = \mathbf{v}_0 + \boldsymbol{\omega} \wedge \mathbf{x}$, $A = A_0 \exp[|\boldsymbol{\omega}|^2 |\mathbf{x}|^2 - (\boldsymbol{\omega} \cdot \mathbf{x})^2]$, and $\beta =$ constant (where $\mathbf{v}_0$ and $\boldsymbol{\omega}$ are constant vectors, and A_0 is a constant scalar). (Hint: $\log M$ is a linear combination of the collision invariants and satisfies the same equation as M. This immediately leads to a system of twelve partial differential equations for the coefficients a, $\mathbf{b}$, c of the polynomial. Six of these are trivial and give that a and c are constant; the remaining six by a suitable manipulation imply that $\mathbf{b}$ is linear in $\mathbf{x}$ and, by substitution...(See Ref. 3.)

1.8.2 Prove the statement that in Eq. (1.8.3) the parameters must be constant in an equilibrium state.

1.8.3 Check that the general solution of Eq. (1.8.4) is of the form $f = f(\mathbf{x} - \boldsymbol{\xi}t, \mathbf{x} \wedge \boldsymbol{\xi}, \boldsymbol{\xi})$.

1.8.4 Find all the Maxwellians that are solutions of the Boltzmann equation (see Refs. 3, 11, and 12).

1.9. Model Equations

When trying to solve the Boltzmann equation for practical problems, one of the major shortcomings is the complicated structure of the collision term, Eq. (1.4.2). When one is not interested in fine details, it is possible to obtain reasonable results by replacing the collision integral by a so-called collision model, a simpler expression $J(f)$ that retains only the qualitative and average properties of the collision term $Q(f, f)$. The equation for the distribution function is then called a kinetic model or a model equation.

The most widely known collision model is usually called the Bhatnagar, Gross, and Krook (BGK) model, although Welander proposed it independently at about the same time as the above mentioned authors.[13,14] It reads as follows:

$$J(f) = \nu[\Phi(\boldsymbol{\xi}) - f(\boldsymbol{\xi})], \tag{1.9.1}$$

where the collision frequency ν is independent of $\boldsymbol{\xi}$ (but depends on the density ρ and the temperature T) and Φ denotes the local Maxwellian, that is, the (unique) Maxwellian having the same density, bulk velocity, and temperature as f:

$$f = \rho(2\pi RT)^{-3/2} \exp[-|\boldsymbol{\xi} - \mathbf{v}|^2/(2RT)]. \tag{1.9.2}$$

Here ρ, $\mathbf{v}$, and T are chosen is such a way that for any collision invariant ψ we have

$$\int_{R^3} \psi(\boldsymbol{\xi})\Phi(\boldsymbol{\xi})d\boldsymbol{\xi} = \int_{R^3} \psi(\boldsymbol{\xi}) f(\boldsymbol{\xi})d\boldsymbol{\xi}. \tag{1.9.3}$$

It is easily checked that, thanks to Eq. (1.9.3):

a) f and Φ have the same density, bulk velocity, and temperature (Problem 1.9.1);
b) $J(f)$ satisfies conservation of mass, momentum, and energy; that is, for any collision invariant:

$$\int_{R^3} \psi(\boldsymbol{\xi})J(f)d\boldsymbol{\xi} = 0; \tag{1.9.4}$$

c) $J(f)$ satisfies the Boltzmann inequality

$$\int_{R^3} \log f J(f)d\boldsymbol{\xi} \leq 0, \tag{1.9.5}$$

where the equality sign holds if and only if f is a Maxwellian (Problem 1.9.3).

It should be remarked that the nonlinearity of the BGK collision model, Eq. (1.9.1), is much worse than the nonlinearity in $Q(f, f)$; in fact the latter is simply quadratic in f, while the former contains f in both the numerator and denominator of an exponential, because $\mathbf{v}$ and T are functionals of f, defined by Eqs. (1.6.2) and (1.6.27).

The main advantage in the use of the BGK model is that for any given problem one can deduce integral equations for ρ, $\mathbf{v}$, and T that can be solved with moderate effort on a computer. Another advantage of the BGK model is offered by its linearized form, as will become clear in the applications to be discussed in this book (see Chapters 3 and 4).

The BGK model has the same basic properties as the Boltzmann collision integral, but it has some shortcomings. Some of them can be avoided by suitable modifications, at the expense, however, of the simplicity of the model. A first modification can be introduced to allow the collision frequency ν to depend on the molecular velocity, more precisely on the magnitude of the random velocity $\mathbf{c}$ (defined by Eq. (1.6.5)), while requiring that Eq. (1.9.4) still holds. All the basic properties, including Eq. (1.9.5), are retained, but the density, velocity, and temperature appearing in Φ are not the local ones of the gas but some fictitious local parameters related to five functionals of f different from ρ, $\mathbf{v}$, and T; this follows from the fact that Eq. (1.9.3) must now be replaced by

$$\int_{R^3} \nu(|\mathbf{c}|)\psi(\boldsymbol{\xi})\Phi(\boldsymbol{\xi})d\boldsymbol{\xi} = \int_{R^3} \nu(|\mathbf{c}|)\psi(\boldsymbol{\xi})f(\boldsymbol{\xi})d\boldsymbol{\xi}. \tag{1.9.6}$$

A different kind of correction to the BGK model is obtained when a complete agreement with the compressible Navier–Stokes equations is required for large values of the collision frequency. In fact the BGK model has only one parameter (at a fixed space point and time instant): the collision frequency ν; the latter can be adjusted to give a correct value for either the viscosity μ or the heat conductivity κ, but not for both. This is shown by the fact that the Prandtl number $\mathrm{Pr} = \mu/c_p\kappa$ (where c_p is the specific heat at constant pressure) turns out[3,6] to be unity for the BGK model, whereas it is about 2/3 for a monatomic gas (according to both experimental data and the Boltzmann equation). In order to have a correct value for the Prandtl number, one is led[15,16] to replacing the local Maxwellian in Eq. (1.9.1) by

$$\Phi(\boldsymbol{\xi}) = \rho(\pi)^{-3/2}(\det \mathsf{A})^{1/2} \exp\left(-(\boldsymbol{\xi} - \mathbf{v}) \cdot [\mathsf{A}(\boldsymbol{\xi} - \mathbf{v})]\right), \tag{1.9.7}$$

where A is the inverse of the matrix

$$\mathsf{A}^{-1} = (2RT/\mathrm{Pr})\mathsf{I} - 2(1 - \mathrm{Pr})\mathsf{p}/(\rho\,\mathrm{Pr}), \tag{1.9.8}$$

where I is the identity and p the stress matrix. If we let $\Pr = 1$, we recover the BGK model.

Only recently[17] has this model (called ellipsoidal statistical (ES) model) been shown to possess the property expressed by Eq. (1.9.5) (Problems 1.9.3 and 1.9.4). Hence the H-theorem holds for the ES model.

Other models with different choices of Φ have been proposed[18,6] but they are not so interesting, except for linearized problems (see Chapters 3 and 4).

Another model is the integro-differential model proposed by Lebowitz, Frisch, and Helfand,[19] which is similar to the Fokker–Planck equation used in the theory of Brownian motion. This model reads as follows:

$$J(f) = D\sum_{k=1}^{3}\left[\frac{\partial^2 f}{\partial \xi_k^2} + \frac{1}{RT}\frac{\partial}{\partial \xi_k}[(\xi_k - v_k)f]\right], \tag{1.9.9}$$

where D is a function of the local density ρ and the local temperature T. If we take D proportional to the pressure $p = \rho RT$, Eq. (1.9.9) has the same kind of nonlinearity (i.e., quadratic) as the true Boltzmann equation.

The idea of kinetic models can be naturally extended to mixtures and polyatomic gases[18,20,21] as we shall see in Chapter 6.

Problems

1.9.1 Prove that if Eq. (1.9.3) holds for any collision invariant ψ, then the local Maxwellian (1.9.2) has the density, bulk velocity, and temperature of the distribution function f.

1.9.2 Prove Eq. (1.9.4).

1.9.3 Prove the Boltzmann inequality for the BGK model, Eq. (1.9.5). (Hint: Use Eq. (1.5.5) with $z = \Phi$ and $y = f$.)

1.9.4 Let B, with elements B_{ik}, be the symmetric matrix A^{-1} defined in Eq. (1.9.8). Prove that the minimum of the H-function ($H(g) = \int g \log g\, d\boldsymbol{\xi}$) on the (nonempty) set $\mathcal{X} = \{g \geq 0, (1 + |v|^2)g$ integrable, $\int g d\boldsymbol{\xi} = \rho, \int \boldsymbol{\xi} g d\boldsymbol{\xi} = \rho v, \int \xi_i \xi_k g d\boldsymbol{\xi} = \rho v_i v_k + \rho B_{ik}\}$ is reached when g equals the anisotropic Gaussian Φ given by (1.9.7), provided that $\Pr \geq 2/3$. (Hint: Diagonalize p with elements p_i; then B is also diagonal and the diagonal elements are, letting $\Pr = (1-\nu)^{-1}$ $(-1/2 \leq \nu \leq 1)$:$[(1+2\nu)p_1 + (1-\nu)p_2 + (1-\nu)p_3]/3$, $[(1+2\nu)p_2 + (1-\nu)p_1 + (1-\nu)p_3]/3$ $[(1+2\nu)p_1 + (1-\nu)p_1 + (1-\nu)p_2]/3$. These numbers are obviously positive when $0 \leq \nu \leq 1$. For negative values of ν, the worst case occurs when two eigenvalues p_i (say with $i = 1, 2$) vanish, yielding for the other eigenvalue of B, $(1+2\nu)p_3/3$. Then B is positive definite. Then we can use the convexity of the H function and the method of Lagrange multipliers without any problem. See Ref. 17.)

1.9.5 Prove that $H(\Phi) = \int \Phi \log \Phi \, d\boldsymbol{\xi}$, where Φ is the anisotropic Gaussian given by (1.9.7), is not larger than the H-functional evaluated at the local Maxwellian, that is, the same Φ with Pr = 1, provided that Pr $\geq$ 2/3. (Hint: Compute the difference between the two H-functions and show that it equals $(\rho^2)\log[\det(\mathrm{p})\det(\rho\mathrm{A})]$. In order to prove that the second determinant is larger then the first, diagonalize p; then the two determinants are $p_1 p_2 p_3$ and, letting Pr $= (1-\nu)^{-1}$ $(-1/2 \leq \nu \leq 1)$, $[(1+2\nu)p_1 + (1-\nu)p_2 + (1-\nu)p_3][(1+2\nu)p_2 + (1-\nu)p_1 + (1-\nu)p_3][(1+2\nu)p_1 + (1-\nu)p_1 + (1-\nu)p_2]/9$. The latter expression can be shown, either directly or by convexity arguments, to have a maximum at $\nu = 1$. See Ref. 17.)

1.10. The Linearized Collision Operator

On several occasions we shall meet the so-called linearized collision operator, related to the bilinear operator defined in Eq. (1.4.3) by

$$Lh = 2M^{-1}Q(Mh, M), \tag{1.10.1}$$

where M is a Maxwellian distribution, usually with zero bulk velocity. When we want to emphasize the fact that we linearize with respect to a given Maxwellian, we write L_M instead of just L.

A more explicit expression of Lh reads as follows:

$$Lh = \int_{\Re^3}\int_{\mathcal{B}_+} M_*(h' + h'_* - h_* - h)\mathcal{B}(\theta, V)d\boldsymbol{\xi}_* d\mathbf{n}, \tag{1.10.2}$$

where we have taken into account that $M'M'_* = MM_*$. Because of Eq. (1.4.10) (with Mh in place of f, M in place of g, and g in place of ϕ), we have the identity

$$\int_{\Re^3} MgLhd\boldsymbol{\xi} = -\frac{1}{4}\int_{\Re^3}\int_{\Re^3}\int_{\mathcal{B}_+}(h' + h'_* - h - h_*)$$
$$\times(g' + g'_* - g - g_*)\mathcal{B}(\theta, V)d\boldsymbol{\xi}_* d\boldsymbol{\xi} d\mathbf{n}. \tag{1.10.3}$$

This relation expresses a basic property of the linearized collision term. In order to make it clear, let us introduce a bilinear expression, the scalar product in the

Hilbert space of square summable functions of $\boldsymbol{\xi}$ endowed with a scalar product weighted with M:

$$(g, h) = \int_{\Re^3} \bar{g} h M d\boldsymbol{\xi}, \tag{1.10.4}$$

where the bar denotes complex conjugation. Then Eq. (1.1.7) (with $\bar{g}$ in place of g) gives (thanks to the symmetry of the expression in the right-hand side of Eq. (1.10.3) with respect to the interchange $g \Leftrightarrow h$)

$$(g, Lh) = (Lg, h). \tag{1.10.5}$$

Further,

$$(h, Lh) \leq 0 \tag{1.10.6}$$

and the equality sign holds if and only if

$$h' + h'_* - h - h_* = 0 \tag{1.10.7}$$

that is, if and only if h is a collision invariant.

Equations (1.10.5) and (1.10.6) indicate that the operator L is symmetric and nonpositive in the aforementioned Hilbert space.

Problems

1.10.1 Prove that (1.10.2) holds. (See Refs. 3 and 6.)

1.10.2 Prove (1.10.3). (See Refs. 3 and 6.)

1.10.3 Prove (1.10.5) and (1.10.6). (See Refs. 3 and 6.)

1.10.4 Prove that the equality sign holds in (1.10.6) if and only if (1.10.7) holds (almost everywhere). (See Refs. 3 and 6.)

1.11. Boundary Conditions

The Boltzmann equation must be accompanied by boundary conditions, which describe the interaction of the gas molecules with the solid walls. It is to this interaction that one can trace the origin of the drag and lift exerted by the gas on the body and the heat transfer between the gas and the solid boundary.

The study of gas–surface interaction may be regarded as a bridge between the kinetic theory of gases and solid state physics and is an area of research by itself. The difficulties of a theoretical investigation are due, mainly, to our lack of knowledge of the structure of surface layers of solid bodies and hence

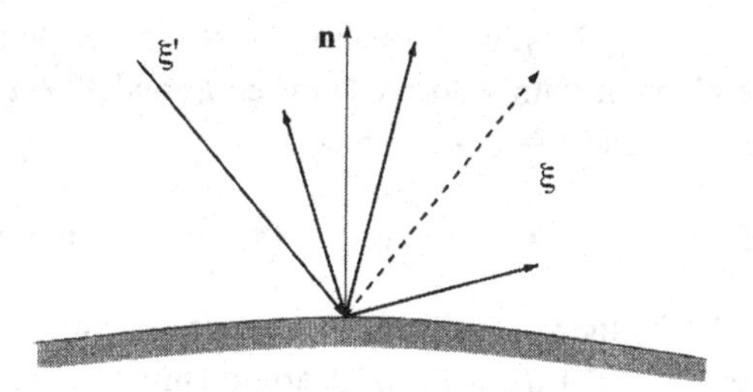

Figure 1.3. The velocity $\boldsymbol{\xi}$ of a reemerging molecule is not uniquely determined by the velocity possessed by the same molecule before hitting the wall, unless specular reflection applies (dashed line).

of the effective interaction potential of the gas molecules with the wall. When a molecule impinges upon a surface, it is adsorbed and may form chemical bonds, dissociate, become ionized, or displace surface molecules. Its interaction with the solid surface depends on the surface finish, the cleanliness of the surface, its temperature, etc. It may also vary with time because of outgassing from the surface. Preliminary heating of a surface also promotes purification of the surface through emission of adsorbed molecules. In general, adsorbed layers may be present; in this case, the interaction of a given molecule with the surface may also depend on the distribution of molecules impinging on a surface element. For a more detailed discussion the reader should consult Refs. 3, 22, and 23.

In general, a molecule striking a surface with a velocity $\boldsymbol{\xi}'$ reemerges from it with a velocity $\boldsymbol{\xi}$ that is strictly determined only if the path of the molecule within the wall can be computed exactly. This computation is a very difficult one, because it depends upon a great number of details, such as the locations and velocities of all the molecules of the wall and an accurate knowledge of the interaction potential. Hence it is more convenient to think in terms of a probability density $R(\boldsymbol{\xi}' \to \boldsymbol{\xi}; \mathbf{x}, t; \tau)$ that a molecule striking the surface with velocity between $\boldsymbol{\xi}'$ and $\boldsymbol{\xi}' + d\boldsymbol{\xi}'$ at the point $\mathbf{x}$ and time t will reemerge at practically the same point with velocity between $\boldsymbol{\xi}$ and $\boldsymbol{\xi} + d\boldsymbol{\xi}$ (Fig. 1.3) after a time interval τ (adsorption or sitting time). If R is known, then we can easily write down the boundary condition for the distribution function $f(\mathbf{x}, \boldsymbol{\xi}, t)$. To simplify the discussion, the surface will be assumed to be at rest.

The mass of molecules emerging with velocity between $\boldsymbol{\xi}$ and $\boldsymbol{\xi} + d\boldsymbol{\xi}$ from a surface element dA about $\mathbf{x}$ in the time interval between t and $t + dt$ is

$$d^*\mathcal{M} = f(\mathbf{x}, \boldsymbol{\xi}, t)|\boldsymbol{\xi} \cdot \mathbf{n}|dt\,dA\,d\boldsymbol{\xi} \qquad (\mathbf{x} \in \partial\Omega, \boldsymbol{\xi} \cdot \mathbf{n} > 0), \qquad (1.11.1)$$

where $\mathbf{n}$ is the unit vector normal to the surface $\partial\Omega$ at $\mathbf{x}$ and directed from the wall into the gas. Analogously, the probability that a molecule impinges upon the same surface element with velocity between $\boldsymbol{\xi}'$ and $\boldsymbol{\xi}' + d\boldsymbol{\xi}'$ in the time interval between $t - \tau$ and $t - \tau + dt$ $(\tau > 0)$ is

$$d^*\mathcal{M}' = f(\mathbf{x}, \boldsymbol{\xi}', t-\tau)|\boldsymbol{\xi}' \cdot \mathbf{n}|dtdAd\boldsymbol{\xi}' \qquad (\mathbf{x} \in \partial\Omega, \boldsymbol{\xi}' \cdot \mathbf{n} < 0). \qquad (1.11.2)$$

If we multiply $d^*\mathcal{M}'$ by the probability of a scattering event from velocity $\boldsymbol{\xi}'$ to a velocity between $\boldsymbol{\xi}$ and $\boldsymbol{\xi} + d\boldsymbol{\xi}$ with an adsorption time between τ and $\tau + d\tau$ (i.e., $R(\boldsymbol{\xi}' \to \boldsymbol{\xi}; \mathbf{x}, t; \tau)d\boldsymbol{\xi}d\tau$) and integrate over all the possible values of $\boldsymbol{\xi}'$ and τ, we must obtain $d^*\mathcal{M}$ (here we assume that each molecule reemerges from the surface element into which it entered, which is not so realistic when τ is large):

$$d^*\mathcal{M} = d\boldsymbol{\xi} \int_0^\infty d\tau \int_{\boldsymbol{\xi}' \cdot \mathbf{n} < 0} R(\boldsymbol{\xi}' \to \boldsymbol{\xi}; \mathbf{x}, t; \tau)\, d^*\mathcal{M}' \qquad (\mathbf{x} \in \partial\Omega, \boldsymbol{\xi} \cdot \mathbf{n} > 0). \qquad (1.11.3)$$

Equating the expressions in Eqs. (1.11.1) and (1.11.3) and canceling the common factor $dAd\boldsymbol{\xi}dt$, we obtain

$$f(\mathbf{x}, \boldsymbol{\xi}, t)|\boldsymbol{\xi} \cdot \mathbf{n}| = \int_0^\infty d\tau \int_{\boldsymbol{\xi}' \cdot \mathbf{n} < 0} R(\boldsymbol{\xi}' \to \boldsymbol{\xi}; \mathbf{x}, t; \tau) f(\mathbf{x}, \boldsymbol{\xi}', t-\tau)|\boldsymbol{\xi}' \cdot \mathbf{n}|d\boldsymbol{\xi}' \qquad (\mathbf{x} \in \partial\Omega, \boldsymbol{\xi} \cdot \mathbf{n} > 0). \qquad (1.11.4)$$

The kernel R can be assumed to be independent of f under suitable conditions, which we shall not detail here.[3,22,23] If, in addition, the effective adsorption time is small compared to any characteristic time of interest in the evolution of f, we can let $\tau = 0$ in the argument of f appearing in the right-hand side of Eq. (1.2.4); in this case the latter becomes

$$f(\mathbf{x}, \boldsymbol{\xi}, t)|\boldsymbol{\xi} \cdot \mathbf{n}| = \int_{\boldsymbol{\xi}' \cdot \mathbf{n} < 0} R(\boldsymbol{\xi}' \to \boldsymbol{\xi}; \mathbf{x}, t) f(\mathbf{x}, \boldsymbol{\xi}', t)|\boldsymbol{\xi}' \cdot \mathbf{n}|d\boldsymbol{\xi} \qquad (\mathbf{x} \in \Omega, \boldsymbol{\xi} \cdot \mathbf{n} > 0), \qquad (1.11.5)$$

where

$$R(\boldsymbol{\xi}' \to \boldsymbol{\xi}; \mathbf{x}, t) = \int_0^\infty d\tau R(\boldsymbol{\xi}' \to \boldsymbol{\xi}; \mathbf{x}, t; \tau). \qquad (1.11.6)$$

Equation (1.2.5) is, in particular, valid for steady problems.

Although the idea of a scattering kernel had appeared before, it was only at the end of 1960s that a systematic study of the properties of this kernel appeared in the scientific literature.[6,22,23] In particular, the following properties were pointed out:[3,6,22–28]

1) Nonnegativeness; that is, R cannot take negative values:

$$R(\boldsymbol{\xi}' \to \boldsymbol{\xi}; \mathbf{x}, t; \tau) \geq 0 \tag{1.11.7}$$

and, as a consequence,

$$R(\boldsymbol{\xi}' \to \boldsymbol{\xi}; \mathbf{x}, t) \geq 0. \tag{1.11.8}$$

2) Normalization, if permanent adsorption is excluded; that is, R, as a probability density for the totality of events, must integrate to unity:

$$\int_0^\infty d\tau \int_{\boldsymbol{\xi}\cdot\mathbf{n}\geq 0} R(\boldsymbol{\xi}' \to \boldsymbol{\xi}; \mathbf{x}, t; \tau) d\boldsymbol{\xi} = 1 \tag{1.11.9}$$

and, as a consequence,

$$\int_{\boldsymbol{\xi}\cdot\mathbf{n}\geq 0} R(\boldsymbol{\xi}' \to \boldsymbol{\xi}; \mathbf{x}, t) d\boldsymbol{\xi} = 1. \tag{1.11.10}$$

3) Reciprocity; this is a subtler property that follows from the circumstance that the microscopic dynamics is time reversible and the wall is assumed to be in a local equilibrium state, not significantly disturbed by the impinging molecule. It reads as follows:

$$|\boldsymbol{\xi}' \cdot \mathbf{n}| M_w(\boldsymbol{\xi}') R(\boldsymbol{\xi}' \to \boldsymbol{\xi}; \mathbf{x}, t; \tau) = |\boldsymbol{\xi} \cdot \mathbf{n}| M_w(\boldsymbol{\xi}) R(-\boldsymbol{\xi} \to -\boldsymbol{\xi}'; \mathbf{x}, t; \tau) \tag{1.11.11}$$

and, as a consequence,

$$|\boldsymbol{\xi}' \cdot \mathbf{n}| M_w(\boldsymbol{\xi}') R(\boldsymbol{\xi}' \to \boldsymbol{\xi}; \mathbf{x}, t) = |\boldsymbol{\xi} \cdot \mathbf{n}| M_w(\boldsymbol{\xi}) R(-\boldsymbol{\xi} \to -\boldsymbol{\xi}'; \mathbf{x}, t). \tag{1.11.12}$$

Here M_w is a (nondrifting) Maxwellian distribution having the temperature of the wall, which is uniquely identified apart from a factor.

We remark that the reciprocity and the normalization relations imply another property:

3′) Preservation of equilibrium; that is, the Maxwellian M_w must satisfy the boundary condition (1.11.4)

$$M_w(\boldsymbol{\xi})|\boldsymbol{\xi} \cdot \mathbf{n}| = \int_0^\infty d\tau \int_{\boldsymbol{\xi}'\cdot\mathbf{n}<0} R(\boldsymbol{\xi}' \to \boldsymbol{\xi}; \mathbf{x}, t; \tau) M_w(\boldsymbol{\xi}')|\boldsymbol{\xi}' \cdot \mathbf{n}| d\boldsymbol{\xi}', \tag{1.11.13}$$

which is equivalent to

$$M_w(\boldsymbol{\xi})|\boldsymbol{\xi} \cdot \mathbf{n}| = \int_{\boldsymbol{\xi}'\cdot\mathbf{n}<0} R(\boldsymbol{\xi}' \to \boldsymbol{\xi}; \mathbf{x}, t) M_w(\boldsymbol{\xi}')|\boldsymbol{\xi}' \cdot \mathbf{n}| d\boldsymbol{\xi}'. \tag{1.11.14}$$

In order to obtain Eq. (1.11.13) it is sufficient to integrate Eq. (1.11.11) with respect to $\boldsymbol{\xi}'$ and τ, taking into account Eq. (1.11.9) (with $-\boldsymbol{\xi}$ and $-\boldsymbol{\xi}'$ in place of $\boldsymbol{\xi}'$ and $\boldsymbol{\xi}$, respectively). We remark that one frequently assumes Eq. (1.11.13) (or (1.11.14)), without mentioning Eq. (1.11.11) (or (1.11.12)); although this is enough for many purposes, reciprocity is very important when constructing mathematical models, because it places a strong restriction on the possible choices. A detailed discussion of the physical conditions under which reciprocity holds has been given by Bärwinkel and Schippers.[29]

The scattering kernel is a fundamental concept in gas–surface interaction, by means of which other quantities should be defined. Frequently its use is avoided by using the so-called accommodation coefficients, with the consequence of lack of clarity, misinterpretation of experiments, bad definitions of terms, and misunderstanding of concepts. The basic information on gas–surface interaction, which should be in principle obtained from a detailed calculation based on a physical model, is summarized in a scattering kernel. The further reduction to a small set of accommodation coefficients can be advocated for practical purposes, provided this concept is firmly related to the scattering kernel.

To describe the accommodation coefficients in a systematic way, it is convenient to introduce, for any pair of functions ϕ and ψ, the following notations:

$$(\psi, \phi)_+ = \int_{\boldsymbol{\xi}\cdot\mathbf{n}>0} \psi(\boldsymbol{\xi})\phi(\boldsymbol{\xi})M_w(\boldsymbol{\xi})|\boldsymbol{\xi}\cdot\mathbf{n}|d\boldsymbol{\xi}, \tag{1.11.15}$$

$$(\psi, \phi)_- = \int_{\boldsymbol{\xi}\cdot\mathbf{n}<0} \psi(\boldsymbol{\xi})\phi(\boldsymbol{\xi})M_w(\boldsymbol{\xi})|\boldsymbol{\xi}\cdot\mathbf{n}|d\boldsymbol{\xi}. \tag{1.11.16}$$

Now, if we factor M_w out of the distribution function f and write

$$f = M_w\phi, \tag{1.11.17}$$

we can define the accommodation coefficient for the quantity ψ when the distribution function at the wall is $M_w\phi$, in the following way:

$$\alpha(\psi, \phi) = [(\psi, \phi)_- - (\psi, \phi)_+]/[(\psi, \phi)_- - (\psi, \iota)_+], \tag{1.11.18}$$

where ι denotes a constant function, such that

$$(\iota, \iota)_+ = (\iota, \phi)_+. \tag{1.11.19}$$

Physically the numerator in Eq. (1.11.18) is the difference between the impinging flow and emerging flow of the quantity, whose density is ψ, when the distribution is $M_w\phi$; the denominator is the same thing when the restriction of f

to $\boldsymbol{\xi} \cdot \mathbf{n} > 0$ is replaced by the wall Maxwellian, normalized in such a way as to give the same entering flow rate as f. In particular, if we let $\psi = \boldsymbol{\xi} \cdot \mathbf{n}$, we obtain the accommodation coefficient for normal momentum; if we let $\psi = \boldsymbol{\xi} \cdot \mathbf{t}$, we obtain the accommodation coefficient for tangential momentum (in the direction of the unit vector $\mathbf{t}$, tangent to the wall); if we let $\psi = |\boldsymbol{\xi}|^2$, we obtain the energy accommodation coefficient. It is convenient to restrict the definition in Eq. (1.1.18) to functions enjoying the property $\psi(\boldsymbol{\xi}) = \psi(\boldsymbol{\xi} - 2\mathbf{n}(\mathbf{n} \cdot \boldsymbol{\xi}))$, which are even functions of $\boldsymbol{\xi} \cdot \mathbf{n}$. This condition is not satisfied by $\psi = \boldsymbol{\xi} \cdot \mathbf{n}$; accordingly, if one wants to define an accommodation coefficient for normal momentum, one has to take $\psi = |\boldsymbol{\xi} \cdot \mathbf{n}|$.

In general, $\alpha(\psi, \phi)$ turns out to depend upon the distribution function of the impinging molecules; accordingly the definition (1.11.18) is not so useful, in general. It becomes more useful, if one selects[3,21,22] a particular class of functions ϕ.

It is clear that there is a relation between the accommodation coefficients and the scattering kernel $R(\boldsymbol{\xi}' \to \boldsymbol{\xi})$ (we omit indicating the space and time arguments), but what this particular relation is depends on the set of functions from which ϕ and ψ are chosen.[3,21,22]

In view of the difficulty of computing the kernel $R(\boldsymbol{\xi}' \to \boldsymbol{\xi})$ from a physical model of the wall, we shall presently discuss a different procedure, which is less physical in nature. The idea is to construct a mathematical model in the form of a kernel $R(\boldsymbol{\xi}' \to \boldsymbol{\xi})$ that satisfies the basic physical requirements expressed by Eqs. (1.11.8), (1.11.10), and (1.11.12) and is not otherwise restricted except by the condition of not being too complicated.

One of the simplest kernels is

$$R(\boldsymbol{\xi}' \to \boldsymbol{\xi}) = \alpha M_w(\boldsymbol{\xi})|\boldsymbol{\xi} \cdot \mathbf{n}| + (1 - \alpha)\delta(\boldsymbol{\xi} - \boldsymbol{\xi}' + 2\mathbf{n}(\boldsymbol{\xi}' \cdot \boldsymbol{\xi})). \quad (1.11.20)$$

This is the kernel corresponding to Maxwell's model,[30] according to which a fraction $(1 - \alpha)$ of molecules undergoes a specular reflection, while the remaining fraction α is diffused with the Maxwellian distribution of the wall M_w. This is the only model for the scattering kernel that appeared in the literature before the late 1960s. Since this model was felt to be somehow inadequate to represent the gas–surface interaction, Nocilla[31] proposed to assume that the molecules are reemitted according to a drifting Maxwellian with a temperature that is, in general, different from the temperature of the wall. While this model is useful as a tool to represent experimental data and has been used in actual calculations, expecially in free-molecular flow,[32] when interpreted at the light of later developments, it does not appear to be tenable, unless its flexibility is severely reduced.[26,33,34] Although the idea of a model such as Nocilla's can be traced back to Knudsen,[35] the full development of these ideas led to the so-called Cercignani–Lampis (CL) model[26], which reads as follows:

$$R(\boldsymbol{\xi}' \to \boldsymbol{\xi}) = \frac{[\alpha_n \alpha_t (2-\alpha_t)]}{\pi/2} \beta_w^2 \xi_n \exp\left[-\beta_w \frac{\xi_n^2 + (1-\alpha_n)\xi_n'^2}{\alpha_n} - \beta_w \frac{|\boldsymbol{\xi}_t - (1-\alpha_t)\boldsymbol{\xi}_t'|^2}{\alpha_t(2-\alpha_t)} \right] I_0\left(\beta_w \frac{(1-\alpha_n)^{1/2}\xi_n \xi_n'}{\alpha_n}\right), \quad (1.11.21)$$

where I_0 denotes the modified Bessel function of first kind and zeroth order defined by

$$I_0(y) = (2\pi)^{-1} \int_0^{2\pi} \exp(y \cos\phi) d\phi. \quad (1.11.22)$$

The two parameters α_t and α_n have a simple meaning; the first one is the accommodation coefficient for the tangential components of momentum, the second one is the accommodation coefficient for ξ_n^2, hence for the part of kinetic energy associated with the normal motion. This model, which was obtained by a series expansion of the kernel under suitable assumptions on its eigenfunctions and eigenvalues in Ref. 26, became rather popular because it was found by other methods as well: through an analogy with Brownian motion by I. Kuščer et al.,[34] under a special mathematical assumption by T. G. Cowling,[36] and through an analogy with the scattering of electromagnetic waves from a surface by M. M. R. Williams.[37] Finally, it followed from the solution of the steady Fokker–Planck equation describing a (somewhat artificial) physical model of the wall.[28] Also, comparisons with the data from beam scattering experiments were quite encouraging.[26,38,3] In spite of this, it must be clearly stated that there are no physical reasons why this model should be considered better than others. In particular we remark that any linear combination of scattering kernels with positive coefficients adding to unity is again a kernel that satisfies all the basic properties. Thus, from a kernel with two parameters, such as the CL model, we can construct a general model, containing an arbitrary function of those parameters.[3] Another method to produce scattering kernels in a simple way is described in Refs. 3 and 23.

There are various systematic procedures to produce scattering kernels. One of them was introduced by the author[3] and has been investigated recently in an effort to reproduce experimental data.[22,39,40] One first constructs on the basis of physical intuition, experimental information, and mathematical simplicity a kernel R_0 that satisfies all the properties except normalization. Then a new kernel that satisfies all the properties can be constructed by means of the following rule:

$$R(\boldsymbol{\xi}' \to \boldsymbol{\xi}) = R_0(\boldsymbol{\xi}' \to \boldsymbol{\xi}) + \xi_n f_0(\boldsymbol{\xi})(1 - H(-\boldsymbol{\xi}'))(1 - H(\boldsymbol{\xi}))/I, \qquad (1.11.23)$$

where $H(\boldsymbol{\xi}')$ and I are obtained from

$$H(\boldsymbol{\xi}') = \int_{\xi_n > 0} R_0(-\boldsymbol{\xi}' \to \boldsymbol{\xi}) d\boldsymbol{\xi}, \qquad (1.11.24)$$

$$I = \int_{\xi_n > 0} \xi_n f_0(\boldsymbol{\xi})(1 - H(\boldsymbol{\xi})) d\boldsymbol{\xi}. \qquad (1.11.25)$$

We refer to the aforementioned papers[22,39,40] for examples and applications of this idea.

All the models discussed above contain pure diffusion according to a non-drifting Maxwellian as a limiting case (Eq. 1.11.20 with $\alpha = 1$). The use of this particular model is justified for low-velocity flows over technical surfaces, but it is inaccurate for flows with orbital velocity. In fact, the elaboration of the measurements of lift-to-drag ratio for the Shuttle Orbiter in the free-molecular regime[41] implies a significant departure from diffuse, fully equilibrated re-emission of molecules at the wall.

It is remarkable that, for any scattering kernel satisfying the three properties of normalization, positivity, and preservation of equilibrium, a simple inequality holds. The latter was stated by Darrozès and Guiraud[42] who also sketched a proof. More details were given later.[28,3] It reads as follows:

$$\mathcal{J}_n = \int_{R^3} \boldsymbol{\xi} \cdot \mathbf{n} f \log f \, d\boldsymbol{\xi} \leq -(2RT_w)^{-1} \int_{R^3} \boldsymbol{\xi} \cdot \mathbf{n} |\boldsymbol{\xi}|^2 f d\boldsymbol{\xi}. \qquad (\mathbf{x} \in \partial\Omega). \qquad (1.11.26)$$

Equality holds if and only if f coincides with M_w (the wall Maxwellian) on $\partial\Omega$ (unless the kernel in Eq. (1.11.5) is a delta function). We remark that if the gas does not slip upon the wall, the right-hand side of Eq. (1.11.26) equals $-q_n/(RT_w)$, where q_n is the heat flow along the normal, according to its definition given in Section 1.5. If the gas slips on the wall, then one must add the power of the stresses $\mathbf{p}_n \cdot \mathbf{v}$ to q_n. In any case, however, the right-hand side equals $q_n^{(w)}$, where $q_n^{(w)}$ is the heat flow in the solid at the interface, because the normal energy flow must be continuous through the wall and stresses have vanishing power in the solid, because the latter is at rest. If we identify the function H introduced in Section 1.7 with $-\eta/R$ (where η is the entropy of the gas), the inequality in Eq. (1.11.26) is exactly what one would expect from the second law of thermodynamics.

Problem

1.11.1 Show that, if Eqs. (1.11.5) and (1.11.10) apply, then the normal component of the bulk velocity of the gas, as defined in Section 1.5, vanishes at the wall.

References

[1] C. Cercignani, "On the Boltzmann equation for rigid spheres," *Transp. Theory Stat. Phys.* **2**, 211–225 (1972).
[2] O. Lanford, III, "The evolution of large classical systems," in *Dynamical Systems, Theory and Applications*, J. Moser, ed., **LNP 35**, 1–111, Springer-Verlag, Berlin (1975).
[3] C. Cercignani, *The Boltzmann Equation and Its Applications*, Springer-Verlag, New York (1988).
[4] R. Illner and M. Pulvirenti, "Global validity of the Boltzmann equation for a two-dimensional rare gas in a vacuum," *Comm. Math. Phys.* **105**, 189–203 (1986)
[5] R. Illner and M. Pulvirenti, "Global validity of the Boltzmann equation for two- and three-dimensional rare gas in vacuum: Erratum and improved result," *Comm. Math. Phys.* **121**, 143–146 (1989).
[6] C. Cercignani, *Mathematical Methods in Kinetic Theory*, Plenum Press, New York (1969; revised edition 1990).
[7] C. Cercignani, *Ludwig Boltzmann. The Man Who Trusted Atoms*, Oxford University Press, Oxford, (1998).
[8] L. Arkeryd, "On the Boltzmann equation. Part II: The full initial value problem," *Arch. Rat. Mech. Anal.* **45**, 17–34 (1972).
[9] C. Cercignani, "Are there more than five linearly independent collision invariants for the Boltzmann equation?," *J. Statistical Phys.* **58**, 817–824 (1990).
[10] L. Arkeryd and C. Cercignani, "On a functional equation arising in the kinetic theory of gases," *Rend. Mat. Acc. Lincei* s.9, **11**, 139–149 (1990).
[11] L. Boltzmann, "Über die Aufstellung und Integration der Gleichungen, welche die Molekular-bewegungen in Gasen bestimmen," *Sitzunsberichte der Akademie der Wissenschaften, Wien* **74**, 503–552 (1876).
[12] C. Truesdell and R. G. Muncaster, *Fundamentals of Maxwell's Kinetic Theory of a Simple Monatomic Gas,* Academic Press, New York (1980).
[13] P. L. Bhatnagar, E. P. Gross, and M. Krook, "A model for collision processes in gases. Small amplitude processes in charged and neutral one-component systems," *Phys. Rev.* **94**, 511–525 (1954).
[14] P. Welander, "On the temperature jump in a rarefied gas," *Arkiv Fysik* **7**, 507–553 (1954).
[15] L. H. Holway, Jr., "Approximation procedures for kinetic theory," Ph.D. Thesis, Harvard (1963).
[16] C. Cercignani and G. Tironi, "Nonlinear heat transfer between two parallel plates at large temperature ratios," in *Rarefied Gas Dynamics*, C. L. Brundin, ed.,Vol. I, 441–453, Academic Press, New York (1967).
[17] P. Andries, P. Le Tallec, J. P. Perlat, and B. Perthame, "The Gaussian–BGK model of Boltzmann equation with small Prandtl number," to appear in *Euro. J. Mechanics B* (1999).
[18] L. Sirovich, "Kinetic modeling of gas mixtures," *Phys. Fluids* **5**, 908–918 (1962).
[19] J. L. Lebowitz, H. L. Frisch and E. Helfand, "Nonequilibrium distribution functions in a fluid," *Phys. Fluids* **3**, 325–338 (1960).

[20] T. F. Morse, "Kinetic model equations in a fluid," *Phys. Fluids* **7**, 2012–2013 (1964).
[21] F. B. Hanson and T. F. Morse, "Kinetic models for a gas with internal structure," *Phys. Fluids* **10**, 345–353 (1967).
[22] I. Kuščer, "Phenomenological aspects of gas–surface interaction," in *Fundamental Problems in Statistical Mechanics* **IV**, E.G.D. Cohen and W. Fiszdon, eds., 441–467, Ossolineum, Warsaw (1978).
[23] C. Cercignani, "Scattering kernels for gas–surface interaction," *Proceedings of the Workshop on Hypersonic Flows for Reentry Problems*, Vol. I, 9–29, 1990, INRIA, Antibes.
[24] C. Cercignani, "Boundary value problems in linearized kinetic theory," in *Transport Theory*, R. Bellman, G. Birkhoff, and I. Abu-Shumays, eds., 249–268, AMS, Providence, RI. (1969).
[25] I. Kuščer, in *Transport Theory Conference*, AEC Report ORO-3588-1, Blacksburgh, VA. (1969).
[26] C. Cercignani and M. Lampis, "Kinetic models for gas–surface interactions," *Transport Theory Stat. Phys.* **1**, 101–114 (1971).
[27] I. Kuščer, "Reciprocity in scattering of gas molecules by surfaces", *Surface Science* **25**, 225–237 (1971).
[28] C. Cercignani, "Scattering kernels for gas–surface interactions", *Transp. Theory Stat. Phys.* **2**, 27–53 (1972).
[29] K. Bärwinkel and S. Schippers, "Nonreciprocity in noble-gas metal-surface scattering," in *Rarefied Gas Dynamics: Space-Related Studies*, E. P. Muntz, D. P. Weaver, and D. H. Campbell, eds., 487–501, AIAA, Washington (1989).
[30] J. C. Maxwell, "On stresses in rarified gases arising from inequalities of temperature," *Phil. Trans. Roy. Soc.* **170**, 231–256, Appendix (1879).
[31] S. Nocilla, "On the interaction between stream and body in free-molecule flow," in *Rarefied Gas Dynamics*, L. Talbot, ed., 169–208, Academic Press, New York (1961).
[32] F. Hurlbut and F.S. Sherman, "Application of the Nocilla wall reflection model to free-molecule kinetic theory," *Phys. Fluids* **11**, 486–496 (1968).
[33] M. N. Kogan, *Rarefied Gas Dynamics*, Plenum Press, New York (1969).
[34] I. Kuščer, J. Možina, and F. Krizanic, "The Knudsen model of thermal accommodation," in *Rarefied Gas Dynamics*, Dini et al., eds., Vol. I, 97–108, Editrice Tecnico-Scientifica, Pisa (1974).
[35] M. Knudsen, *The Kinetic Theory of Gases*, Methuen, London (1950).
[36] T. G. Cowling, "On the Cercignani–Lampis formula for gas–surface inte actions," *J. Phys. D. Appl. Phys.* **7**, 781–785 (1974).
[37] M. M. R. Williams, "A phenomenological study of gas–surface interactions," *J. Phys. D. Appl. Phys.* **4**, 1315–1319 (1971).
[38] C. Cercignani, "Models for gas–surface interactions: Comparison between theory and experiment," in *Rarefied Gas Dynamics*, D. Dini et al., eds., vol. I, 75–96, Editrice Tecnico- Scientifica, Pisa (1974).
[39] Cercignani, C., Lampis, M., and Lentati, A., "A new scattering kernel in kinetic theory of gases," *Transp. Theory Stat. Phys.* **24**, 1319–1336 (1995).
[40] C. Cercignani and M. Lampis, "New scattering kernel for gas–surface interaction," *AIAA Journal* **35**, 1000–1001 (1997).
[41] R. C. Blanchard, "Rarefied flow lift to drag measurement of the Shuttle Orbiter," Paper No. ICAS 86–2.10.1, 15th ICAS Congress, London (September 1986).
[42] J.-S. Darrozès and J.-P. Guiraud, "Généralisation formelle du théorème H en présence de parois. Applications," *C.R.A.S.* (Paris) **A262**, 1368–1371.

2

Problems for a Gas in a Slab: General Aspects and Preliminary Examples

2.1. Introduction

This chapter deals with problems in which the gas molecules can be imagined to have a distribution function that depends on just one Cartesian coordinate (as well as on molecular velocity and, possibly, time). Thus the gas will be assumed to occupy a slab (the limiting case of a half-space will be treated in the next chapter). These problems, although somehow oversimplified, are interesting in several respects. They have a conceptual value, since they are next to equilibria in a classification based on mathematical simplicity. Thus they lend themselves to the introduction of the kinds of flow regimes that one meets in rarefied gas dynamics. They are also suitable to introduce the various approximation techniques that have been proposed to solve the Boltzmann equation. Finally, the half-space problems will be shown to provide important information on the Knudsen layers, or kinetic boundary layers, which occur near solid boundaries and are of paramount importance in the so-called slip regime.

Let us first consider steady problems. Here, of course, the role of boundary conditions is of paramount importance. In this section we shall not restrict the solutions to depend on just one space variable, since we shall discuss properties that have a wider validity.

The Boltzmann equation in the steady case may be written as follows:

$$\boldsymbol{\xi} \cdot \partial_{\mathbf{x}} f = Q(f, f)(\mathbf{x}, \boldsymbol{\xi}), \tag{2.1.1}$$

where $\mathbf{x}$ varies in some region Ω of ordinary space and $\boldsymbol{\xi}$ in the entire velocity space.

Among the possible boundary conditions discussed in the previous chapter there are only two that are physically acceptable and deterministic for the molecular velocity since the kernel R reduces to a delta function. They are the condition of specular reflection (the normal component of ξ is reversed at the boundary) and the reverse reflection or bounce-back boundary condition (the velocity vector is reversed).

We use these to show that for certain kinds of boundary conditions steady solutions may be trivial or even nonexistent.

As we discussed in the previous chapter (see also Refs. 1 and 2), a steady Maxwellian distribution describes a gas in a (statistical) equilibrium state with any reasonable definition of the latter concept. In particular, we saw that (in a general domain not rotationally invariant) a Maxwellian with constant parameters is the most general solution of the Boltzmann equation with boundary conditions of specular reflection.

The boundary conditions in Chapter 1 have been written assuming the walls to be at rest; these conditions can be immediately extended to moving walls by replacing everywhere the molecular velocity by the relative velocity with respect to the wall. The boundary condition of specular reflection feels this change only if the wall has a movement in the direction normal to the wall itself, but it is completely insensitive to any motion tangential to the wall. Thus we cannot have a flow between two parallel plates, one of which moves in its own plane with respect to the other. This is the so-called (plane) Couette flow to be discussed in detail in this and the next chapter.

The boundary condition of reverse reflection appears to be more realistic because it allows a momentum transfer between the walls and the gas. Thus one can consider the Couette flow in this case. But there is high price to pay. There is no steady solution. In fact energy exchanges are still not allowed and the energy supplied by the motion of the walls is partly transformed into random motions (i.e., heat) that cannot leave the gas. Thus the temperature grows exponentially in time.

This is true no matter what kind of geometry we consider, provided the walls move in such a way as to transmit a motion that leads to creation of entropy. Since a lack of entropy production corresponds to Maxwellian distributions, it is a very general conclusion that the bounce-back boundary condition permits only steady solutions that are Maxwellians. If these are not compatible with the boundary conditions then an unsteady solution is the only possibility.

Here we must pause for a while to discuss the concept of mean free path. The mean free path is the average distance covered by a molecule, that is, (to simplify), a hard sphere whose diameter we shall denote by σ, between two subsequent collisions. We can consider a moving molecule that strikes any of the other molecules, which are assumed to be motionless (this introduces no

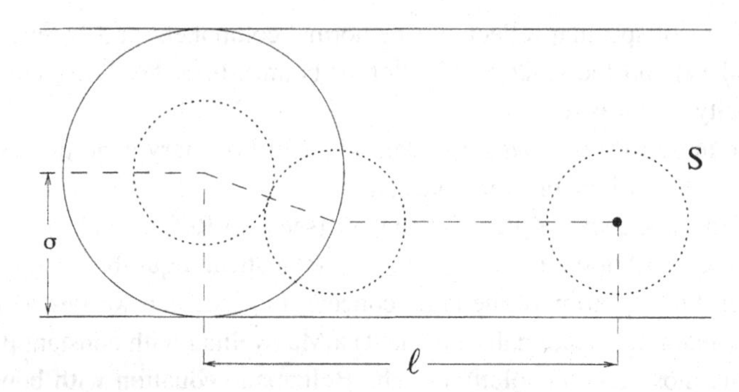

Figure 2.1. The free path of a molecule between two collisions. The moving molecule S is represented as a point, the other molecules as spheres with double radius (the distance between the centers at the point of contact). The dotted circles represent the actual spheres.

essential difference and simplifies the explanation). A collision occurs when the center of the moving sphere reaches a distance equal to σ from the center of a sphere at rest. Then, to compute the length of the path between two collisions, we can consider the moving sphere S as a point and the sphere at rest as having a doubled radius, σ (Fig. 2.1). Then if S travels a distance ℓ on average between two impacts, this means that there is only one molecule (i.e., S) in a cylinder of base $\pi\sigma^2$ and height ℓ or $n\pi\sigma^2\ell \cong 1$, where $n = N/V$ is the number of molecules per unit volume. Hence $\ell \cong 1/(n\pi\sigma^2)$. This simple argument shows that ℓ equals $1/(n\pi\sigma^2)$, except for a numerical factor that depends on the way we perform the average implied by the use of the word "mean." For molecules other than hard spheres the concept becomes a little fuzzy although it retains its usefulness as a typical length preferred by a molecule between two (significant) interactions. Thus the concept of mean free path is a qualitative one; it can be made precise, as we shall see, in several different ways. The different ways produce slightly different values for ℓ; thus it is advisable to check the precise definition adopted by each author. Whenever we adopt the standard definition in this book (to be given in Section 2.5) we shall use the letter λ rather than ℓ.

Both direct and indirect methods give a value for ℓ (or λ) at room temperature and atmospheric pressure of about a millionth of a centimeter. The mean free path becomes smaller at lower densities, such as those encountered by an object flying in the upper atmosphere.

In the next section we shall investigate the case of Couette flow between parallel plates for bounce-back boundary conditions.

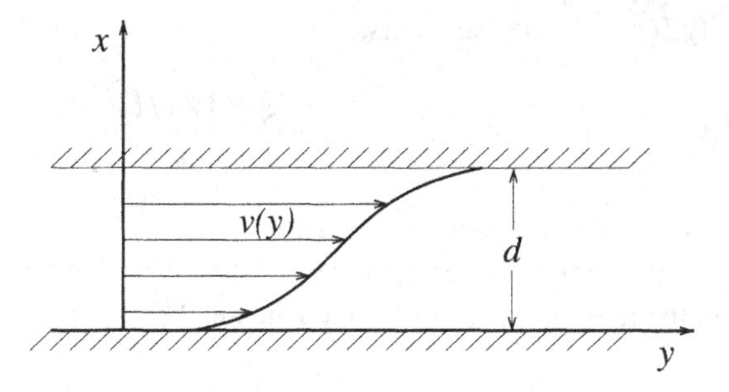

Figure 2.2. Geometry of a Couette flow.

2.2. Couette Flow for Bounce-Back Boundary Conditions

Let us consider a monatomic rarefied gas with average density ρ_0, in motion between two parallel plates located at $x = 0$ and $x = L$, respectively (see Fig. 2.2). The upper plate moves with velocity V, whereas the lower one is at rest.

To orient ourselves in a new field and chart the phenomena that can be expected, it is very useful to consider dimensional analysis and take a look at the typical nondimensional parameters associated with the equations that rule these phenomena. Rarefied gas dynamics is no exception: It makes use of two basic nondimensional numbers, the Knudsen number Kn_L, based on the distance between the plates L ($\mathrm{Kn}_L = \lambda_0/L$), and the speed ratio $S = U/V_{\mathrm{th}}$ (where $V_{\mathrm{th}} = (2RT_0)^{1/2}$ is the thermal speed and T_0 the temperature of the gas at $t = 0$), related to the Mach number Ma (the ratio between V and the speed of sound for a monatomic gas in continuum mechanics) by $\mathrm{Ma} = (6/5)^{1/2}\, S = 1.095\, S$. λ_0 is the mean free path at $t = 0$.

We are going to look for a solution that is self-similar, in the sense that if we cut the slab at $x = L'$ and imagine putting a wall there, the solution in the slab between $x = 0$ and $x = L'$ remains the same, provided we give the plate at $x = L'$ the velocity $V' = VL'/L$. Then the basic parameter must depend on the ratio V/L and should be the product of the previous two parameters. In other words it must coincide with the nondimensional velocity gradient

$$\mathcal{T} = \frac{V\lambda_0}{L(RT_0)^{1/2}}. \tag{2.2.1}$$

We complete our formulation with the following initial and boundary conditions:

a) At time zero we assume a Maxwellian distribution with temperature T_0 and bulk velocity Ux/L along the y axis:

$$f(0, x, \boldsymbol{\xi}) = \rho_0 \pi^{-3/2} V_{\text{th}}^{-3} \exp\left(-\frac{|\boldsymbol{\xi} - \mathbf{j}Vx/L|^2}{V_{\text{th}}}\right), \tag{2.2.2}$$

where $\mathbf{j}$ is the unit vector along the y axis.

b) At the plates we assume that the molecules satisfy the bounce-back boundary condition in the reference frame of each plate and hence write

$$f(t, 0, y, \boldsymbol{\xi}) = f(t, 0, y, -\boldsymbol{\xi}), \tag{2.2.3}$$

$$f(t, L, y, \boldsymbol{\xi}) = f(t, L, y, 2V\mathbf{j} - \boldsymbol{\xi}). \tag{2.2.4}$$

Following a paper by the author,[3] we shall look for solutions such that the variable $\mathbf{x}$ appears in f only through the bulk velocity

$$f = f(\mathbf{c}, t), \tag{2.2.5}$$

where $\mathbf{c} = \boldsymbol{\xi} - \mathbf{v}$ is the random velocity.

Although in this book we are interested in computing solutions and not in just proving their existence, we mention that solutions of the Boltzmann equation with this property exist. They are called homoenergetic affine flows, because the thermal energy per unit volume of the gas is space independent and the bulk velocity $\mathbf{v}$ is an affine (=linear inhomogeneous) function of position $\mathbf{x}$, given by

$$\mathbf{v} = \mathsf{K}(t)\mathbf{x} + \mathbf{v}_0(t), \tag{2.2.6}$$

where K is a linear operator (3×3 matrix) in R^3.

In particular one can show[3] that $f(\mathbf{c}, t)$ must satisfy the following equation:

$$\frac{\partial f}{\partial t} - \frac{\partial f}{\partial \mathbf{c}} \cdot \mathsf{K}\mathbf{c} = Q(f, f) \tag{2.2.7}$$

and prove that Eq. (2.2.7) admits the following:

Existence Theorem.[3] *There exists a solution f of Eq. (2.1), where the kernel $B(\theta, |\boldsymbol{\xi} - \boldsymbol{\xi}_*|)$ of the collision term $Q(f, f)$ does not grow more than quadratically in $\boldsymbol{\xi}, \boldsymbol{\xi}_*$ and the initial mass density, energy density, and H-functional ($= \int f \log f d\boldsymbol{\xi}$) are finite at time 0. These quantities remain bounded for $0 \le t \le \bar{t}$. The time $\bar{t}$ is arbitrary provided $\mathsf{K}(0)$ has no negative eigenvalues. If $\mathsf{K}(0)$ possesses negative eigenvalues and t_0^{-1} is their largest absolute value, then $\bar{t}$ must be smaller than t_0.*

It is convenient to introduce the nth-order moments of the distribution function:

$$m_{i_1 i_2 \ldots i_n} = \int_{R^3} \xi_{i_1} \xi_{i_2} \ldots \xi_{i_n} f \, d\boldsymbol{\xi} \qquad (i_k = 1, 2, 3). \tag{2.2.8}$$

Among the moments $m_{i_1 i_2 \ldots i_n}$ we find for $n = 2$ the momentum flow m_{ij}. Its trace is $\rho|\mathbf{v}|^2 + 2\rho e$, where ρ is the density and e the thermal energy per unit mass.

Frequently it turns out that central moments are more convenient. They are analogous to the moments $m_{i_1 i_2 \ldots i_n}$ except for the fact that we replace the components of the molecular velocity $\boldsymbol{\xi}$ by the components of the peculiar velocity $\mathbf{c}$. Thus we have

$$p_{i_1 i_2 \ldots i_n} = \int_{R^3} c_{i_1} c_{i_2} \ldots c_{i_n} f \, d\boldsymbol{\xi} \qquad (i_k = 1, 2, 3). \tag{2.2.9}$$

Among the moments $p_{i_1 i_2 \ldots i_n}$ we find for $n = 2$ the stress tensor p_{ij}. Its trace is $3p = 3\rho RT = 2\rho e$, where p is the pressure, T the absolute temperature, and R the gas constant. For $n = 3$, by summing over two indices, we find that $\sum_{j=1}^{3} p_{ijj} = 2q_i$, where q_i is the ith component of the heat flow vector.

These moments are space-homogeneous if f has the form Eq. (2.2.5); this applies in particular to the thermal energy per unit volume of the gas.

The moments of the distribution function are particularly useful if we assume Maxwell molecules. Then we have

$$\mathcal{B}(\theta, |\boldsymbol{\xi} - \boldsymbol{\xi}_*|) = \mathcal{B}(\theta), \tag{2.2.10}$$

and the mean free time is independent of temperature and thus constant if the density remains constant. The reason why the moments are more useful when one assumes Maxwell molecules is that when we try to form the equations satisfied by the moments, the contribution from the collision term contains a finite number of moments of order not higher than the order of the moment arising from the time derivative term in the Boltzmann equation. This property was discovered by Maxwell and is characteristic of the molecules that are now named after him.

Homoenergetic affine solutions for the moments of a Maxwell gas were first found by Galkin and turned out to be homoenergetic dilatations.[4,5] The book by Truesdell and Muncaster[5] gives a unified discussion of homoenergetic affine flows for a general medium. The defining properties are the following:

a) The body force (per unit mass) **X** acting on the molecules is constant:

$$\mathbf{X} = \text{const.} \tag{2.2.11}$$

b) The central moments are space-homogeneous.
c) The bulk velocity **v** is an affine function of position **x**:

$$\mathbf{v} = \mathsf{K}(t)\mathbf{x} + \mathbf{v}_0(t). \tag{2.2.12}$$

An analysis of the momentum balance equation based on a), b), and c) immediately leads to the following restrictions on K and $\mathbf{v}_o$:

$$\begin{cases} \dot{\mathsf{K}} + \mathsf{K}^2 = 0, \\ \dot{\mathsf{K}}\mathbf{v}_0 + \mathsf{K}\mathbf{v}_0 = \mathbf{X}. \end{cases} \tag{2.2.13}$$

The general solution of this system is

$$\begin{cases} \mathsf{K}(t) = [\mathsf{I} + t\mathsf{K}(0)]^{-1}\mathsf{K}(0), \\ \mathbf{u}_0(t) = [\mathsf{I} + t\mathsf{K}(0)]^{-1}\left[\mathbf{v}_0(0) + t\mathbf{X} + \frac{1}{2}t\mathsf{K}(0)\mathbf{X}\right], \end{cases} \tag{2.2.14}$$

where I is the 3×3 identity matrix. This solution exists globally for $t > 0$ if the eigenvalues of $\mathsf{K}(0)$ are nonnegative; otherwise the solution ceases to exist for $t = t_0$, where $-t_0^{-1}$ is the largest, in absolute value, among the negative eigenvalues of $\mathsf{K}(0)$.

In particular, if

$$[\mathsf{K}(0)]^2 = 0, \tag{2.2.15}$$

then $[\mathsf{I} + t\mathsf{K}(0)]^{-1} = \mathsf{I} - t\mathsf{K}(0)$ and therefore $\mathsf{K}(t)$ is independent of time. The velocity **v** is then steady if and only if

$$\mathsf{K}(0)\mathbf{X} = \mathbf{0}, \tag{2.2.16}$$

and $\mathbf{v}_0(0)$ is chosen in such a way that

$$\mathsf{K}(0)\mathbf{v}_0(0) = \mathbf{X}. \tag{2.2.17}$$

In particular, this is always possible if $\mathbf{X} = \mathbf{0}$.

Equation (2.2.15) is satisfied if and only if a coordinate system exists for which the matrix representation of $\mathsf{K}(0)$ is given by

$$((K_{ij})) = \begin{pmatrix} 0 & 0 & 0 \\ K & 0 & 0 \\ 0 & 0 & 0 \end{pmatrix}. \tag{2.2.18}$$

For a simple proof of this, see Problem 2.2.1. When (2.2.15) applies one talks, for obvious reasons, of a homoenergetic shear flow. The density is not only space-homogeneous but also constant for this flow.

As shown by Galkin[4,6–8] and Truesdell,[9] the second-order moment equations for a Maxwell gas, associated with a homoenergetic affine flow, are decoupled from those of higher order and can be solved explicitly. This result generalizes the result mentioned above for homoenergetic dilatations. In order to obtain the moment equations, one multiplies Eq. (2.2.7) by the appropriate monomial $c_{i_1} c_{i_2} \dots c_{i_n}$ and integrates over the velocity space. The term with the derivative with respect to the velocity variables must be handled by partial integration; the collision term is complicated unless we adopt Maxwell molecules, as we shall do.

The most interesting moment equations are those for $p_{22} = \int c_1^2 f \, d\boldsymbol{\xi}$, $p_{12} = \int c_1 c_2 f \, d\boldsymbol{\xi}$, $p = \frac{1}{3} \int |\mathbf{c}|^2 f \, d\boldsymbol{\xi}$ (see Problem 2.2.2):

$$\tau \dot{p} + \frac{2}{3} \mathcal{T} p_{12} = 0,$$

$$\tau \dot{p}_{12} + p_{12} + \mathcal{T} p_{11} = 0,$$

$$\tau \dot{p}_{11} - p + p_{11} = 0. \tag{2.2.19}$$

Here τ is a relaxation time, given by

$$\tau = \frac{1}{A\rho}, \tag{2.2.20}$$

where A is a well-defined constant that only depends on the molecular mass m and on the constant giving the strength of the intermolecular force (see Problem 2.2.2).

It turns out that τ is related to the viscosity coefficient μ of the gas when the Boltzmann equation agrees with the Navier–Stokes equations (see next section) by $\tau = p\mu$. $\mathcal{T}$ is the nondimensional number introduced in Eq. (2.2.1) and takes on the form:

$$\mathcal{T} = \tau K = \frac{\mu}{p} K. \tag{2.2.21}$$

Equation (2.2.19) is a system of three linear, homogeneous, first-order differential equations that possesses a set of linearly independent solutions of the form:

$$p = p^0 \exp(\chi t/\tau), \quad p_{12} = p_{12}^0 \exp(\chi t/\tau), \quad p_{11} = p_{11}^0 \exp(\chi t/\tau); \tag{2.2.22}$$

here χ is one of the roots of the third-degree equation

$$\chi(\chi+1)^2 = \frac{2}{3}\mathcal{T}^2. \tag{2.2.23}$$

This equation can be solved explicitly by means of the formula found by Scipione del Ferro, Tartaglia, and Cardano. It is enough here to remark that it certainly has a real root r; the other two roots can be easily expressed in terms of r and turn out to be complex with real part $-1 - r/2$ and imaginary part $\pm[r(1+3r/4)]^{1/2}$. This immediately shows that r must be positive, because the product of the roots must equal the right-hand side of Eq. (2.2.23). Thus, in general, the solution of system (2.2.19) will possess a part that grows exponentially in time and another part that oscillates with an exponentially decaying amplitude. The asymptotically dominating part agrees with the Navier–Stokes equations with an error $O(\mathcal{T}^2)$. The exponential growth of the dominating part can be easily explained by the circumstance that the solution under consideration forbids any heat diffusion; thus the work done by the tangential stress heats up the gas and increases the pressure, which in turn increases the stress components p_{11} and p_{12} because viscosity increases with temperature, and hence with pressure, with the density being constant. Condition b) above holds for the solutions obtained by Truesdell[9] and Galkin.[4,6–8]

The homoenergetic affine solutions satisfy the Boltzmann equation in the entire space and as such they have been discussed in the papers mentioned so far. Boundary conditions for the homoenergetic shear flow in a slab were first discussed in a more recent paper.[10] It is, in fact, easy to check that solutions of the form $f = f(\mathbf{c}, t)$ satisfy the boundary conditions (2.2.3) and (2.2.4) provided the initial distribution is even in $\mathbf{c}$ and $\mathbf{v}(x=0)=0$ and $\mathbf{u}(x=L)=V\mathbf{j}$. The second condition is immediately satisfied by letting $\mathbf{v}=\mathbf{j}Vx/L$.

Thus we have checked that in our case the solution has a steady bulk velocity, whereas the stress components and the temperature grow exponentially in time. Although the explicit calculation applies only to Maxwell molecules, this behavior must be considered as typical for other models as well.

We remark that although we have found explicitly the first few moments of the distribution function, we cannot compute the distribution function itself. This can be done for the BGK model (see Chapter 1, Section 1.9); in this case in fact the equation can be easily solved if the first five moments are known (see Problem 2.2.3)

It is interesting to look for the asymptotic behavior of the solution. Since the complex roots of Eq. (2.2.23) have a negative real part, the corresponding contributions will decay in time and the second-order moments will asymptotically behave as pure exponentials. Although the coefficients depend on the initial data in a way that one may easily calculate (see Problem 2.2.4), the ratios of p_{12} and p_{11} to p do not (in this asymptotic expression) and they can be easily calculated by replacing $\dot{p}$, $\dot{p}_{12}$, and $\dot{p}_{11}$ in (2.2.19) by rp, rp_{12}, and rp_{11} respectively, where r is the real root of Eq. (2.2.23). In particular, we obtain

$$p_{12} = -\frac{3}{2}\frac{r}{\mathcal{T}}p = -\mu K\frac{3r}{2\mathcal{T}^2}. \tag{2.2.24}$$

We remark that, according to Navier–Stokes equations, one should obtain

$$p_{12} = -\mu K. \tag{2.2.25}$$

Let us look for the behavior of r for small values of $\mathcal{T}$. It is clear from Eq. (2.2.23) that we can find r for small values of $\mathcal{T}$ by using the iteration scheme (see Problem 2.2.5)

$$r_n = \frac{2}{3}\mathcal{T}^2 - 2r_{n-1}^2 - r_{n-1}^3 \qquad (n = 1, 2, \ldots), \quad r_0 = 0; \tag{2.2.26}$$

here the sequence r_n approaches r when $n \to \infty$. Thus we obtain

$$r = \frac{2}{3}\mathcal{T}^2 - \frac{8}{9}\mathcal{T}^4 + O(\mathcal{T}^6) \qquad (\text{as } \mathcal{T} \to 0). \tag{2.2.27}$$

Inserting this result into Eq. (2.2.24) shows that the Navier–Stokes constitutive relation, Eq. (2.2.25), is indeed satisfied to lowest order in $\mathcal{T}$, but corrections arise when the latter parameter is no longer negligible with respect to 1. Thus the relation between the stresses and the strain rate is not linear, according to the Boltzmann equation, but higher order contributions are also present. Therefore we should expect the Navier–Stokes equations not to be valid for high strain rates.

Problems

2.2.1 Show that Eq. (2.2.15) is satisfied if and only if a coordinate system exists for which the matrix representation of $\mathsf{K}(0)$ is given by Eq. (2.2.18). (Hint: See Ref. 3.)

2.2.2 Show that Eqs. (2.2.19) hold. (Hint: Multiply the left-hand side of Eq. (2.2.7), with K given by Eq. (2.2.18), by the appropriate functions of $\mathbf{c}$ and perform a partial integration with respect to $\mathbf{c}$ in the right-hand side. This supplies the terms with time derivatives and those containing $\mathcal{T}$, without the factor τ. In the first equation the right-hand side vanishes, because of conservation of energy. As for the other two equations, change the variables in such a way that $\mathbf{c}'$ and $\mathbf{c}'_*$ appear explicitly and not as arguments of f. Then perform the integral with respect to $\boldsymbol{\xi}$ and $\boldsymbol{\xi}_*$ in terms of the moments to obtain a constant times ρ times either $-p_{12}$ or $p - p_{11}$. See Ref. 5.)

2.2.3 Show that one can compute the distribution function for the problem discussed in the text, if the BGK model is used. (Hint: Assume the collision frequency in the collision term of the BGK model, Eq. (2.9.1), to be proportional to the density ρ and hence constant. We thus have a linear partial differential equation for f, since Φ is explicitly known, the temperature being $p/(R\rho)$. The equation can be integrated with the method of characteristics, to yield f in terms of a suitable integral.)

2.2.4 Solve the system of Eqs. (2.2.19) for assigned initial data, assuming the real root r of Eq. (2.2.23) to be known.

2.2.5 Show that the iteration scheme in Eq. (2.2.26) converges if $\mathcal{T}^2$ is sufficiently small (e.g., smaller that $7^{1/2}/2 - 1$) (Hint: Use contraction mapping and the fact that the first iterate is the largest.)

2.3. Couette Flow at Small Mean Free Paths

The solution discussed in the previous section must be considered exceptional for two reasons. First of all, closed-form solutions of the Boltzmann equation (or even of the corresponding moment equations) are very rare when the solution depends on space coordinates. Second, the solution does not represent what one would expect from an experiment on plane Couette flow, where the gas should be able to exchange heat with the walls and reach a steady state.

The main fact is that general boundary conditions for the Boltzmann equation have a half-range character: We assign the distribution function of the molecules leaving a wall in terms of the distribution of those arriving at the wall. This situation is unsymmetric and can be made symmetric only in the case of deterministic boundary conditions, those of specular and reverse reflection. We have seen, however, that the latter conditions are of theoretical but not of practical interest.

If we pass to the easiest among the more realistic boundary conditions, that of diffusion according to a Maxwellian distribution (Eq. (1.11.20) with $\alpha = 1$), we are unable to find any accurate information on the solution, unless we introduce approximations or resort to numerical methods. This is related to the fact that, because of the half-range nature of the boundary conditions, something unusual must happen near a wall: The molecules arriving there "do not know" that there is a wall and have a distribution function that reflects the presence of a boundary only indirectly (because of the collisions they suffer with the molecules coming from the wall).

It is clear then that, near a wall, there must be a layer, of the order of a few mean free paths, where the solution is widely different from that prevailing in the remaining part of the slab. Layers of this kind are called *Knudsen layers* after the great experimentalist Martin Knudsen who made basic investigations on the flow of rarefied gases in the early part of the twentieth century. The Knudsen layer is, of course, meaningful only if the region occupied by the gas is significantly larger than the mean free path.

Hence, in order to understand what happens, we look at the case when the mean free path is small with respect to the other typical lengths occurring in a given problem.

Let us write the Boltzmann equation (2.1.1) in the following form:

$$\xi_1 \partial_x f = \frac{1}{\epsilon} Q(f, f)(x, \boldsymbol{\xi}). \tag{2.3.1}$$

The factor $1/\epsilon$ indicates that the mean free path is small and can be obtained from Eq. (2.1.1) (rewritten in the case when f depends on just one space coordinate x) by the replacement $x \Rightarrow x/\epsilon$. Thus ϵ can be considered as the mean free path or, better, its nondimensional form, the Knudsen number. Here and henceforth we shall simply write ∂_x for partial derivatives with respect to x.

In spite of the fact that we face a singular perturbation problem (the small parameter multiplies the only derivative in our equation), the great mathematician Hilbert[5] proposed the approach to be presently described. He worked on the general case, but the particular case we are considering is worth some attention.

Let us try to find, as Hilbert suggested, a solution of our problem for the Boltzmann equation in the form

$$f = \sum_{n=0}^{\infty} \epsilon^n f_n. \tag{2.3.2}$$

By inserting this formal series into Eq. (2.3.1) and matching the various orders in ϵ, we obtain equations that one can hope to solve recursively (see Problem 2.3.1):

$$Q(f_0, f_0) = 0, \qquad (2.3.3)_0$$

$$2Q(f_1, f_0) = \xi_1 \partial_x f_0 \equiv S_0, \qquad (2.3.3)_1$$

$$\ldots,$$

$$2(Q(f_j, f_0) = \xi_1 \partial_x f_{j-1} - \sum_{i=1}^{j-1} Q(f_i, f_{j-i}) \equiv S_{j-1}, \qquad (2.3.3)_j$$

$$\ldots,$$

where $Q(f, g)$ denotes the symmetrized collision operator and the sum is empty for $j = 1$. The first equation, namely Eq. $(2.3.3)_0$, gives

$$f_0 = M \tag{2.3.4}$$

with the five parameters $(\rho_0, \mathbf{v}_0, T_0)$ still unknown.

Equation $(2.3.3)_1$ can be written as

$$L_M h_1 = s_0, \tag{2.3.5}$$

where L_M denotes the linearized collision operator about the Maxwellian M (i.e., $L_M h = 2Q(M, Mh)/M$), $h_j = f_j/M$, and $s_j = S_j/M$. Note that by the Fredholm alternative for integral equations, this equation has a solution if and only if the integrals of the product of $S_0 = \xi_1 \partial_x M$ by the collision invariants vanish. Therefore, the solvability condition is

$$\int_{R^3} \psi_\alpha \xi_1 \partial_x M d\boldsymbol{\xi} = 0 \quad (\alpha = 0, 1, 2, 3, 4),$$

where the ψ_α are the collision invariants.

The above five equations tell us something about the first five moments of the distribution function f_0, though surprisingly little. To simplify matters, we assume that we already know that u, the x components of the bulk velocity, is zero; this follows from conservation of mass, which tells us that ρu must be constant and, since the first component of the bulk velocity, u, is zero at the (nonporous) walls, it follows that u must vanish everywhere. Then we assume the walls to move in the y direction and hence the third component of the bulk velocity, w, is also assumed to be zero. This leaves us with just three macroscopic fields, in the Maxwellian, (ρ_0, v_0, T_0). In terms of these, the mass, momentum, and energy conservation equations are trivially satisfied except for the x component of the momentum equation, which reduces to

$$\partial_x p_0 = 0, \tag{2.3.6}$$

that is, the pressure p_0 is constant.

In addition, f_1 is not completely known. In fact, if we write

$$f_1 = f_1^O + f_1^F \tag{2.3.7}$$

to denote the decomposition of f_1 into a part that gives zero integral when multiplied by the five collision invariants and a linear combination of the latter, multiplied by M (this decomposition is known to be unique), then f_1^O is determined by Eq. (2.3.5). The property $\int \psi_\alpha f_1^O d\boldsymbol{\xi} = 0$ ($\alpha = 0, 1, 2, 3, 4$) is usually expressed by saying that f_1^O is orthogonal to the collision invariants.

Let us next analyze Eq. (2.3.3)$_2$. If we write the right-hand side as $S_1 = Ms_1$ then

$$L_M h_2 = s_1 \tag{2.3.8}$$

is solvable if

$$\int S_1 \psi_\alpha d\boldsymbol{\xi} = 0.$$

Since $\int \psi_\alpha Q(f, g) d\boldsymbol{\xi} = 0$, this is equivalent to

$$\int \psi \xi_1 \partial_x f_1 d\boldsymbol{\xi} = 0. \tag{2.3.9}$$

By Eq. (2.3.8),

$$f_1^O = ML_M^{-1} s_0, \tag{2.3.10}$$

where L_M^{-1} is the inverse of L_M in the subset of functions orthogonal to the collision invariants. By Eq. (2.3.9),

$$\int_{R^3} \psi \xi_1 \partial_x f_1^O d\boldsymbol{\xi} = -\int_{R^3} \psi \xi_1 \partial_x f_1^F d\boldsymbol{\xi}. \tag{2.3.11}$$

Now something interesting comes out of the left-hand side of Eq. (2.3.11) (the right-hand side gives just a pressure gradient as before). The interesting equations are the second component of the momentum balance and the energy equation. They read as follows:

$$\begin{aligned} \partial_x(\mu_0 \partial_x v_0) &= 0, \\ \partial_x(\kappa_0 \partial_x T_0 + v_0 \mu_0 \partial_x v_0) &= 0. \end{aligned} \tag{2.3.12}$$

Here μ_0 and κ_0 are two functions of T_0 given by

$$\begin{aligned} \mu_0 &= -(RT_0)^{-1} \int_{R^3} M c_1 c_2 L_M^{-1}(c_1 c_2)\, d\mathbf{c}, \\ \kappa_0 &= -(4RT_0^2)^{-1} \int_{R^3} M c_1 |\mathbf{c}|^2 L_M^{-1}(c_1(|\mathbf{c}|^2 - 5RT_0)) d\mathbf{c}. \end{aligned} \tag{2.3.13}$$

In other words we have discovered that the component v_0 of the bulk velocity and the temperature T_0 are given, to the lowest order, by the Navier–Stokes equations for a compressible fluid with viscosity and heat conductivity coefficients determined by the collision operator (i.e., by the molecular interaction). For power-law potentials it is not hard to show that these transport coefficients are proportional to a power of the temperature T_0. In fact, if we change the integration variable from $\mathbf{c}$ to $\mathbf{C} = \mathbf{c}(2RT_0)^{-1/2}$, the operator L_M factorizes into the product of an operator independent of temperature times the power $(T_0)^{\alpha}$, where $\alpha = (n-5)/[2(n-1)]$ for an intermolecular force inversely proportional to the nth power of the distance ($\alpha = 1/2$ for hard sphere molecules). Collecting the other factors coming from powers of T_0 explicitly appearing in the expressions of μ_0 and κ_0 after the replacement of $\mathbf{c}$ by $\mathbf{C}(2RT_0)^{1/2}$, we find that the transport coefficients are both proportional to ${T_0}^{1-\alpha}$. This explains why the Prandtl number for gases, $\mathrm{Pr} = c_p\mu/\kappa$, where c_p is the specific heat at constant pressure, is only slightly dependent on temperature; it would be exactly independent for molecules interacting with (any!) power-law potential. We incidentally remark that the transport coefficients are also independent of the gas density; in fact, ρ_0 appears as a factor in both the Maxwellian distribution and the linearized collision operator; hence it cancels out in the above expressions for the transport coefficients. In the case of Maxwell molecules we can carry the computation a step further[1] and show that the Prandtl number is exactly equal to $2/3$.

The results that we have just found are of interest because they confirm the intuitive idea that, when the mean free path is small, the gas behaves as a continuum described by the Navier–Stokes equations. Several comments are, however, in order. First of all, we have said nothing about the boundary conditions: How should we solve (2.3.12)? Second: What is the price that we paid, that is, what have we lost when expanding in powers of ϵ? We shall discuss the first question in the next chapter. As for the second, we have lost the possibility of describing the phenomena occurring on the scale of ϵ (i.e., in regions having the thickness of a mean free path). Thus we must expect two Knudsen layers near the walls, where the Navier–Stokes equations, and our expansion, do not hold.

Before proceeding to investigate the solution of the Navier–Stokes equations, we remark that the Hilbert expansion may not be convergent. In spite of that, it can be an asymptotic expansion and hence provide a good approximation for small values of ϵ.

We rewrite the Navier–Stokes equations (2.3.12), suppressing the label 0 since we shall not go beyond the lowest order in our expansion. Then we have

$$\partial_x(\mu\partial_x v) = 0,$$

$$\partial_x(\kappa\partial_x T + v\mu\partial_x v) = 0. \tag{2.3.14}$$

The first equation simply says that

$$\mu \partial_x v = \tau_w, \tag{2.3.15}$$

where τ_w is a constant. The notation has been chosen in agreement with standard boundary layer theory; τ_w has the meaning of the opposite (with our conventions) of the stress component p_{12} at both walls.

Were μ independent of temperature, or the latter constant, we would obtain the linear velocity profile considered in the previous section (we remark that this was the case in the previous section; T was there a function of t and not of x and hence that profile satisfies Eq. (2.3.15) provided we allow μ and τ_w to be time dependent, for consistency).

In a similar way the second equation of our system, the energy equation, gives

$$\kappa \partial_x T + v\mu \partial_x v = -q_w, \tag{2.3.16}$$

where q_w is a constant giving the value of the heat flow at a wall, per unit area and unit time, if the velocity v vanishes there. Since we assumed the wall at $x = 0$ to be at rest, v actually vanishes there, if the gas does not slip over the wall. Even if there is a slip, q_w would retain the meaning of heat supplied to the gas by a wall, per unit area and unit time, since the energy flow must be continuous through a steady wall and the stress does not exert any work on the wall if the latter does not move.

Equation (2.3.15) shows the convenience of using a variable s defined by

$$s = \int_0^x \frac{dx}{\mu(T)}. \tag{2.3.17}$$

In terms of this variable, we have a linear profile, because equation (2.3.15) becomes

$$\partial_s v = \tau_w. \tag{2.3.18}$$

The same variable is useful in the energy equation (2.3.16), provided we assume the Prandtl number $\mathrm{Pr} = c_p\mu/\kappa$ to be independent of temperature, as is the case for power-law potentials and hard spheres and, as we said above, approximately true for other molecular interactions as well. If we make this assumption, we obtain for the energy equation:

$$\frac{c_p}{\mathrm{Pr}} \partial_s T + v\partial_s v = -q_w. \tag{2.3.19}$$

We can now easily integrate the last two equations to give

$$v = \tau_w s + v_w, \tag{2.3.20$_1$}$$

$$c_p T + \frac{1}{2} \mathrm{Pr}\, v^2 = -\, \mathrm{Pr}\, q_w s + c_p T_w. \tag{2.3.20$_2$}$$

Here v_w and T_w are two new integration constants. v_w is the velocity of the gas at the wall located at $x = 0$, and T_w can be interpreted as the temperature of the gas there only if the gas does not slip on the wall ($v_w = 0$).

If we eliminate v from the $(2.3.20)_2$ by means of $(2.3.20)_1$, we obtain

$$c_p T = -\frac{1}{2} \mathrm{Pr}\, (\tau_w s + v_w)^2 - \mathrm{Pr}\, q_w s + c_p T_w. \tag{2.3.21}$$

We have thus obtained the general solution of the system (2.3.14) in terms of the variable s. We remark that $c_p = (5/2) R$ is just a constant; in fact, from the general thermodynamic definitions and properties, we obtain for a monatomic Boltzmann gas:

$$c_p = R + c_v = R + \left(\frac{\partial e}{\partial T} \right)_\rho = \frac{5}{2} R,$$

where c_v is the specific heat at a constant volume.

Our solution still has a drawback to be presently eliminated. The convenient variable s was defined implicitly in Eq. (2.3.17). Now, knowing T as a function of s, we obtain

$$x = \int_0^s \mu(T(s)) ds, \tag{2.3.22}$$

that is, the relation between the variables x and s, provided we are able to perform the integral in (2.3.22). For power-law potentials μ varies as a power of T. Two typical cases in which the integral can be explicitly performed are the cases $\mu = \mu_w (T/T_w)$ (Maxwell molecules) and $\mu = \mu_w (T/T_w)^{1/2}$ (hard spheres). The first case is the easiest and yields

$$\begin{aligned} x &= \frac{\mu_w}{c_p T_w} \int_0^s \left(-\frac{1}{2} \mathrm{Pr}\, (\tau_w s + v_w)^2 - \mathrm{Pr}\, q_w s + c_p T_w \right) ds \\ &= \frac{\mu_w}{c_p T_w} \left(-\frac{1}{2} \mathrm{Pr} \left(\frac{1}{3} \tau_w^2 s^3 + v_w \tau_w s^2 + v_w^2 s \right) - \mathrm{Pr}\, \frac{1}{2} q_w s^2 + c_p T_w s \right). \end{aligned} \tag{2.3.23}$$

The relation between x and s is cubic and can, in principle, be inverted explicitly. We remark that one can obtain a relation between v and x without effort. In

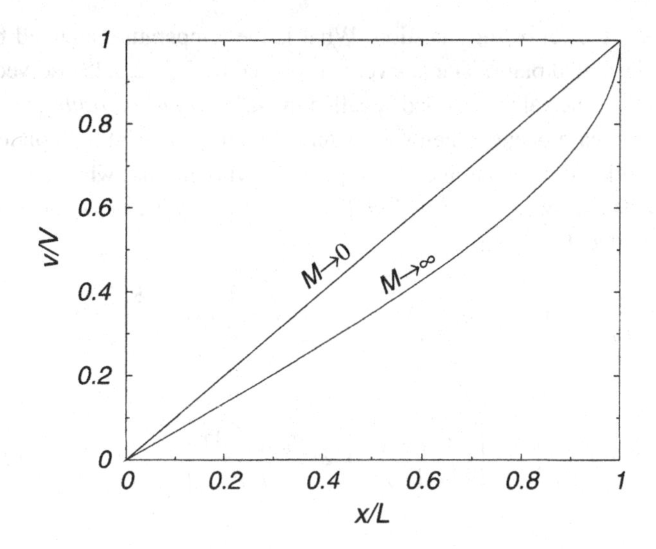

Figure 2.3. Limiting velocity profiles for plane Couette flow.

fact, $(2.3.20)_1$ gives $s = (v - v_w)/\tau_w$, which can be inserted in Eq. (2.3.23) to obtain the desired relation.

It is remarkable that we have proceeded so far without using boundary conditions. These must be specified in order to compute the integration constants. In continuum mechanics, one usually assumes that the gas does not slip on the wall. This gives $v_w = 0$ and a relation to determine τ_w:

$$\tau_w L = \frac{\mu_w}{c_p T_w}\left(-\mathrm{Pr}\frac{1}{6}V^3 - \mathrm{Pr}\frac{1}{2\tau_w}q_w V^2 + c_p T_w V\right). \tag{2.3.24}$$

The profile changes considerably with the value of the Mach number. The limiting cases Ma→ 0 and Ma→ ∞ are shown in Fig. 2.3.

The above relation is a bit clumsy but it simplifies a little if we assume the plate at $x = 0$ to be thermally insulated. Then $q_w = 0$ and, if $\overline{T}$ is the temperature at $x = L$, Eq. $(2.3.20)_1$ gives

$$c_p\overline{T} = -\frac{1}{2}\mathrm{Pr}\,(V)^2 + c_p T_w. \tag{2.3.25}$$

Hence we can replace T_w in terms of $\overline{T}$ in Eq. (2.3.24) to obtain

$$\tau_w L = \frac{\overline{\mu}}{c_p\overline{T}}\left(-\frac{5}{3}\mathrm{Pr}\,V^3 + c_p T_w V\right). \tag{2.3.26}$$

One can ask the following question: What is the temperature attained by the gas at the insulated plate? The answer is provided by Eq. (2.3.25), solved with respect to T_w. The value obtained is called the *recovery temperature*.

A different case occurs when we assign the temperatures at both plates. We shall not work out the general case but just deal with the case when both plates are at rest. The answer is clearly $T = T_w - q_w s \mathrm{Pr}/c_p$, where s is obtained (for Maxwell molecules) from

$$x = \frac{\mu_w}{T_w} \int_0^s \left(T_w - q_w \frac{\mathrm{Pr}}{c_p} s \right) ds = \mu_w s - \frac{1}{2} \mu_w q_w \frac{\mathrm{Pr}}{c_p T_w} s^2 \tag{2.3.27}$$

or

$$s = \frac{c_p T_w}{q_w \mathrm{Pr}} - \left[\left(\frac{c_p T_w}{q_w \mathrm{Pr}} \right)^2 + 2x \frac{c_p T_w q_w \mathrm{Pr}}{\mu_w} \right]^{1/2}. \tag{2.3.28}$$

Hence

$$T = \left(T_w^2 + 2x \frac{T_w q_w}{\kappa_w} \right)^{1/2} \tag{2.3.29}$$

and, by letting $x = L$ and $T = \overline{T}$,

$$q_w = \frac{\kappa_w (\overline{T}^2 - T_w^2)}{2 L T_w}. \tag{2.3.30}$$

This formula looks unsymmetric, because when exchanging the temperatures of the two walls, T_w and $\overline{T}$, one should obtain just a change of sign in the right-hand side. Since, however, κ/T is independent of temperature by assumption, the asymmetry is only apparent.

Problem

2.3.1 Check that Eqs. (2.3.3)$_j$ actually follow when the expansion (2.3.2) is inserted into Eq. (2.3.1) and the coefficients of the different powers in the left and right sides are equated (Hint: Use the quadratic nature of $Q(.,.)$ and Cauchy's rule for the product of two series.)

2.4. Couette Flow at Large Mean Free Paths

After seeing the behavior of a gas when it behaves like a continuum, in this section we consider the opposite case in which the small parameter is the Knudsen number (or the inverse of the mean free path).

In keeping with the traditional convention that ϵ denotes a small number, we shall replace it by $1/\epsilon$. The new ϵ is *inversely* proportional to the mean free path:

$$\xi_1 \partial_x f = \epsilon Q(f, f). \tag{2.4.1}$$

By analogy with what we did in the previous section, we might be tempted to use a series expansion of the form (2.3.2), albeit with a different meaning of the expansion parameter. This, however, does not work, for a reason to be presently explained. The factor ξ_1 in Eq. (2.4.1) takes all possible values and hence also values of order ϵ; thus we should expect troubles from the molecules traveling almost parallel to the wall or with low speed, because then the left-hand side can become smaller than the right-hand side, in spite of the small factor ϵ. This is confirmed by an actual calculation. If we try the expansion, we obtain

$$\xi_1 \partial_x f_0 = 0, \tag{2.4.2$_0$}$$

$$\xi_1 \partial_x f_1 = Q(f_0, f_0), \tag{2.4.2$_1$}$$

$$\dots,$$

$$\xi_1 \partial_x f_j = \sum_{i=0}^{j-1} Q(f_i, f_{j-i-1}), \tag{2.4.2$_j$}$$

$$\dots.$$

Now, it is possible to solve Eq. (2.4.2)$_0$ (we shall do it soon). If we try, however, to solve Eq. (2.4.2)$_1$ we encounter problems with the right-hand side. The latter, after the zeroth approximation f_0 has been computed, is a known function and we can try to solve our equation by dividing by ξ_1 and integrating with respect to x. This simple method meets with the aforementioned difficulty. In fact we are confronted with an integrand with a vanishing denominator and the integral does not converge unless the term $Q(f_0, f_0)$ vanishes at ξ_1 (which it does not). A further nuisance is that f_0 is discontinuous, as we shall see, at the singular point.

Let us now solve the first equation in our perturbation scheme, which, by itself, does not pose many problems. The solution has a physical meaning: It describes the behavior of a gas with no molecular collisions (Knudsen gas). We exclude the possibility of a constant concentration of molecules traveling parallel to the wall (if there are no molecular collisions it is always possible to add such a parallel beam, without disturbing the other molecules, which only collide with the walls). Under the said assumption, we can drop the factor ξ_1 in Eq. (2.4.2)$_0$ and obtain that f_0 is constant with respect to x for any $\boldsymbol{\xi}$. In other words, f_0 is an arbitrary function of $\boldsymbol{\xi}$. This function, as usual, is determined by the boundary conditions. If we adopt the most general boundary conditions, the problem becomes complicated and requires solving an integral equation. Then we restrict ourselves to the simplest case, that of Maxwell's boundary conditions, Eq. (1.10.20). Actually, we shall consider the case of diffuse reflection ($\alpha = 1$ in Eq. (1.10.20)). Then we find that f_0 equals two different Maxwellians M_0 and M_L, according to whether ξ_1 is positive or negative. The two Maxwellians will differ in the values of their tangential bulk velocity and temperature. Thus,

$$f_0 = \begin{cases} A_0 \exp(-\beta_0 |\boldsymbol{\xi}|^2) & \text{if } \xi_1 > 0, \\ A_L \exp(-\beta_L |\boldsymbol{\xi} - V\mathbf{j}|^2) & \text{if } \xi_1 < 0, \end{cases} \tag{2.4.3}$$

where β_0 and β_L are inversely proportional to T_0 and T_L, respectively. The two constants A_0 and A_L must be presently determined. We remark that the equations and the boundary conditions do not determine the solution uniquely: We must also assign the amount of gas, or equivalently, since everything is constant with respect to x, the density ρ. The other condition is provided by the boundary conditions, because, if the wall is assumed to be nonporous, the mass flow at each wall must be zero. Since the mass flow ρv is independent of x (mass conservation!), we need only impose the condition at one wall. Thus we have two conditions to determine A_0 and A_L; if the walls have equal temperatures and hence $\beta_0 = \beta_L = \beta$, then the two constants are both equal to $\rho/2(\beta/\pi)^{3/2}$. Otherwise we have

$$A_0 \left(\frac{\pi}{\beta_0}\right)^{3/2} + A_L \left(\frac{\pi}{\beta_L}\right)^{3/2} = \rho, \tag{2.4.4}$$

$$A_0 \left(\frac{\pi}{2\beta_0^2}\right) = A_L \left(\frac{\pi}{2\beta_L^2}\right). \tag{2.4.5}$$

The second equation shows that

$$A_0 = A\beta_0^2, \quad A_L = A\beta_L^2, \tag{2.4.6}$$

where A is a constant to be determined through the first equation, and direct substitution shows that

$$A = \frac{\rho}{\pi^{3/2}\left(\beta_0^{1/2} + \beta_L^{1/2}\right)}. \tag{2.4.7}$$

Thus we have a complete solution. We remark that all the moments of the solution, including the velocity and temperature, are constant between the plates. In particular the y component of the bulk velocity is

$$v = V\frac{(T_L)^{1/2}}{(T_L)^{1/2} + (T_0)^{1/2}}. \tag{2.4.8}$$

It is clear that the gas slips over both walls. If the temperatures are equal, $v = V/2$ and the amount of slip is equal at both plates: one half of the speed of the moving plate.

The temperature of the gas is given by

$$T = (T_L T_0)^{1/2} + \frac{V^2}{3k_B}\frac{(T_L T_0)^{1/2}}{[(T_L)^{1/2} + (T_0)^{1/2}]^2}. \tag{2.4.9}$$

The density of the molecules ρ has been assigned, but it is convenient to introduce the densities ρ^+, ρ^- of molecules traveling with positive, respectively negative values of ξ_1. They are implicitly determined by the previous calculation of the constants $A_\pm$. Obviously $\rho^- + \rho^- = \rho$; in addition mass conservation at the walls implies $\rho^+(T_0)^{1/2} = \rho^-(T_L)^{1/2}$. Hence

$$\rho^+ = \rho\frac{(T_L)^{1/2}}{(T_L)^{1/2} + (T_0)^{1/2}}; \quad \rho^- = \rho\frac{(T_0)^{1/2}}{(T_L)^{1/2} + (T_0)^{1/2}}. \tag{2.4.10}$$

It is also easy to compute the stress $p_{12} = -\tau_w$. We obtain

$$p_{12} = -\frac{V}{2\pi^{1/2}}\left[\rho^-\left(\frac{2k_B T_L}{m}\right)^{1/2} + \rho^+\left(\frac{2k_B T_0}{m}\right)^{1/2}\right]. \tag{2.4.11}$$

Thus there is a drag exerted on the plates, which depends on the density but not on the distance between the plates. Finally, we can compute the heat flow $q_1 = q_w$. We obtain

$$q_1 = \frac{1}{\pi^{1/2}}\left[\rho^+\left(\frac{2k_B T_0}{m}\right)^{3/2} - \rho^-\left(\frac{2k_B T_0}{m}\right)^{3/2}\right] + \frac{V^2}{4\pi^{1/2}}\left[\rho^+\left(\frac{2k_B T_0}{m}\right)^{1/2} + \rho^-\left(\frac{2k_B T_0}{m}\right)^{1/2}\right]\frac{(T_L)^{1/2}}{(T_L)^{1/2} + (T_0)^{1/2}}. \tag{2.4.12}$$

One can repeat the calculations for Maxwell's boundary conditions (1.10.20) with $\alpha \neq 1$. They are easy but cumbersome, especially because the molecules scattered from each of the walls do not have the same temperature as the wall. Once these two temperatures have been computed, the rest of the calculation proceeds in much the same way as that for $\alpha = 1$, the results of which have just been stated.

Let us see now what happens at the next order in the inverse Knudsen number ϵ. When we substitute our free-molecular solution in $(2.4.2)_1$, we find something that, generally speaking, is not finite at $\xi_1 = 0$. Thus there is no hope in proceeding with the expansion. In spite of this, we can introduce a method (called *collision iteration*), first introduced by Willis[11–15] to compute the solution in the nearly free molecular regime. We split the collision term into gain and loss terms (this requires some cutoff for grazing collisions for intermolecular potentials with infinite range) and write as a first step in an iteration scheme:

$$\xi_1 \partial_x f_1 + \epsilon f_1 R(f_0) = \epsilon G(f_0, f_0). \tag{2.4.13}$$

The splitting may look arbitrary but can be justified. In fact the gain term and the factor $R(f)$ in the loss term tend to smooth out the singularities, whereas the f in the loss term retains, of course, any singularity. Now, we can solve for f_1 without problems, because f_0 is constant with respect to x. Hence

$$f_1 = \begin{cases} f_0 \exp\left(-\dfrac{\epsilon R(f_0)x}{\xi_1}\right) + \dfrac{G(f_0, f_0)}{R(f_0)}\left[1 - \exp\left(-\dfrac{\epsilon R(f_0)x}{\xi_1}\right)\right] & (\xi_1 > 0), \\ f_0 \exp\left(-\dfrac{\epsilon R(f_0)(L-x)}{|\xi_1|}\right) + \dfrac{G(f_0, f_0)}{R(f_0)}\left[1 - \exp\left(-\dfrac{\epsilon R(f_0)(L-x))}{|\xi_1|}\right)\right] & (\xi_1 < 0). \end{cases} \tag{2.4.14}$$

We remark that the solution remains regular even at ξ_1. If we try to expand into a power series in ϵ, the singularity arises again. It can be shown, however, that the error of the approximate solution we have found is of order higher than ϵ. In particular the solution does not satisfy the conservation equations, but the error is of the order just mentioned.

We remark that our approximate solution, in addition to being explicit, supplies some ideas on the behavior of the exact solution.

Let us look now at the dependence of the solution on ϵ. We shall consider Maxwell molecules with an angular cutoff, to avoid unnecessary complications, and use the trivial identity

$$F(\xi_1, \xi_2, \xi_3) = [F(\xi_1, \xi_2, \xi_3) - F(0, \xi_2, \xi_3)g(\xi_1)] + F(0, \xi_2, \xi_3)g(\xi_1), \tag{2.4.15}$$

which holds for any pair of functions F and g. We want to apply it to $F = F_0 = G(f_0, f_0)/R(f_0)$ and a suitably chosen g. Since we shall apply it for either positive or negative values of ξ_1, by $F(0)$ we shall understand the limit from the right and from the left, respectively, since F may be discontinuous at $\xi_1 = 0$.

As for the function g, we cannot take $g = 1$ (the easiest choice), because it cannot be integrated on a half line. We shall take for g a one-dimensional Maxwellian $g = \exp(-\xi_1^2/(2RT))$, where T is the temperature computed above. Any function with the same qualitative properties would lead to the same result.

If we now compute some moments of the distribution f_1, which do not contain ξ_1 as a factor in the integrand (e.g., the density or the second component of the bulk velocity), we meet integrals of the form

$$I = \int_0^\infty \exp\left(-\beta\xi_1^2 - \frac{\epsilon R(f_0)x}{\xi_1}\right) d\xi_1. \tag{2.4.16}$$

In the case of Maxwell molecules $R(f_0)$ is a constant with respect to ξ_1 (essentially the density ρ computed above). If we introduce the so-called Abramowitz functions:[16,17]

$$T_n(s) = \int_0^\infty t^n \exp\left(-t^2 - \frac{s}{t}\right) dt \qquad (n = \ldots -1, 0, 1, 2, \ldots), \tag{2.4.17}$$

we obtain

$$I = \beta^{-1/2} T_0(\epsilon R(f_0)x\beta^{1/2}). \tag{2.4.18}$$

The functions $T_n(s)$ do not possess a powers series expansion in s but are sums of two functions, one of which has such power expansion, while the other is the product of an expandable function times $\log s$. In fact, using simple formulas (Problem 2.4.1), we obtain

$$T_0(s) = \frac{\pi^{1/2}}{2} - \left(\frac{5}{2} - \frac{3\gamma}{2}\right) s + s\log s + O(s^2 \log s), \tag{2.4.19}$$

where $\gamma = 0.5772156649\ldots$ is the Euler–Mascheroni constant. It is important to underline that, for practical purposes, the terms containing a logarithm multiplying a power s^n are comparable with terms containing just s^n. In fact, a small number, in typical engineering practice, is 10^{-2} or 10^{-3} (i.e., something which will produce a natural logarithm close to 5). This explains why, in common practice, terms of order $s^n \log s$ and s^n are collected together, though, formally, the former is dominant. This is, of course, justified, if the quantity we are considering is not vanishing at $\epsilon = 0$.

We have thus discovered why the simple-minded power expansion did not work. We need logarithms as well. Once this is clarified, the computations can be carried out. We remark that, had we considered moments containing a power of ξ_1, we would have found Abramowitz functions of order n higher than zero. For these functions, we also meet logarithms but only later in the expansion. Thus the stress p_{12} has a contribution of order ϵ, but not one of order $\epsilon \log \epsilon$. The same applies to the heat flow q_1.

The logarithmic singularity has been found under the assumption that we can separate the gain and loss terms. Unless the molecules are hard spheres, this involves some kind of cutoff. In the non-cutoff case, a detailed analysis shows that an expansion of the solution for large values of the Knudsen number produces a correction of the order of $\mathrm{Kn}^{-(s-1)/(s+1)}$ if the intermolecular force varies according to the inverse sth power of the distance.[14]

Problem

2.4.1 Show that if $T_0(s)$ is defined by Eq. (2.4.17) with $n = 0$, then (2.4.19) applies, at least for positive values of s. (Hint: First subdivide the integral giving $T_0(s) - T_0(0)$ into two parts, with lower and upper limit 1, respectively. Then subtract $\exp(-s/t)$ in the second integral and add the integral of the latter function between 0 and 1. We have now three integrals. The first two can be expanded in powers of s up to a remainder of order s^2. The coefficient of s is evaluated by recalling that one of the expressions of γ is $\gamma = \int_0^1 [1 - \exp(-t)]dt - \int_1^\infty t^{-1} \exp(-t)dt$. The third integral is dealt with by replacing t by $1/t$. A partial integration produces a term of order s^2 minus another integral of the form $\int_{1/s}^\infty t^{-1} \exp(-t)dt$. The latter by a further splitting produces a term $-\log s$ and a linear term and a remainder.)

2.5. Rarefaction Regimes

We have seen that, using perturbation methods, we can, in principle, compute the behavior of a rarefied gas in the limiting cases of large and small mean free paths. We have also uncovered several important features of rarefied flows. The most remarkable one is that Navier–Stokes equations do not apply, unless the mean free path is sufficiently small. Actually, to see where the usual description of a fluid breaks down, it is sufficient to remark that the definition of the components of the stress tensor in Eq. (1.6.10) indicates that they form a nonnegative definite matrix and hence each nondiagonal component is certainly less than one half of the trace ($= 3p$). Hence the validity of the Navier–Stokes constitutive relation would imply, for a simple shear flow, such as plane Couette flow ($v = v_2$ being the y component of the velocity vector $\mathbf{v}$ and μ the viscosity coefficient):

$$|\mu \partial v/\partial x| \leq 3p/2. \tag{2.5.1}$$

The significance of (2.5.2) is clear if we introduce the mean free path, which was discussed in Section 2.3, which turns out to be related to the viscosity by the arguments discussed there. Rather than sticking to the qualitative concept of Sections 2.2 and 2.3, we introduce a precise value for the mean free path, denoted by λ, and related to μ in the following way:

$$\lambda = \mu(2RT)^{1/2}/p, \tag{2.5.2}$$

so that Eq. (2.5.2) becomes

$$|\partial v/\partial x| \leq 3(2RT)^{1/2}/\lambda. \tag{2.5.3}$$

To appreciate this simple result, we remark that the mean free path λ is about 1 meter at about 130 km. In any case, Eq. (2.5.3) indicates that for rarefied gases and/or high speeds, one cannot rely on the usual Navier–Stokes equations for a compressible fluid and one must resort to kinetic theory.

Thus the Boltzmann equation became a practical tool for the aerospace engineers, when they started to remark that flight in the upper atmosphere must face the problem of decreasing ambient density with increasing height. This density reduction would alleviate the aerodynamic forces and heat fluxes that a flying vehicle would have to withstand. However, for virtually all missions, the increase of altitude is accompanied by an increase in speed; thus it is not uncommon for a spacecraft to experience its peak heating at considerable altitudes, such as, for example, 70 km. When the density of a gas decreases, there is, of course, a reduction of the number of molecules in a given volume and, what is more important, an increase in the distance between two subsequent collisions of a given molecule, till one may well question the validity of the Euler and Navier–Stokes equations, which are usually introduced on the basis of a continuum model, which does not take into account the molecular nature of a gas. It is to be remarked that, according to the analysis of Section 2.3, the use of those equations can also be based on the kinetic theory of gases, which justifies them as asymptotically useful models when the mean free path is negligible.

In the area of environmental problems, the Boltzmann equation is also required. Understanding and controlling the formation, motion, reactions, and evolution of particles of varying composition and shapes, ranging from a diameter of the order of .001 μm to 50 μm, as well as their space–time distribution under gradients of concentration, pressure, temperature, and the action of radiation, has grown in importance, because of the increasing awareness of the local and global problems related to the emission of particles from electric power plants, chemical plants, and vehicles as well as of the role played by small particles in the formation of fog and clouds, in the release of radioactivity from nuclear reactor accidents, and in the problems arising from the exhaust streams of aerosol reactors, such as those used to produce optical fibers, catalysts, ceramics, silicon chips, and carbon whiskers.

One cubic centimeter of atmospheric air at ground level contains approximately 2.5×10^{19} molecules. About a thousand of them may be charged (ions). A typical molecular diameter is 3×10^{-10} m (3×10^{-4} μm) and the average distance between the molecules is about ten times as much. The mean free path is of the order of 10^{-8} m, or 10^{-2} μm. In addition to molecules and ions one cubic centimeter of air also contains a significant number of particles varying in size, as indicated above. In relatively clean air, these particles can number 10^5 or more, and include pollen, bacteria, dust, and industrial emissions. They can be both beneficial and detrimental, and they arise from a number of natural sources as well as from the activities of all living organisms, especially humans. The particles can have complex chemical compositions and shapes and may even be toxic or radioactive.

A suspension of particles in a gas is known as an aerosol. Atmospheric aerosols are of global interest and have important impact on our lives. Aerosols are also of great interest in numerous scientific and engineering applications.[18]

A third area of application of rarefied gas dynamics has emerged in the past few years. Small size machines, called micromachines, are being designed and built. Their typical sizes range from a few microns to a few millimiters. Rarefied flow phenomena that are more or less laboratory curiosities in machines of more usual size can form the basis of important systems in the micromechanical domain.

One should not forget the design and simulation of aerosol reactors, used to produce optical fibers, catalysts, ceramics, silicon chips, and carbon whiskers, which have been mentioned above as sources of air pollution.

A further area of interest occurs in the vacuum industry. Although this area existed for a long time, the expense of the early computations with kinetic theory precluded applications of numerical methods. The latter could develop only in the context of the aerospace industry, because until recently the big budgets required were available only there.

We have met already the basic parameter Kn measuring the degree of rarefaction of a gas. Of course, one can consider several Knudsen numbers, based on different characteristic lengths, exactly as one does for the Reynolds number.

Thus, in the flow past a body, there are two important macroscopic lengths: the local radius of curvature and the thickness of the viscous boundary layer δ, and one can consider Knudsen numbers based on either length. Usually the second one ($\mathrm{Kn}_\delta = \lambda/\delta$) gives the most severe restriction to the use of Navier–Stokes equations in aerospace applications.

In the case of Couette flow, the basic length is the distance between the plates. When Kn is larger than (say) 0.01, we have already remarked (Section 2.3) that the presence of a thin layer near the wall, of thickness of the order λ (Knudsen layer), influences the viscous profile in a significant way.

This and other regimes to be presently described are of interest in both high altitude flight and aerosol science; in particular they are all met by a shuttle when returning to Earth. The case of Couette flow is important because it exhibits all these regimes in a very simple way. The only phenomenon of importance, that does not show up in our simple example is the formation of shock waves, which we shall discuss in Chapter 4.

When the mean free path increases, one witnesses a thickening of the shock waves, whose thickness is of the order of 6λ. The bow shock in front of a body merges with the viscous boundary layer; that is why this regime is sometimes called the merged layer regime by aerodynamicists. We shall use the other frequently used name of *transition regime*.

When Kn is large (few collisions), phenomena related to gas–surface interaction play an important role. We discussed this case in the previous section. One distinguishes between free molecular and nearly free molecular regimes. In the first case the molecular collisions are completely negligible, while in the second they can be treated as a perturbation. We gave example of both calculations in the case of Couette flow.

It is clear that the transition regime is the typical regime where the Boltzmann equation, in its full complexity, is needed. Thus we shall devote the next two sections and the next chapter to methods that have been devised to deal with the transition regime.

2.6. Moment Methods for Plane Couette Flow

To deal with the transition regime, one can try to improve upon the perturbation methods used for either large or mean free paths by computing further terms. There are, however, several drawbacks. In addition to the fact that the calculations become increasingly prohibitive, the two methods are not suitable for two different reasons, to be presently explained.

We might expect that the Hilbert expansion is not convergent, but only asymptotic; this means that we cannot improve accuracy by enlarging the number of terms but only by reducing the Knudsen number. In other words, we can improve upon the Navier–Stokes calculations only when they are already good. These general statements should not be taken as applying to all solutions: There might be problems in which, because of certain simplifications or importance of certain terms with respect to others, the Hilbert expansion, or suitable modifications of it, might be useful. We shall return on this point in Chapters 5 and 7.

As for the collision iteration, one can conjecture that it converges, but its convergence will be rather slow and thus not suitable to reach Knudsen numbers of order unity.

A method that looks promising is to consider the moment equations of the Boltzmann equation. Even in the simplest case, Maxwell molecules, a finite number of equations, however, does not constitute a closed system, and a closure method must be introduced. There are, of course, exceptions to this general statement. Space-homogeneous problems and homoenergetic affine flows belong to the family of exceptions (for Maxwell molecules).

The first few moment equations are the conservation equations. In the case of Couette flow, under the assumption, justified in Section 2.3, that $v_1 = u = 0$ and $v_3 = 0$, they reduce to

$$\begin{aligned} \partial_x p_{11} &= 0, \\ \partial_x p_{12} &= 0, \\ \partial_x (q_1 + v p_{12}) &= 0. \end{aligned} \tag{2.6.1}$$

Let us consider the next few equations. We can consider those for the moments associated with the monomials ξ_1^2 $\xi_1\xi_2$ $\xi_1|\boldsymbol{\xi}|^2/2$: We obtain (Problem 2.6.3)

$$\partial_x p_{111} = -\frac{p}{\mu}(p_{11} - p),$$

$$\partial_x (p_{121} + p_{11} v) = -\frac{p}{\mu} p_{12}, \qquad (2.6.2)_1$$

$$\partial_x \left(q_{11} + v p_{121} + \frac{1}{2} p_{11} v^2 \right) = -\frac{p}{\mu} \left(p_{12} v + \frac{2}{3} q_1 \right), \qquad (2.6.2)_2$$

where

$$q_{11} = \frac{1}{2} \int_{R^3} \xi_1^2 |\boldsymbol{\xi}|^2 f \, d\boldsymbol{\xi} \tag{2.6.3}$$

is one half of the trace of p_{11ij}. μ is, of course, the viscosity coefficient. Equations $(2.6.2)_1$ and $(2.6.2)_2$ are exact only for Maxwell molecules, because, otherwise, infinitely many moments arise in any single moment equation from the collision term. Equation $(2.6.2)_2$ is frequently rewritten in the form

$$\partial_x q_{11} + p_{121}\partial_x v = -\frac{2}{3}\frac{p}{\mu}q_1. \qquad (2.6.2)_{2'}$$

In fact, since p_{11} is constant because of (2.6.1), Eq. $(2.6.2)_1$ gives

$$\frac{1}{2}\partial_x(p_{11}v^2) = v\partial_x(p_{11}v) = -v\partial_x(p_{121}) - \frac{p}{\mu}p_{12}v,$$

which can be used to transform Eq. $(2.6.2)_2$ into Eq. $(2.6.2)_2$. A third form in which Eq. $(2.6.2)_2$ is written is

$$\partial_x(q_{11} + vp_{121}) = -\frac{5pR}{2\kappa}q_1, \qquad (2.6.2)_{2''}$$

where κ is the heat conductivity related to μ by $\mathrm{Pr} = c_p\mu/\kappa = 5R\mu/(2\kappa)$ and $\mathrm{Pr} = 2/3$ (exactly) for Maxwell molecules.

It is immediately obvious that there are more unknowns than equations. The extra unknowns are p_{111}, p_{121}, and q_{11}. This is unavoidable and constitutes the main drawback of the moment equations. Due to the presence of the factor ξ_1 in front of the space derivative, moments of higher order appear in each equation and we face a system that is not closed. To avoid the necessity of dealing with infinitely many equations, we must introduce a closure recipe, dictated by physics and simplicity.

The simplest *recipe* would be to assume that the higher order moments are related to those of low order by the same relation holding for a Maxwellian distribution. This simplification is, however, too drastic. In fact, to be consistent, we should also let the heat flow be zero. A better approach is to assume an approximate form for the distribution function depending on just a finite number of moments and compute those of higher order in terms of those appearing in the approximation. Early versions of this approach can be found in Maxwell's papers. A systematic procedure was introduced by Grad in 1949.[19] This approach goes under the name of "thirteen moment equations," because exactly thirteen moments were chosen to appear in the approximation to the distribution function for a general problem (the number of required moments may reduce for a particular problem, such as plane Couette flow). The choice is dictated by the fact that these moments are those which have a recognized status in continuum mechanics, being the density, the three components of the bulk velocity, the six components of the stress tensor, and the three components of the heat flow vector. Other important physical quantities, such as pressure p, thermal energy per unit volume e, and temperature T are related to the components of the stress tensor by well-known relations.

Grad's *ansatz* for the distribution function reads as follows:

$$f = M\left[1 + \frac{3}{4pe}\sum_{ij}(p_{ij} - p\delta_{ij})c_i c_j + \frac{3}{2pe}\sum_i q_i\left(\frac{3|\mathbf{c}|^2}{10e} - 1\right)\right], \tag{2.6.4}$$

where M is the local Maxwellian distribution.

Equation (2.6.4) provides the required closure solutions for the moments:

$$p_{ijk} = \frac{2}{5}(q_i\delta_{jk} + q_j\delta_{ik} + q_k\delta_{ij}), \tag{2.6.5}$$

$$\begin{aligned} p_{ijkr} &= RT(p_{ij}\delta_{kr} + p_{ik}\delta_{jr} + p_{ir}\delta_{jk} + p_{jr}\delta_{ik} + p_{kr}\delta_{ij} + p_{jk}\delta_{ir}) \\ &\quad + \frac{pRT}{3}(\delta_{ij}\delta_{kr} + \delta_{ik}\delta_{jr} + \delta_{ir}\delta_{jk}). \end{aligned} \tag{2.6.6}$$

In order to close the system of equations for Couette flow, we specialize these relations to yield

$$\begin{aligned} p_{111} &= \frac{6}{5}q_1, \\ p_{121} &= \frac{2}{5}q_2, \\ q_{11} = \frac{1}{2}\sum_k p_{11kk} &= \frac{7RT}{2}p_{11} + \frac{4pRT}{3}. \end{aligned} \tag{2.6.7}$$

The second of these relations brings in a component of the heat flow vector, which was not present in the Navier–Stokes approximation. We cannot use geometrical symmetry arguments to exclude immediately this component. Even if we assume it to be zero we see that the set of equations is remarkably complicated.

The method of thirteen moments has the advantage of producing a number of partial differential equations in ordinary space, in place of the integro-differential Boltzmann equation in phase space. These equations are notably more complicated than the Navier–Stokes equations, but this is not the main reason why they have not become popular in rarefied gas dynamics. One reason is related to the fact that the distribution function behaves in a peculiar way in the neighborhood of a boundary, because of the presence of Knudsen layers, and its behavior is poorly approximated by a polynomial.

The problem of boundaries also arises in connection with the problem of obtaining the boundary conditions for the thirteen moment equations. We know that the boundary conditions for the Boltzmann equation have a half-range character. Thus we should form, roughly speaking, just one half of the moments of the distribution at a boundary. Otherwise, we would obtain too many boundary conditions. In the problem under consideration, the boundary conditions should essentially involve four quantities (v, p_{12}, T, q_1) and thus we should expect two boundary conditions at each plate (the differential equations are of first order). A reasonable approach, which can be traced back to Maxwell, is to take the moments corresponding to flows in a direction normal to the plate. The reason why this procedure appears to be reasonable is that the conservation equations tell us that some of these flows (those corresponding to conserved quantities) are unchanged through the slab. Thus the value beyond the Knudsen layer coincides with that at the wall. Further, the molecules arriving at the wall suffer few, if any, collisions in the Knudsen layer and thus it is reasonable to think of approximating their distribution with a polynomial. Once theirs is known, the distribution function of the molecules leaving the wall is given by the boundary conditions. It is, of course, an approximation, the limitations of which will be investigated in the following chapters.

Let us consider the simplest case, when the molecules are diffused according to a Maxwellian distribution M_w at the boundary $x=0$. We first remark that, by equating the mass flow in the x direction of the molecules traveling away from the wall, as given by the distribution (2.6.4), to the same quantity evaluated with the Maxwellian of the wall, we obtain that the density appearing in front of the wall Maxwellian in the boundary conditions can be assumed to be the same as that of the gas except for a term proportional to $p_{11} - p$, which can be considered of higher order (see Problem 2.6.4). To compute the first boundary condition, we can essentially equate the momentum flow of the half-range wall Maxwellian (which is zero) to the momentum flow arising from the distribution (2.6.4) restricted to positive values of ξ_1 (see Problem 2.6.5) and write

$$0 = \frac{1}{2}p_{12} + \int_{\xi_1>0} \xi_1\xi_2 M d\boldsymbol{\xi} = \frac{1}{2}p_{12} + \frac{\rho v}{2}\left(\frac{2RT}{\pi}\right)^{1/2}. \tag{2.6.8}$$

The calculation of the relation between the temperature and the heat flow at the same wall is slightly more complicated. In fact the contribution of the wall

Maxwellian to the energy flow is not zero. We obtain

$$\int_{\xi_1>0} \xi_1 |\mathbf{c}|^2 M_w \, d\boldsymbol{\xi} = \frac{\rho}{4}\left(\frac{2RT_w}{\pi}\right)^{3/2} = \frac{\rho}{4}\left(\frac{2RT}{\pi}\right)^{3/2} + \frac{1}{2} q_1. \qquad (2.6.9)$$

These relations show that the tangential velocity and the temperature of the gas are not the same as those of the wall; there is a velocity slip and a temperature jump. Thus the approximate argument we have just given shows that the simplest effects produced by the presence of the Knudsen layers are certainly compatible with the method of thirteen moments.

At the other wall we shall have similar boundary conditions. The only changes will be changes of sign, because the velocity component ξ_1 are negative in the integrals, and the fact that the left-hand side of Eq. (2.6.8) will be nonzero. It will equal $-(\rho/2)(2RT_w/\pi)^{1/2}V$, where V is the speed of the wall.

The fact that the behavior of the gas near a wall has a half-range nature suggests a different ansatz for the distribution function. We can take it to equal two different Maxwellians $M_\pm$ according to whether $\pm\xi_1 > 0$:

$$\begin{aligned} M_+ &= A_+(x)\exp(-\beta_+(x)|\boldsymbol{\xi} - v_+(x)\mathbf{j}|^2) \quad \text{if} \quad \xi_1 > 0, \\ M_- &= A_-(x)\exp(-\beta_-(x)|\boldsymbol{\xi} - v_-(x)\mathbf{j}|^2) \quad \text{if} \quad \xi_1 < 0, \end{aligned} \qquad (2.6.10)$$

where $\beta_+(x)$ and $\beta_-(x)$ are inversely proportional to fictitious temperatures $T_+(x)$ and $T_-(x)$, respectively, and $A_\pm(x) = \rho_\pm(x)[2RT_\pm(x)]^{-3/2}$.

This method was introduced by Liu and Lees.[20,21] It has several advantages, including the fact that we can satisfy completely diffusive boundary conditions exactly and it becomes exact in the free-molecular limit. The main disadvantage is the complication in the calculation of the terms arising from the collision integral for molecules different from Maxwell's. But this is a feature shared by other moment methods. In addition, whenever quantities are sensitive to a detailed effect of collisions, the method becomes inadequate. A simple example of this failure is provided by plane Poiseuille flow (see Chapter 5, Section 5.8). The method can become valuable, when the ansatz on the distribution function is suitably modified to take into account specific features of the flow under consideration. An example will be provided in Chapter 8 when the method will be applied to evaporation and condensation problems.

All the moments are expressed, in general, in terms of ten quantities, $\rho^\pm$, $v_i^\pm$ $(i = 1, 2, 3)$, and $T^\pm$. In the case of Couette flow we can suppress the first and third components of the velocity in both Maxwellians, $v_1^\pm$ and $v_3^\pm$. Thus we have essentially six unknowns.

The governing equations are the moment equations provided by the four conservation equations:

$$\rho u_1 = 0,$$
$$p_{11} = \text{const.},$$
$$p_{12} = \text{const.},$$
$$q_1 + v p_{12} = \text{const.} \tag{2.6.11}$$

and the two moment equations for the balance of tangential stress and normal heat flow.

If we use the two-stream Maxwellian defined above as a distribution function, the system of moment equations turns out to have the form:

$$\rho_+(T_+)^{1/2} - \rho_-(T_-)^{1/2} = 0,$$
$$\rho_+T_+ + \rho_-T_- = \alpha_1,$$
$$\rho_+(T_+)^{1/2}(v_- - v_+) = \alpha_2,$$
$$\rho_+(T_+)^{1/2}\left[T_- - T_+ + \frac{1}{4R}(v_-^2 - v_+^2)\right] = \alpha_3,$$
$$\frac{d}{dx}(\rho_+u_+T_+ + \rho_-u_-T_-) = -A\alpha_2(\rho_+ + \rho_-),$$
$$\frac{d}{dx}\left[\rho_+T_+^2 + \rho_-T_-^2 + \frac{1}{5R}(\rho_+v_+^2T_+ + \rho_-v_-^2T_-)\right]$$
$$= -\frac{2}{15\,R}A\alpha_2(\rho_+v_+ + \rho_-v_-) - \frac{8}{15}\alpha_3(\rho_+ + \rho_-), \tag{2.6.12}$$

where

$$A = \frac{(R)^{1/2}T}{\mu(T)(2\pi)^{1/2}}$$

is a constant, which can be evaluated at any value of T (for Maxwell's molecules the viscosity coefficient is proportional to T).

Clearly one can reduce the solution to solving a system of two ordinary differential equations. One of these can be integrated immediately. When the result is inserted into the other equation, the variables can be separated and the second equation can be solved in closed form as well.

Liu and Lees[21] discussed the above system in detail. They first concentrated on the low Mach number case. Then they extended their analysis to the case of an arbitrary Mach number. For large Mach numbers and high Knudsen numbers they were not able to find a solution; this indicates that even for a simple flow such as Couette flow, an a priori imposed distribution cannot capture all the features of the Boltzmann equation. A qualitative understanding of the basic features is first needed.

Problems

2.6.1 Show that we may relate the third- and fourth-order moments (defined in Section 2.2, Eq. (2.2.8)) to the corresponding central moments (defined in Section 2.2, Eq. (2.2.9)) in the following way:

$$m_{ijk} = \rho v_i v_j v_k + p_{ij} v_k + p_{jk} v_i + p_{ik} v_j + p_{ijk},$$

$$\begin{aligned} m_{ijkl} = {} & p_{ijkl} + p_{ijk} v_l + p_{jkl} v_i + p_{kli} v_j + p_{lij} v_k + p_{ij} v_k v_l + p_{jk} v_l v_i \\ & + p_{kl} v_i v_j + p_{li} v_j v_k + p_{ik} v_j v_l + p_{jl} v_i v_k + \rho v_i v_j v_k v_l. \end{aligned}$$

2.6.2 Show that in the case of Couette flow ($\mathbf{v} = (0, v, 0)$) we have

$$m_{111} = p_{111}, \quad m_{121} = p_{121} + p_{11} v,$$

$$\sum_{k=1}^{3} m_{11kk} = \sum_{k=1}^{3} p_{11kk} + 2p_{112} v + p_{11} |\mathbf{v}|^2.$$

2.6.3 Show that Eqs. (2.6.2) hold. (Hint: Proceed in a way analogous to that suggested for Problem 2.2.2 with obvious modifications. Use also the results of the previous problem when convenient.)

2.6.4 Show that equating the mass flow of the molecules traveling away from the wall, as given by the distribution (2.6.4), to the same quantity based on the Maxwellian M_w normalized to some density ρ_w, we obtain that the density ρ of the gas equals ρ_w except for a term proportional to $p_{11} - p$. (Hint: Recall that the normal component of the bulk velocity is zero at the wall.)

2.6.5 Show that equating the momentum flow of the molecules traveling away from the wall, as given by the distribution (2.6.4), to the same quantity based on the Maxwellian M_w normalized to some density ρ_w, we obtain Eq. (2.6.8).

2.6.6 Show that Eqs. (2.6.12) hold (Hint: Use the equations for moments and not for central moments.)

2.6.7 Solve the system (2.6.12) in a closed form and discuss the solution (see Ref. 21).

2.7. Perturbations of Equilibria

The first steady solutions other than Maxwellian to be investigated were perturbations of the latter. Although we are interested in actual calculations and not in the theorems proved about the exact solutions, we mention here an early result of existence. A paper of the author dealing with problems in a slab for the Boltzmann equation linearized about a Maxwellian in 1967[22] was immediately used by Y.-P. Pao[23] to prove that one could use that existence result to prove that the nonlinear Boltzmann equation has a solution provided the Maxwellians associated with the two boundaries are sufficiently close in some sense.

The method of perturbation of equilibria is different from the Hilbert method because the small parameter is not contained in the Boltzmann equation but in auxiliary conditions, such as boundary or initial conditions. In Couette flow this parameter may be chosen to be proportional to the relative speed of the plates divided by the thermal speed and their temperature difference divided by the average temperature; both parameters must be small for the method to apply. Let us try to find a solution of our problem for the Boltzmann equation in the form

$$f = \sum_{n=0}^{\infty} \epsilon^n f_n, \tag{2.7.1}$$

where at variance with previous expansions ϵ is a parameter that *does not* appear in the Boltzmann equation. In addition f_0 is assumed from the start to be a Maxwellian distribution.

By inserting this formal series into Eq. (2.3.1) and matching the various orders in ϵ, we obtain equations that one can hope to solve recursively:

$$\xi_1 \partial_x f_1 = 2Q(f_1, f_0), \qquad (2.7.2)_1$$

$$\ldots,$$

$$\xi_1 \partial_x f_j = 2(Q(f_j, f_0) + \sum_{i=1}^{j-1} Q(f_i, f_{j-i}), \qquad (2.7.2)_j$$

$$\ldots,$$

where, as in Section 2.3, $Q(f, g)$ denotes the symmetrized collision operator and the sum is empty for $j = 1$.

Although in principle one can solve the subsequent equations by recursion, in practice one solves only the first equation, which is called the *linearized Boltzmann equation*. This equation can be rewritten as follows:

$$\xi_1 \partial_x h = L_M h, \tag{2.7.3}$$

where L_M denotes the linearized collision operator about the Maxwellian M, that is, $L_M h = 2Q(M, Mh)/M$, with $h = f_1/M$ (see Chapter 1, Section 1.10). We shall assume, as is usually done with little loss of generality, that the bulk velocity in the Maxwellian is zero and we shall denote the unperturbed density and temperature by ρ_0 and T_0.

Although the equation is now linear, and hence all the weapons of linear analysis are available, it is far from easy to solve for a given boundary value problem, such as Couette flow. Yet it is possible to gain an insight on the behavior of the general solution of Eq. (2.7.3).

This solution can be found by looking for solutions of exponential form in x, familiar from the solution of ordinary differential equations:

$$h_\gamma = g_\gamma(\boldsymbol{\xi}) \exp(\gamma x). \tag{2.7.4}$$

Inserting this expression into Eq. (2.7.3), we find that the equation is satisfied, provided $g_\gamma(\boldsymbol{\xi})$ is a solution of

$$L_M g_\gamma = \gamma \xi_1 g_\gamma. \tag{2.7.5}$$

The study of this equation is not so simple; explicit solutions are available only if models such as BGK are used (see next chapter). It is possible however to study the set of values of γ for which Eq. (2.7.5) has a solution. The reader is invited to consult Refs. 1 and 2 for details.

The first important result is that no complex values are admitted for γ. In addition the set of admitted values is symmetric with respect to the origin, and at least two cases may occur; either the set is made up of the two half-lines $\gamma \leq -\overline{\gamma}$ and $\gamma \geq \overline{\gamma}$, where $\overline{\gamma}$ is some nonnegative number, plus discrete values between $-\overline{\gamma}$ and $\overline{\gamma}$, or it coincides with the entire real axis. The first case occurs for hard sphere molecules, the second for molecules interacting at a distance with an angular cutoff.

Among the possible values there is always $\gamma = 0$ (in the first of the two cases above, it may be the only discrete value between $-\overline{\gamma}$ and $\overline{\gamma}$). The case $\gamma = 0$ is special for two reasons: First, it leads to solutions that do not have an exponential behavior in x; second, the value $\gamma = 0$ is degenerate and offers some surprises. If we let $\gamma = 0$ in Eq. (2.7.5), we see that there are five possible solutions, the collision invariants; thus the eigenvalue $\gamma = 0$ is at least fivefold degenerate. It turns out that it is eightfold degenerate, but the additional degeneracy does not show up in terms of more functions satisfying Eq. (2.7.5). The situation is similar to the case of the ordinary differential equation $f'' = 0$. If we look for exponential solutions of the form $f = C\exp(\gamma x)$, for a nonzero constant $C \neq 0$, we obtain $\gamma^2 = 0$. In addition to $f = C$, we know that we have another, linearly independent, solution of the form $f = Dx$, where D is an arbitrary constant. In the same way, it turns out that, to obtain the general solution of Eq. (2.7.3), it is not enough to assume a linear combination of functions of the form (2.7.4); we must also add solutions that are linear in x. Thus we must look for this kind of solution. If we insert $h = A(\boldsymbol{\xi})x + B(\boldsymbol{\xi})$ in Eq. (2.7.3), we obtain, equating terms of zeroth and first degree in x,

$$\xi_1 A = L_M B, \qquad 0 = L_M A. \tag{2.7.6}$$

Thus A must be a collision invariant ψ_α, but, for the equation for B to be solvable, $\int \xi_1 A \psi_\beta d\boldsymbol{\xi}$ must be zero for any collision invariant ψ_β ($\beta = 0, ..., 4$). This condition is automatically satisfied when either A or ψ_β does not contain a term proportional to ξ_1. When A is proportional to ξ_1, then the solvability condition is not satisfied (to see it, it is sufficient to let $\psi_\beta = 1$). However, if $\psi_\beta = \xi_1$, among the remaining collision invariants ψ_α, ψ_2, and ψ_3 satisfy the solvability condition, whereas ψ_0 does not and $\psi_4 = |\boldsymbol{\xi}|^2$ does not either; there is, however, a linear combination of ψ_0 and ψ_4 that satisfies the condition and hence, in the rest of this section, we let ψ_4 denote this linear combination, $\psi_4 = |\boldsymbol{\xi}|^2 - 5RT_0$ (Problem 2.7.1). Then we have just three solutions linear in x (this justifies the previous statement that the eigenvalue $\gamma = 0$ is eightfold degenerate), and we can write the general solution of Eq. (2.7.3) in the following form:

$$\begin{aligned} h = & \sum_{\alpha=0}^{4} A_\alpha \psi_\alpha + \sum_{\alpha=2}^{4} B_\alpha [\psi_\alpha x + L^{-1}(\xi_1 \psi_\alpha)] + \int_{-\infty}^{-\overline{\gamma}} C(\gamma) g_\gamma(\boldsymbol{\xi}) \exp(\gamma x) d\gamma \\ & + \int_{\overline{\gamma}}^{\infty} C(\gamma) g_\gamma(\boldsymbol{\xi}) \exp(\gamma x) d\gamma, \end{aligned} \tag{2.7.7}$$

where $\overline{\gamma}$ is nonnegative. In the case $\overline{\gamma} \neq 0$, there might be discrete exponential terms, but we have omitted these for simplicity. Further, since the solution of Eq. (2.7.5) is determined up to a constant factor depending upon γ, we have written the contribution as the product of a fixed factor $g_\gamma(\boldsymbol{\xi})$, normalized in some way, and an arbitrary weight $C(\gamma)$, which changes from one solution to another. This weight and, likewise, the coefficients A_α and B_α are determined by the boundary conditions. It can be shown (see Refs. 24 and 1) that Eq. (2.7.7) really provides the general solution of Eq. (2.7.3).

The above expression contains important information on the behavior of the solution. There are terms that have exponential behavior and terms that are polynomials of at most first degree in x.

Let us consider the exponential terms first. A feature that strikes the eye is that both increasing and decreasing exponentials are present; this may look odd, if one compares these terms with similar solutions in which time replaces the space variable x. In that case, in fact, the increasing exponentials are usually absent. A moment's thought will however show that this feature is absolutely required, because the increasing exponentials become decreasing, if we look at decreasing values of x. Thus, if we have two boundaries, located, say, at $x = 0$ and $x = L$, the terms with positive and negative values of γ will describe terms that decrease when moving away from the boundaries in the interval $[0, L]$. To see this in an explicit fashion, we exhibit a strikingly simple example, which embodies the essential features of Eq. (2.7.3) as far as the exponential terms are concerned. Let us consider the equation

$$\xi_1 \partial_x h + \nu h = 0, \tag{2.7.8}$$

where ν is a constant. The general solution is obviously

$$h = A(\xi_1) \exp\left(-\frac{\nu x}{\xi_1}\right) \tag{2.7.9}$$

where the constant of integration A depends on ξ_1 as well as on other parameters not explicitly appearing in Eq. (2.7.8), such as ξ_2 and ξ_3. If we assign the values of h at $x = 0$ for $\xi_1 > 0$, $h_+(\xi_1)$, and at $x = L$ for $\xi_1 < 0$, $h_-(\xi_1)$, we have the solution

$$h = \begin{cases} h_+(\xi_1) \exp\left(-\dfrac{\nu x}{\xi_1}\right) & \text{for } \xi_1 > 0, \\ h_-(\xi_1) \exp\left(-\dfrac{\nu(L-x)}{|\xi_1|}\right) & \text{for } \xi_1 < 0, \end{cases} \tag{2.7.10}$$

where we have written $-|\xi_1|$ in place of ξ_1 when the latter is negative, in order to emphasize the signs in the exponentials. We see that the solution exhibits a decreasing exponential for $\xi_1 > 0$ and an increasing one for $\xi_1 < 0$. However, they are both bounded in x and one of them decreases away from one of the boundaries, $x = 0$ or $x = L$ respectively, in the interval $[0, L]$. The term that does not decrease is exponentially small, of order $\exp(-\nu L/|\xi_1|)$ near the corresponding boundary.

Thus the exponential terms describe the Knudsen layers and play no role away from the boundaries, provided that the latter are sufficiently away from each other. We can see how far away by remarking that γ has the dimensions of an inverse length; though γ takes infinitely many values, we can surmise that a typical magnitude of γ in a significant contribution to the solution will be of the order of the inverse of the mean free path. Thus, provided the boundaries are several mean free paths apart, the exponential terms will give a significant contribution just in thin layers near the boundaries (the Knudsen layers).

Let us turn now to the other terms in Eq. (2.7.7). It is easy to compute the contributions of these terms to the first few moments of the distribution function (Problem 2.7.2). Here is the result, where the superscript F denotes that the contributions from the exponential terms do not appear (though the contribution from the unperturbed Maxwellian is present):

$$\rho^F = \rho_0[1 + A_0 - 2RT_0(A_4 + B_4x)],$$

$$v_1^F = A_1RT_0,$$

$$v_i^F = A_iRT_0 + B_iRT_0x \quad (i = 2, 3),$$

$$T^F = T_0[1 + 2RT_0(A_4 + B_4x)],$$

$$p_{ij}^F = RT_0\rho_0(1 + A_0)\delta_{ij} - \mu_0(B_i\delta_{1j} + B_j\delta_{1i})RT_0 \quad (i, j = 1, 2, 3)\ (B_1 = 0),$$

$$q_1^F = -2\kappa_0B_4RT_0^2,$$

$$q_i^F = 0\ (i = 2, 3), \tag{2.7.11}$$

where μ_0 and κ_0 are given by Eq. (2.3.13). We thus see that the stress tensor and the heat flow vector satisfy the Navier–Stokes constitutive relations; further, it is easy to see (Problem 2.7.3) that ρ^F, $\mathbf{v}^F$ and T^F satisfy the Navier–Stokes equations linearized about $\rho = \rho_0$, $\mathbf{v} = 0$, and $T = T_0$.

Thus we have the following picture: Provided the plates are sufficiently far apart (several mean free paths), there are two Knudsen layers near the boundaries, where the behavior of the solution is strongly dependent on the boundary conditions, and a central core (a few mean free paths away form the plates), where the solution of the Navier–Stokes equations holds (with a slight reminiscence of the boundary conditions). If the plates are close in terms of the mean free path, then all the terms in the general solution are simultaneously important. If the mean free path is much larger than the distance between the plates, the exponential terms dominate and the nearly free solution discussed in Section 2.4 applies.

Although we have given evidence for the above statements just in the case of the linearized Boltzmann equation, there is strong evidence that this qualitative picture applies to nonlinear flows as well, with a major exception. In general, compressible flows develop shock waves at large speeds (see Chapter 4) and these do not appear in Eq. (2.7.7). These shocks are not surfaces of discontinuity as for an ideal fluid, governed by the Euler equations, but are layers of rapid change of the solution (on the scale of the mean free path). One can obtain solutions for flows containing shocks from the Navier–Stokes equations, but, since they change significantly on the scale of the mean free path, they are inaccurate. Other regions where this picture is inaccurate are the zones of high rarefaction, where nearly free-molecular conditions may prevail, even if the rest of the flow is reasonably described in terms of Navier–Stokes equations, Knudsen layers, and shock layers.

Subtler phenomena may occur if the solutions depend on more than one space coordinate. We shall discuss this aspect in Chapters 5 and 7.

Problems

2.7.1 Show that there is a linear combination of $|\boldsymbol{\xi}|^2 + c$ of ψ_4 and ψ_0 such that when multiplied by $\xi_1 M$ and integrated gives a vanishing result. Compute c and show that it equals $-5RT_0$.

2.7.2 Show that Eqs. (2.7.11) hold.

2.7.3 Linearize the Navier–Stokes equations about a constant density ρ_0, a constant temperature T_0, and a vanishing bulk velocity. Show that the expressions of ρ^F, $\mathbf{v}^F$, and T^F given by Eqs. (2.7.11) hold.

References

[1] C. Cercignani, *The Boltzmann Equation and Its Applications*, Springer-Verlag, New York (1988).
[2] C. Cercignani, *Mathematical Methods in Kinetic Theory*, Plenum Press, New York (1990).
[3] C. Cercignani, "Existence of homoenergetic affine flows for the Boltzmann equation," *Archive for Rational Mechanics and Analysis* **105**, 377–387 (1989).
[4] V. S. Galkin, "Exact solutions of the kinetic moment equations of a mixture of monatomic gases," *Fluid Dynamics* **1**, 29–34 (1966).
[5] C. Truesdell and R. G. Muncaster, *Fundamentals of Maxwell's Kinetic Theory of a Simple Monatomic Gas*, Academic Press, New York (1980).
[6] V. S. Galkin, "On a solution of the kinetic equation," *PMM* (in Russian) **20**, 445–446 (1956).
[7] V. S. Galkin, "On a class of solutions of Grad's moment equations," *PMM* **22**, 532–536 (1958).
[8] V. S. Galkin, "One-dimensional unsteady solutions of the equations for the kinetic moments of a monatomic gas," *PMM* **28**, 226–229 (1964).
[9] C. Truesdell, "On the pressures and the flux of energy in a gas according to Maxwell's kinetic theory, II," *J. Rational Mech. Analysis* **5**, 55–128 (1956).
[10] C. Cercignani and S. Cortese, "Validation of a Monte Carlo simulation of the plane Couette flow of a rarefied gas," *J. Stat. Phys.* **75**, 817–838 (1994).
[11] D. R. Willis, "A study of some nearly free molecular flow problems," Princeton University Aero. Engineering Lab. Report no. 440 (1960).
[12] D. R. Willis, "Theoretical solutions to some nearly free molecular problems," in *Rarefied Gas Dynamics*, M. Devienne, ed., 246–257, Pergamon Press, New York (1960).

[13] D. R. Willis, General Electric Co. Report TIS 60SD399 (1960).
[14] D. R. Willis, "The effect of the molecular model on rarefied gas flows," Rand Corporation Memorandum TM-4638-PR (1965).
[15] M. Abramowitz, "Evaluation of the integral $\int_0^\infty e^{-u^2-x/u}du$," *J. Math. Phys.* **32**, 188–192 (1953).
[16] M. Abramowitz and I. A. Stegun, *Handbook of Mathematical Functions*, Applied Mathematical Series, National Bureau of Standards, Washington (1964).
[17] M. M. R. Williams and S. K. Loyalka, *Aerosol Science. Theory & Practice*, Pergamon Press, Oxford (1991).
[18] H. Grad, "On the kinetic theory of rarified gases," *Comm. Pure Appl. Math.* **2**, 331–407 (1949).
[19] L. Lees, "A kinetic theory description of rarefied gases," GALCIT Hypersonic Research Project Memo. No. 51, California Institute of Technology (1959).
[20] C. Liu and L. Lees, "Kinetic theory description of plane compressible Couette flow," in *Rarefied Gas Dynamics*, L. Talbot, ed., 391–428, Academic Press, New York (1961).
[21] C. Cercignani, "Existence and uniqueness in the large for boundary value problems in kinetic theory," *J. Math. Phys.* **8**, 1653–1656 (1967).
[22] Y.-P. Pao, "Boundary-value problems for the linearized and weakly nonlinear Boltzmann equation," *J. Math. Phys.* **8**, 1893–1898 (1967).
[23] C. Cercignani, "On the general solution of the steady linearized Boltzmann equation," in *Rarefied Gas Dynamics*, M. Becker and M. Fiebig, eds., Vol. I, A.9-1-11, DFLVR Press, Porz-Wahn (1974).

3

Problems for a Gas in a Slab or a Half-Space: Discussion of Some Solutions

3.1. Use of Models

To get further insight into the solutions of the Boltzmann equation, it is useful to adopt a simplified model, such as the BGK model mentioned in Chapter 1, Section 1.9. The machineries developed in the last section of Chapter 2 also apply to them with some modifications. In particular, the viscosity coefficient and the heat conductivity turn out to be simply related to the collision frequency, and the Prandtl number turns out to be exactly unity, as opposed to the value 2/3, which is exact for Maxwell molecules and reasonably accurate for other molecular models. This shows a rather large error, of the order of 30%. Yet, frequently errors are smaller than this, especially when suitable tricks are used, such as the adjustment of the value of the collision frequency.

The main advantages of the use of models appear when numerical methods or the linearized Boltzmann equation are used. Since we have not discussed numerical solutions as yet, we shall briefly deal with the use of models for the linearized Boltzmann equation. If we consider the BGK model and linearize about a fixed Maxwellian distribution, M, we have essentially to approximate the local Maxwellian Φ appearing in Eq. (1.9.1) in terms of M and the moments of h. A simple calculation gives the following linearized operator (Problem 3.1.1):

$$L_{BGK} = \nu_0 \left[\int \hat{M}(\boldsymbol{\xi}_*)h(\boldsymbol{\xi}_*)d\boldsymbol{\xi}_* + \frac{\boldsymbol{\xi}}{RT_0} \cdot \int \boldsymbol{\xi}_* \hat{M}(\boldsymbol{\xi}_*)h(\boldsymbol{\xi}_*)d\boldsymbol{\xi}_* + \left(\frac{|\boldsymbol{\xi}|^2}{2RT_0} - \frac{3}{2} \right) \int \left(\frac{|\boldsymbol{\xi}_*|^2}{2RT_0} - \frac{3}{2} \right) \hat{M}(\boldsymbol{\xi}_*)h(\boldsymbol{\xi}_*)d\boldsymbol{\xi}_* - h \right], \quad (3.1.1)$$

where ν_0 is the collision frequency evaluated at the density ρ_0 and temperature T_0 of the unperturbed state, and $\hat{M}$ denotes M/ρ_0. The operator (3.1.1) has a simple structure. It can be used to transform linearized problems into rather simple integral equations or systems of (at most two) such equations (see next section). Another advantage is that the solutions of Eq. (2.7.5) can be computed explicitly and their properties studied.[1–5]

Further, the solution of problems in a slab or a half-space can be transformed to problems involving essentially the variable ξ_1 and not the entire vector $\boldsymbol{\xi}$.[1,5] To this end we remark that by momentum conservation $\int \xi_1 \hat{M} h d\boldsymbol{\xi}$ is a constant C. Then it is sufficient to write h in the following form:

$$h = h_0(x, \xi_1) + \frac{\xi_1 C}{RT_0} + \left(\frac{\xi_2^2 + \xi_2^2}{2RT_0} - 1\right) h_1(x, \xi_1) + 2\xi_2 h_2(x, \xi_1) + 2\xi_3 h_3(x, \xi_1) + h_R(x, \boldsymbol{\xi}), \tag{3.1.2}$$

where the decomposition is uniquely determined by the conditions

$$\int h_R \hat{M} d\boldsymbol{\xi} = 0, \quad \int \xi_i h_R \hat{M} d\boldsymbol{\xi} = 0 \quad (i = 2, 3),$$
$$\int \left(\frac{\xi_2^2 + \xi_2^2}{2RT_0} - 1\right) h_R \hat{M} d\boldsymbol{\xi} = 0. \tag{3.1.3}$$

Then the linearized BGK model is equivalent to the following five equations (Problem 3.1.2):

$$\xi_1 \partial_x h_0 = \nu_0 \left\{ \int \hat{M}_1(\xi_{1*}) h_0(\xi_{1*}) d\xi_{1*} + \frac{1}{6}\left(\frac{\xi_1^2}{RT_0} - 1\right) \left[\int \left(\frac{\xi_{1*}^2}{RT_0} - 1\right) \hat{M}_1(\xi_{1*}) h_0(\xi_{1*}) d\xi_{1*} + \int \hat{M}_1(\xi_{1*}) h_1(\xi_{1*}) d\xi_{1*} \right] - h_0 \right\},$$

$$\xi_1 \partial_x h_1 = \nu_0 \left\{ \frac{1}{3}\left(\frac{\xi_1^2}{RT_0} - 1\right) \left[\int \hat{M}_1(\xi_{1*}) h_0(\xi_{1*}) d\xi_{1*} + \frac{2}{3} \int \hat{M}_1(\xi_{1*}) h_1(\xi_{1*}) d\xi_{1*} \right] - h_1 \right\},$$

$$\xi_1 \partial_x h_i = \nu_0 \left[\int \hat{M}_1(\xi_{1*}) h_i(\xi_{1*}) d\xi_{1*} - h_i \right] \quad (i = 2, 3),$$

$$\xi_1 \partial_x h_R = -\nu_0 h_R, \tag{3.1.4}$$

where $\hat{M}_1$ is the one-dimensional Maxwellian distribution obtained by integrating $\hat{M}$ with respect to ξ_2 and ξ_3.

The last equation in the above system is simple (it is essentially our example (2.7.8)) but usually plays little, if any, role. The previous relation contains two identical equations for $i = 2, 3$ and describes a shear flow when the plates move in the x_i direction. The first two equations are coupled and describe heat transfer problems (which are decoupled from motion in the linearized approximation).

The general solution of the above equations can be obtained in an explicit form.[1–5] In other words we can write the expansion (2.7.7) with an explicit expression for the eigensolutions g_γ (see Problem 3.1.3). Thus one can consider the problem of matching the boundary conditions. Here the half-range nature of the latter comes into play. If we equate the solution at each boundary to the data there, the resulting relations hold for either $\xi_1 > 0$ or $\xi_1 < 0$ and one cannot solve for the weights $C(\gamma)$ explicitly. One can obtain Fredholm equations determining $C(\gamma)$.[6] Actually, one can obtain two different equations, which lend themselves to obtaining converging expansions for either large or small values of the Knudsen number.[6]

We recall here that other, more complicated models can be devised. Although they can be nonlinear, it is only at the linearized level that systematic approaches can be developed, without renouncing any property. The linearized models are essentially of two kinds. The first procedure relies on the use of the eigenfunctions of the linearized collision operator for Maxwell molecules (i.e., polynomials $\psi_\alpha(\boldsymbol{\xi})$ that are orthogonal with respect to a Maxwellian weight); in other words, $(\psi_\alpha, \psi_\beta) = \delta_{\alpha\beta}$, where

$$(g, h) = \int ghM d\boldsymbol{\xi}. \tag{3.1.5}$$

Then we can obtain models of the following form:

$$L_N = \sum_{\alpha=0}^{N} \nu_\alpha \psi_\alpha(\psi_\alpha, h) - \nu_N h, \tag{3.1.6}$$

where $\nu_\alpha = \nu_N$ for $\alpha = 0, 1, 2, 3, 4$. This model generalizes the BGK model (to which it reduces for $N = 4$) in a simple way and is capable of approximating the linearized collision operator for Maxwell molecules. A further generalization along the same lines is

$$L_N = \sum_{\alpha,\beta=0}^{N} \nu_{\alpha\beta} \psi_\alpha(\psi_\beta, h) - \nu_N h. \tag{3.1.7}$$

This model can approximate molecules other than Maxwell's, but some features are lost, especially at large speeds. This remark leads to considering other models having the form

$$L_N = \nu(|\boldsymbol{\xi}|)\left(\sum_{\alpha=0}^{N} \mu_\alpha \phi_\alpha(\nu\phi_\beta, h) - h\right), \tag{3.1.8}$$

where $\phi_\alpha(\boldsymbol{\xi})$ are suitable functions orthogonal with respect to a Maxwellian distribution multiplied by a function $\nu = \nu(|\boldsymbol{\xi}|)$, the collision frequency ($(\phi_\alpha, \nu\phi_\beta) = \delta_{\alpha\beta}$).

These models can be treated in the same way as the BGK model. Of course, the calculations become more complicated, but the qualitative features are the same. We shall not enter into details here and refer the reader to the literature.[4,5]

Problems

3.1.1 Show that the linearized BGK model has the form indicated in Eq. (3.1.1) (Hint: Write the moments appearing in (1.9.1) as the sum of an unperturbed contribution and a perturbed one (zero for the bulk velocity) and expand, neglecting terms of order higher than first in the perturbed quantities.)

3.1.2 Show that the linearized BGK model can be split into the equations of the system (3.1.4). (Hint: Use the definitions (3.1.2) and (3.1.3). See Refs. 1 and 5)

3.1.3 Find all the solutions of the system (3.1.4) in the form (2.7.4). (Hint: The last equation is trivial and was already solved in the previous example (see Eqs. (2.7.8–2.7.9). The two previous equations are identical and it is enough to solve one of them, by normalizing $\int \hat{M} h_i \, d\boldsymbol{\xi}_* = 1$. The solutions are not ordinary functions but distributions. See Refs. 1–5.)

3.2. Transformation of Models into Pure Integral Equations

One of the advantages of the models is that they can be transformed into pure integral equations. We shall just consider the case of isothermal Couette flow to which we can apply the relevant equation of the system (3.1.4). We rewrite the latter abolishing all the subscripts, since no confusion arises:

$$\xi \partial_x h = \nu_0[v(x) - h]. \tag{3.2.1}$$

Here we have replaced the integral by $v(x)$, the component of the bulk velocity along the y axis, because this is the meaning of that integral. We want to solve the above equation with the following boundary conditions:

$$h = 0 \quad \text{for } x = 0 \text{ and } \xi > 0, \qquad h = V \quad \text{for } x = L \text{ and } \xi < 0. \tag{3.2.2}$$

Now we take advantage of the fact that the term $v(x)$ in Eq. (3.2.1) is independent of ξ and assume for a moment that it is known. Then we can write the solution of Eq. (3.2.1) with the boundary condition (3.2.2) as follows:

$$h = \begin{cases} \frac{\nu_0}{\xi}\int_0^x v(x_*)\exp\left(-\nu_0\frac{x-x_*}{\xi}\right)dx_* & \text{for } \xi > 0 \\ V\exp\left(-\nu_0\frac{L-x}{|\xi|}\right) + \frac{\nu_0}{|\xi|}\int_x^L v(x_*)\exp\left(-\nu_0\frac{|x-x_*|}{|\xi|}\right)dx_* & \text{for } \xi < 0. \end{cases} \tag{3.2.3}$$

This equation shows that h is known (at least in principle, since we must perform two integrations) when $v(x)$ is known. We must now obtain an equation for $v(x)$. To this end we take into account the fact that $v(x)$ is given by the integral appearing in Eq. (3.2.1). Then if we insert h as given by Eq. (3.2.3) into the latter integral, and equate the result to $v(x)$, we obtain the equation determining the latter quantity. The result reads as follows (Problem 3.2.1):

$$v(x) = \pi^{-1/2}VT_0\left(\frac{L-x}{\lambda}\right) + (\lambda)^{-1}\pi^{-1/2}\int_0^L T_{-1}\left(\frac{|x-x_*|}{\lambda}\right)v(x_*)dx_*, \tag{3.2.4}$$

where the functions $T_0(s)$ and $T_{-1} = -dT_0/ds$ were defined in Eq. (2.4.17) and $\lambda = (2RT_0)^{1/2}/\nu_0$ is the mean free path introduced in the previous chapter. Equation (3.2.4) can be solved by iteration, since the integral operator is contracting. This is, however, not the best method to solve the above integral equation. Discrete ordinate methods work very well. The general form of a numerical scheme is

$$v_j = S_j + \sum_k \alpha_{jk}v_k, \tag{3.2.5}$$

where two possible choices for the source term S_j and the coefficients are[7]

$$S_j = \pi^{-1/2}VT_0\left(\frac{L-\frac{2j-n+1}{2n}L}{\lambda}\right),$$

$$\alpha_{jk} = (\lambda)^{-1}\pi^{-1/2}\int_{\frac{2k-n}{2n}L}^{\frac{2k-n+2}{2n}L} T_{-1}\left(\frac{\left|\frac{2j-n+1}{2n}L - x_*\right|}{\lambda}\right)dx_*, \tag{3.2.6}$$

$$S_j = n\pi^{-1/2}V\int_{\frac{2j-n}{2n}L}^{\frac{2j-n+2}{2n}L} T_0\left(\frac{L-x}{\lambda}\right)dx,$$

$$\alpha_{jk} = n(2\lambda)^{-1}\pi^{-1/2}\int_{\frac{2j-n}{2n}L}^{\frac{2j-n+2}{2n}L}\int_{\frac{2k-n}{2n}L}^{\frac{2k-n+2}{2n}L} T_{-1}\left(\frac{|x-x_*|}{\lambda}\right)dx\,dx_*. \tag{3.2.7}$$

The integrals appearing here can be explicitly evaluated in terms of the functions T_0 and T_1 (Problem 3.2.3).

As for the system of the first two equations of the system (3.1.4), it is possible, by an analogous procedure, to obtain a system of two integral equations for the perturbations of density and temperature.

Systems of integral equations (of increasing order) can be obtained[1] for the more general models considered at the end of the previous section.

The method of using integral equations was pioneered by Welander[8] for half-space problems (see Section 3.5) and by Willis[9] in the case of a slab. Two schemes of the kind reported above were used by Cercignani and Daneri[7] to study plane Poiseuille flow (see Section 3.4) in 1963. The method of solving numerically the integral equations can be extended to nonlinear problems and was used for several years as the only method that could give reasonably accurate solutions on a computer.[10–12] It is still used to provide solutions with an accuracy that can, in principle, be made as high as one likes in both linearized and nonlinear problems.

Problems

3.2.1 Verify that Eq. (3.2.1) holds, as indicated in the main text.

3.2.2 Prove that the Abromowitz functions satisfy $dT_n/dx = -T_{n-1}$.

3.2.3 Verify that the integral operator in Eq. (3.2.1) is contracting in the space of bounded functions (Hint:

$$(\lambda)^{-1}\pi^{-1/2}\left|\int_0^L T_{-1}\left(\frac{|x-x_*|}{\lambda}\right)v(x_*)dx_*\right|$$
$$\leq 2\pi^{-1/2}\left[T_0(0)-T_0\left(\frac{L}{\lambda}\right)\right]\sup_{0,L}|v(x)|.)$$

3.2.4 Compute the integrals in (3.2.6) and (3.2.7) (see Ref. 7).

3.3. Variational Methods

The integral equation approach lends itself to a variational solution. The main idea of this method (for linear problems) is the following. Suppose that we must solve the equation

$$\mathcal{L}h = S, \tag{3.3.1}$$

where h is the unknown, $\mathcal{L}$ a linear operator, and S a source term. Assume that we can form a bilinear expression $((g, h))$ such that $((\mathcal{L}g, h)) = ((g, \mathcal{L}h))$, for any pair $\{g, h\}$ in the set where we seek a solution. Then the expression (functional)

$$J(\tilde{h}) = ((\tilde{h}, \mathcal{L}\tilde{h})) - 2((S, \tilde{h})) \tag{3.3.2}$$

has the property that if set $\tilde{h} = h + \eta$, then the terms of first degree in η disappear and $J(\tilde{h})$ reduces to $J(h) + ((\eta, \mathcal{L}\eta))$ if and only if h is a solution of Eq. (3.3.1) (Problem 3.3.1). In other words if η is regarded as small (an error), the error in the functional in Eq. (3.3.2) becomes small of second order in the neighborhood of h, if and only if h is a solution of Eq. (3.3.1). Then we say that the solutions of the latter equation satisfy a variational principle or make the functional in Eq. (3.3.2) stationary. Thus a way to look for solutions of Eq. (3.3.1) is to look for solutions that make the functional in Eq. (3.3.2) stationary (variational method).

The method is particularly useful if we know that $((\eta, \mathcal{L}\eta))$ is nonnegative (or nonpositive) because we can then characterize the solutions of Eq. (3.3.1) as maxima or minima of the functional (3.3.2). But, even if this is not the case, the property is useful. First of all, it gives a nonarbitrary recipe to select among approximations to the solution in a given class. Second, if we find that the functional J is related to some physical quantity, we can compute this quantity with high accuracy, even if we have a poor approximation to h. If the error η is of order 10%, then J will be in fact computed with an error of the order of 1%, because the deviation of $J(\tilde{h})$ from $J(h)$ is of order η^2, as we have seen.

The integral formulation of the models lends itself to the application of the variational method.[13] Thus in the case of Eq. (3.2.4) the bilinear expression $((w, v))$ is simply $\int_0^L w(x)v(x)dx$ and the calculations are straightforward (see below). Further, the functional is related to the stress component p_{12}, which is constant and gives the drag exerted by the gas on each plate. Thus this quantity can be computed with high accuracy (see below).

This method can be generalized to other problems and to the more complicated models discussed in the previous section.[1] It can also be used to obtain accurate finite ordinate schemes, by approximating the unknowns by trial functions that are piecewise constant. The second scheme in the previous section was obtained in this way.[7]

In the case of the steady linearized Boltzmann equation, Eq. (3.7.3), a similar method can be used. Let us indicate by Dh the differential part appearing in the left-hand side (for the sake of generality we set $Dh = \xi\partial_x h$ in order to include problems with a more complicated geometry) and assume that there is a source term as well (we shall see an example of a source in Section 3.4) and write our equation in the form

$$Dh - Lh = S. \tag{3.3.3}$$

If we try the simplest possible bilinear expression

$$((g, h)) = \int_0^L \int_{\Re^3} g(\mathbf{x}, \boldsymbol{\xi}) h(\mathbf{x}, \boldsymbol{\xi}) d\mathbf{x} d\boldsymbol{\xi} \tag{3.3.4}$$

and we use it with $\mathcal{L}h = Dh - Lh$ we cannot reproduce the symmetry property $((\mathcal{L}g, h)) = ((g, \mathcal{L}h))$. It works for Lh but not for Dh. There is however a trick[14] that leads to the desired result.

Let us introduce the parity operator in velocity space, P, such that $P[h(\boldsymbol{\xi})] = h(-\boldsymbol{\xi})$. Then we can think of replacing Eq. (3.3.3) by

$$PDh - PLh = PS, \tag{3.3.5}$$

because this is completely equivalent to the original equation. In addition, because of the central symmetry of the molecular interaction, $PLh = LPh$, hence the fact that we had no problems with L is not destroyed by the fact that we use P. However, we have by a partial integration (see Problem 3.3.2):

$$((g, PDh)) = ((PDg, h)) + ((g^+, Ph^-))_B - ((Pg^-, h^+))_B. \tag{3.3.6}$$

Here $g^\pm$ denote the restrictions of a function defined on the boundary to positive, respectively negative, values of $\boldsymbol{\xi} \cdot \mathbf{n}$, where $\mathbf{n}$ is the unit vector normal to the boundary. In addition, we have put

$$((g^\pm, h^\pm))_B = \int_{\partial\Omega} \int_{\pm\boldsymbol{\xi}\cdot\mathbf{n}>0} |\boldsymbol{\xi} \cdot \mathbf{n}| g(\mathbf{x}, \boldsymbol{\xi}) h(\mathbf{x}, \boldsymbol{\xi}) d\boldsymbol{\xi} d\sigma. \tag{3.3.7}$$

In the one-dimensional case, the integration over the boundary $\partial\Omega$ reduces to the sum of the boundary terms at $x = 0$ and $x = L$.

Clearly the last two terms in Eq. (3.3.6) do not fit in our description. We have two ways out of the difficulty. We first recall a property of the boundary conditions, discussed in Section 1.9 of Chapter 1. The boundary conditions must be linearized about the Maxwellian distribution M and this gives them the following form (see below and Problem 3.3.3):

$$h^+ = h_0 + Kh^-. \tag{3.3.8}$$

Because of reciprocity (Eq. (1.11.12)), we have

$$((Pg^-, Kh^-))_B = ((Kg^-, Ph^-))_B. \tag{3.3.9}$$

Hence, if we assume that both g and h satisfy the boundary conditions, we have

$$
\begin{aligned}
&((g^+, Ph^-))_B - ((Pg^-, h^+))_B \\
&\quad = ((h_0, Ph^-))_B - ((Pg^-, h_0))_B + ((Kg^-, Ph^-))_B - ((Pg^-, Kh^-))_B \\
&\quad = ((h_0, Ph^-))_B - ((Pg^-, h_0))_B. \qquad (3.3.10)
\end{aligned}
$$

We remark that we can modify the solution of the problem by adding a combination of the collision invariants with constant coefficients. This does not modify the Boltzmann equation but can be used to modify the boundary conditions. Usually it is possible to dispose of the constant coefficients to make $((h_0, Ph^-))_B = 0$ (and, at the same time, of course, $((h_0, Pg^-))_B = 0$). We assume that this is the case. Using this relation and Eq. (3.3.9), Eq. (3.3.10) reduces to

$$((g^+, Ph^-))_B - ((Pg^-, h^+))_B = 0, \qquad (3.3.11)$$

and the variational principle holds with the operator $\mathcal{L}h = PDh - PLh$ and the source PS.

This variational principle is correct but not particularly useful, because it can be used only with approximations that exactly satisfy the boundary conditions and it can be complicated to construct these approximations. Thus we follow another procedure by incorporating the boundary conditions in the functional. It is enough to consider

$$J(\tilde{h}) = ((\tilde{h}, PD\tilde{h} - PL\tilde{h})) - 2((PS, \tilde{h})) + ((P\tilde{h}^-, \tilde{h}^+ - K\tilde{h}^- - 2h_0))_B. \qquad (3.3.12)$$

In fact, if we let $\tilde{h} = h + \eta$, we find that the terms linear in η disappear from J and the variational principle holds (Problem 3.3.4).

In agreement with what we said before, it is interesting to look at the value attained by J when $\tilde{h} = h$. Equation (3.3.12) becomes

$$J(h) = -((PS, h)) - ((Ph^-, h_0))_B. \qquad (3.3.13)$$

This result acquires its full meaning only when we examine the expressions for h_0 and S. In general $S = 0$ (for an important case in which this is not true, see Section 3.4). If we let $S = 0$, then we must look at the expression of h_0. The boundary source has a special form because it arises from the linearization, about a Maxwellian distribution M, of a boundary condition of the form

$$f_+ = K_w f_-, \qquad (3.3.14)$$

where K_w is an operator, which has several properties, including

$$M_{w+} = K_w M_{w-}. \tag{3.3.15}$$

Now, if let $f = M(1 + h)$ in Eq. (3.3.14), we have

$$h_+ = \frac{K_w M_- h_-}{M_+} + \frac{K_w M_-}{M_+} - 1. \tag{3.3.16}$$

This relation is exact. We can now proceed to neglecting terms of order higher than first in the perturbation parameters. We can replace in K_w the temperature and velocity of the wall by those of the Maxwellian M (i.e., T_0 and 0) and obtain a slightly different operator K_0. Thus we obtain the operator K, which we used before, by letting $K_0 M_- h_- / M_+ = K h_-$. Concerning the source, we have, using Eq. (3.3.15),

$$h_0 = \frac{K_w M_-}{M_+} - 1 = \frac{K_w M_-}{M_+} - \frac{K_w M_{w-}}{M_{w+}}.$$

Since M_w and M differ by terms of first order, we can replace K_w by K_0 because their difference is also of first order and would produce a term of second order in the expression of h_0:

$$h_0 = \frac{K_0 M_-}{M_+} - \frac{K_0 M_{w-}}{M_{w+}} = 1 - \frac{K_0 M_{w-}}{M_{w+}}. \tag{3.3.17}$$

Now, if we neglect terms of higher than first order in the speed of the wall and the temperature difference $T_w - T_0$, we have $M_w = M(1 + \psi)$, where (recalling that M_w is determined up to the density that we can choose to be the same as in M) ψ can be explicitly computed to give (Problem 3.3.5)

$$\psi = \frac{\boldsymbol{\xi} \cdot \mathbf{v}_w}{RT_0} + \left(\frac{|\boldsymbol{\xi}|}{2RT_0} - \frac{3}{2} \right) \frac{T_w - T_0}{T_0} \tag{3.3.18}$$

and, finally, neglecting again terms of order higher than first (Problem 3.3.6),

$$h_0 = \psi_+ - K\psi_-. \tag{3.3.19}$$

Then Eq. (3.3.13) with $S = 0$ gives

$$\begin{aligned} J(h) &= -((Ph^-, \psi_+))_B + ((Ph^-, K\psi_-))_B \\ &= -((Ph^-, \psi_+))_B + ((Kh^-, P\psi_-))_B \\ &= -((Ph^-, \psi_+))_B + ((h^+ - \psi^+ + K\psi^-, P\psi_-))_B \\ &= -((Ph^-, \psi_+))_B + ((h^+ -, P\psi_-))_B + ((K\psi^- - \psi^+, P\psi_-))_B. \end{aligned} \tag{3.3.20}$$

The last term is a known quantity, whereas the first and the second can be combined to give unknown quantities of physical importance. In fact, if we take into account the expression of ψ (3.3.18), we obtain

$$\begin{aligned}
&-((Ph^-, \psi_+))_B + ((h^+-, P\psi_-))_B \\
&= -\int \frac{\boldsymbol{\xi}\cdot\mathbf{v}_w}{RT_0}\boldsymbol{\xi}\cdot\mathbf{n}h(\mathbf{x},\boldsymbol{\xi})Md\boldsymbol{\xi}d\sigma \\
&\quad + \int\left(\frac{|\boldsymbol{\xi}|}{2RT_0}-\frac{3}{2}\right)\frac{T_w-T_0}{T_0}\boldsymbol{\xi}\cdot\mathbf{n}h(\mathbf{x},\boldsymbol{\xi})Md\boldsymbol{\xi}d\sigma \\
&= -\frac{1}{RT_0}\left(\int \mathbf{p_n}\cdot\mathbf{v}_w d\sigma + \int q_\mathbf{n}\frac{T_w-T_0}{T_0}d\sigma\right). \qquad (3.3.21)
\end{aligned}$$

Here $\mathbf{p_n}$ is the normal stress vector and $q_\mathbf{n}$ the normal component of the heat flow. The fact that the mass flow vanishes at the wall has been taken into account.

Because of the linearity of the problem, it is possible and convenient, without loss of generality, to consider separately the two cases $\mathbf{v}_w = \mathbf{0}$ and $T_w = T_0$. Then the two terms in the expression above occur in two different problems. For some typical problems $\mathbf{v}_w$ vanishes on just one part of the boundary, whereas it is a constant on the remaining part of the latter; then the factor multiplying this constant is the drag on the corresponding part of the boundary. Similarly one can consider the case in which the factor in front of q_n vanishes on just one part of the boundary and relate the value of the functional to the heat transfer.

The two variational principles that have been discussed are related to each other.[1]

To give a simple example of the use of the variational method in kinetic theory, let us consider the case of Couette flow for the integral formulation of the BGK model, Eq. (3.2.4). The functional to be considered is

$$\begin{aligned}
J(\tilde{v}) = &\int_0^L [\tilde{v}(x)]^2 - \pi^{-1/2}(\lambda)^{-1}\int_0^L\int_0^L \tilde{v}(x)T_{-1}\left(\frac{|x-x_*|}{\lambda}\right)\tilde{v}(x_*)dx dx_* \\
&- 2\pi^{-1/2}V\int_0^L v(\tilde{x})T_0\left(\frac{L-x}{\lambda}\right)dx. \qquad (3.3.22)
\end{aligned}$$

We remark that

$$J(v) = -\pi^{-1/2}V\int_0^L v(x)T_0\left(\frac{L-x}{\lambda}\right)dx. \qquad (3.3.23)$$

Let us now consider the expression of the stress component p_{12}. To compute it, we have to integrate the expressions of h in (3.2.1) after multiplication by ξ, to obtain

$$p_{12} = (2RT_0)^{1/2} V T_1 \left(\frac{L-x}{\lambda}\right) - (2RT_0)^{1/2}(\lambda)^{-1} \times \int_0^L T_0\left(\frac{|x-x_*|}{\lambda}\right) \operatorname{sgn}(x-x_*) v(x_*) dx_*. \quad (3.3.24)$$

We note that, as we know, p_{12} is constant. To check it, we can differentiate Eq. (3.3.24) with respect to x and use the fact that v satisfies Eq. (3.2.4) (the elementary property of the Abramowitz functions, $dT_n/ds = -T_{n-1}$, which follows from the definition given in Chapter 2, must also be used). Thus we can evaluate p_{12} for any value of x. If we choose $x = L$ and compare with Eq. (3.3.23), we obtain (using $T_1(0) = 1/2$)

$$p_{12} = \left(\frac{RT_0}{2}\right)^{1/2} V + \left(\frac{2RT_0}{V\pi}\right)^{1/2} (\lambda)^{-1} J(v). \quad (3.3.25)$$

Thus the stress has a direct relation to the stationary value (in this case a minimum) of the functional J and we can hope for an accurate expression of p_{12} by taking a rough approximation for v. The simplest choice is $\tilde{v} = V/2 + A(x - L/2)$ where A is a constant to be chosen in such a way as to make the derivative of $J(Ax)$ with respect to A vanish. The reason why we choose the above form for the trial function rather than the simpler Ax is that there is a symmetry about the midpoint between the plates (in the linearized case). We know that a linear trial function is a very poor approximation to the true velocity profile, since it neglects the Knudsen layers completely. Yet, and this is the main advantage of the variational method, we shall obtain very accurate results for the stress tensor

After some simple manipulations based on the properties of the T_n functions, we find[13] the following expression for $J(V/2 + A(x - L/2))$ (see Problem 3.3.7):

$$\begin{aligned} J(V/2 + A(x - L/2)) &= A^2\pi^{-1/2}\left\{\frac{L^2}{2\lambda^2}\left[\frac{1}{2} + T_1\left(\frac{L}{\lambda}\right)\right] + 2\frac{L}{\lambda}T_2\left(\frac{L}{\lambda}\right) - \left[1 - T_3\left(\frac{L}{\lambda}\right)\right]\right\} \\ &\quad - (2\pi RT_0)^{-1/2} V A \left\{\frac{L}{\lambda}\left[2T_1\left(\frac{L}{\lambda}\right) + 1\right] + \left[2T_2\left(\frac{L}{\lambda}\right) - \frac{\pi^{1/2}}{2}\right]\right. \end{aligned} \quad (3.3.26)$$

Table 3.1. *Comparison between the variational*[13] *and numerical*[9] *results for the nondimensional stress in plane Couette flow as a function of the inverse Knudsen number* δ. *The results with the variational method are computed with both a linear (third column) and a cubic trial function (fourth column).*

δ	Willis	Linear	Cubic
20.0	0.0807	0.0805	0.0805
10.0	0.1474	0.1474	0.1474
7.00	0.1964	0.1964	0.1963
5.00	0.2526	0.2524	0.2523
4.00	0.2946	0.2945	0.2943
3.00	0.3539	0.3537	0.3535
2.50	0.3938	0.3935	0.3933
2.00	0.4440	0.4438	0.4437
1.75	0.4745	0.4743	0.4742
1.50	0.5099	0.5097	0.5096
1.25	0.5517	0.5512	0.5511
1.00	0.6008	0.6008	0.6008
0.10	0.9258	0.9258	0.9258

The minimum condition gives the value of A_0:

$$A_0 = \frac{(2RT_0)^{-1/2}V\left\{(L/\lambda)\left[2T_1(L/\lambda)+1\right]+\left[2T_2(L/\lambda)-\frac{1}{2}\pi^{1/2}\right]\right\}}{\left\{(L/\lambda)^2\left[\frac{1}{2}+T_1(L/\lambda)\right]+4L/\lambda T_2(L/\lambda)-2\left[1-T_3(L/\lambda)\right]\right\}}. \tag{3.3.27}$$

If we insert this into $J(V/2 + A_0(x - L/2))$ we can compute an approximate expression for p_{12} given by

$$p_{12} = \rho_0(2RT_0)^{1/2}VT_1(0) + \rho_0\left(\frac{2RT_0}{V\pi}\right)^{1/2}(\lambda)^{-1}J(V/2 + A_0(x - L/2)). \tag{3.3.28}$$

We can use the inverse Knudsen number $\delta = L/\lambda$ as an independent variable and put p_{12} in a nondimensional form π_{12} by dividing by its free-molecular value $p_{12}^{f.m.} = -\rho_0 V(RT_0/2\pi)^{1/2}$. The resulting values are tabulated versus δ in Table 3.1 (third column) and compared with the results obtained by Willis[9] through a numerical solution, mentioned in the previous section. The agreement is excellent, the difference being at most $5 \cdot 10^{-4}$. We can say something more, namely that the results obtained by the variational method are more accurate than Willis's, with one possible exception. As a matter of fact, the variational method gives for π_{12} a value approximated from above, and the values in third column are, with the exception of the first row, never larger than the corresponding ones in the second column.

In order to check the accuracy of the method, we can use a cubic trial function $v = V/2 + A(x - L/2) + B(x - L/2)^3$. In this case the algebra is more formidable but still straightforward.[13] It is clear that the resulting values for π_{12} are only slightly different; the difference between the third and the fourth column is at most $2 \cdot 10^{-4}$. This confirms the accuracy of the method and the fact that the error in both the variational and Willis's results are less than 10^{-3}, with the variational ones being slightly more accurate.

Let us now look at the application of the variational method to the linearized Boltzmann equation in its integro-differential formulation. Here the main difficulty lies in choosing the trial function. In fact the method based on the integral equation has a hidden advantage: No matter how bad is the trial function, we obtain the correct free-molecular limit, because the free streaming is already built into the method. That is why we could obtain such good results with a very simple trial function; the reason was that it was correct in the continuum regime and thus the variational method optimized the interpolation between the two extreme behaviors.

Now we need at least three constants to reproduce the limiting behaviors in a reasonable way and thus (following Ref. 14) we try

$$\tilde{h} = 2\xi_3(\alpha x + \beta\xi_1 + \gamma \operatorname{sgn} \xi_1), \tag{3.3.29}$$

where α, β, and γ are three adjustable constants. The advantages with respect to the previous method are that the calculations are easier (in the case of the BGK model) and that a simple rational approximation to π_{12} is obtained for any molecular model:

$$\pi_{12} = \frac{a + \pi^{1/2}\delta}{a + b\delta + c\delta^2}. \tag{3.3.30}$$

The constants a, b, and c can be computed, in principle, for any molecular model. In order to make the definition of δ unequivocal, we let $\delta = L/\lambda$, where $\lambda = (\mu_0/p_0)(2RT_0)^{1/2}$ (this definition agrees with the previous one in the case of the BGK model). Then $c = 1$. Three cases were considered in Ref. 14, the BGK model, Maxwell molecules, and hard spheres, with the following results:

$$\begin{gathered} a = \tfrac{4-\pi}{\pi-2} \cong 0.7519, \qquad b = \tfrac{\pi^{3/2}}{2(\pi-2)} \cong 2.4388 \quad \text{(BGK)}, \\ a = 0.2225, \qquad b = 2.1400 \quad \text{(Maxwell molecules)}, \\ a = 0.3264, \qquad b = 2.1422 \quad \text{(hard spheres)}. \end{gathered} \tag{3.3.31}$$

Table 3.2. *Comparison between the variational*[14] *and numerical*[98] *results for the drag in plane Couette flow. The last four columns show variational results for different molecular models. The last one, marked with a star, is based on the trial function (3.32); the others are on (3.29).*

δ	Willis	BGK	Maxwell	Hard spheres	Hard spheres$_*$
20	0.0807	0.0805	0.0805	0.0801	0.0807
10	0.1474	0.1476	0.1476	0.1462	0.1483
7	0.1964	0.1969	0.1967	0.1943	0.1980
5	0.2526	0.2534	0.2529	0.2491	0.2550
4	0.2946	0.2958	0.2951	0.2900	0.2980
3	0.3539	0.3556	0.3542	0.3472	0.3583
2	0.4440	0.4462	0.4430	0.4332	0.4496
1	0.6008	0.6024	0.5933	0.5797	0.6051
0.1	0.9258	0.9238	0.8953	0.8994	0.9147

The maximum disagreement between the values of the stress for the BGK model and Willis's result is 0.5%. The results for Maxwell molecules and the BGK model are very close to each other except for high values of the Knudsen number ($\delta < 2$), where a difference of order 3% arises. Surprisingly enough, the results for hard spheres are, in this range, much closer than those for Maxwell molecules to those for the BGK model.

Note the relatively bad behavior of the hard sphere model for small Knudsen numbers, where all the models should agree. The reason is the same as for the inaccuracy of half-range polynomial expansion method, to be discussed in Section 3.9, and the solution is the same:[14] We must take into account the fact that, for molecules other than Maxwell's, the term proportional to $\xi_3\xi_1$ does not give an accurate description for small values of Kn and we must modify Eq. (3.3.29) as follows:

$$\tilde{h} = 2\alpha\xi_3 x + \beta L_M^{-1}(\xi_1\xi_3) + \gamma \operatorname{sgn}\xi_1)\xi_3, \tag{3.3.32}$$

where α, β, and γ are, as before, three adjustable constants to be determined by the variational method and L_M^{-1} is the inverse of usual linearized collision operator. The agreement with the solution obtained by Willis and the variational results for Maxwell molecules, over a wide range of Knudsen numbers, illustrated in Table 3.2, indicates several things, including the accuracy of te BGK model, the usefulness of the variational methods, and the relative insensitivity of the stress in linearized plane Couette flow to the particular molecular model adopted in the calculations.

Variational methods have been applied to the problem of heat transfer between flat plates in the linearized approximation.[15] The heat transfer according

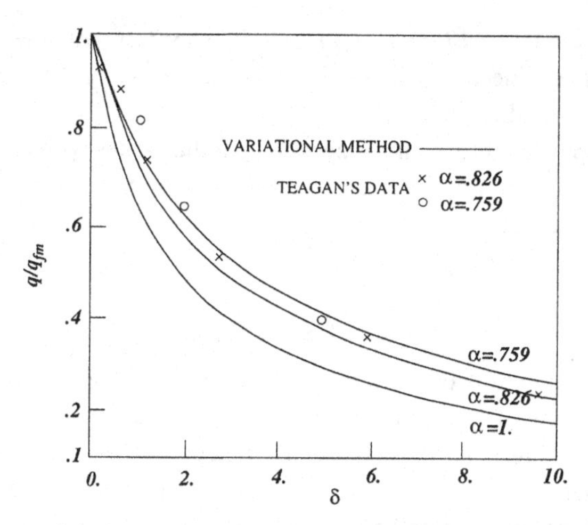

Figure 3.1. Comparison between the variational results for the problem of linearized heat transfer with Teagan and Springer's data. The results for the ratio between the heat flow normal to the walls and its free-molecular value refer to different values of the accommodation coefficient, as adopted in the calculations or given by Teagan and Springer.[16]

to both an accurate numerical and a variational solution of the BGK model[15] are always lower than the experimental data of Teagan and Springer[16] as shown in Fig. 3.1. The same fact occurs for variational results based on different collision models such as hard spheres and Maxwell molecules,[17] and this seems to exclude that the discrepancy is due to the use of the BGK model. A possible explanation might be due to the assumption made in[16] that the chamber pressure and the pressure between the plates are equal.[18] A comparison of the results of the variational results with accurate numerical solutions of the Boltzmann equation will be made later (Section 3.9).

Problems

3.3.1 Verify that if we set $\tilde{h} = h + \eta$, then the terms of first degree in η disappear from $J(\tilde{h})$, which reduces to $J(h) + ((\eta, \mathcal{L}\eta))$ if and only if h is a solution of Eq. (3.3.1).

3.3.2 Verify that Eq. (3.3.6) holds.

3.3.3 Verify that the linearized boundary conditions have the form (3.3.8).

3.3.4 Verify that if we let $\tilde{h} = h + \eta$ in (3.3.12) (where h satisfies the linearized Boltzmann equation and the boundary conditions) we find that the linear terms in η disappear from J and the variational principle holds.

3.3.5 Verify that the difference between the Maxwellians M_w and M_0 is, to first order, $M_0\psi$, where ψ is given by (3.3.18).

3.3.6 Verify that (3.3.17) reduces to (3.3.19) when terms of order higher than first are neglected.

3.3.7 Verify that Eq. (3.3.26) holds.

3.3.8 Compute the heat transfer between parallel plates with the variational method, assuming perfect accommodation and compare the results with Fig. 3.1.

3.4. Poiseuille Flow

Another interesting flow between parallel plates is plane Poiseuille flow. This is a particular case of flow in pipes or channels when a pressure difference is present. Since there is no conceptual difference among different shapes, we start by a generic cross section Σ (a strip in the plane case) in the (x, z) plane and assume that there is a small pressure difference in the y direction (Fig. 3.2). We remark that whereas in the continuum regime this pressure drop might arise from a density or a temperature difference, and the results would not differ in the linearized case, the two cases become more and more distinct when the Knudsen number increases. We shall consider the case of a density gradient, since the difference between the two cases is due to a phenomenon called thermal transpiration, which at low Knudsen number is restricted to the Knudsen layers.

Cylindrical Poiseuille flow (in which Σ is a circle) was carefully investigated by Knudsen in the early part of the twentieth century.[19] He found a striking result. If one fixes the pressure difference between the endpoints of a long, narrow tube (a capillary) and lets the average pressure vary, then the flow rate

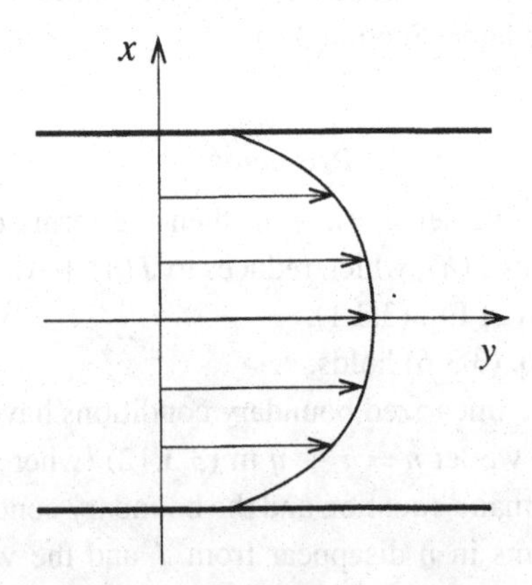

Figure 3.2. Geometry for plane Poiseuille flow. A negative pressure gradient along the y axis causes the gas to move from left to right.

through a cross section of the pipe exhibits a minimum. This phenomenon (known as the Knudsen minimum effect) is somewhat astonishing, because one would expect that the less molecules there are, the smaller will be their flow rate. At the beginning of 1960s, the matter was still debated and most of the experts thought that there was an error in the Knudsen data and the minimum (a rather flat one) was not there.

Let us see how to attack the problem from a theoretical point of view. We assume the boundaries to be at fixed temperature T_0 and the presence of a small density variation along the y axis. Then we can look for a solution linearized about a Maxwellian distribution M with a density $\rho = \rho_0 + \rho_1(y)$, with the second- and higher-order derivatives of smaller order with respect to ρ_1. Then

$$\xi_1 \partial_x h + \xi_2 \partial_y h + \xi_3 \partial_z h + \frac{1}{\rho} \xi_2 \partial_y \rho_1 = L_M h, \tag{3.4.1}$$

where we have neglected h when multiplying $\partial_y \rho_1$, as we have done for all the other second-order terms. Since we assume that the cross section is independent of y, the dependence of h on y is of second order, because of our assumptions. Thus we can assume h to be constant in y and treat $\partial_y \rho_1/\rho = \partial_y \rho/\rho = k$ as a constant.

Then we have the equation

$$\xi_1 \partial_x h + \xi_3 \partial_z h + k\xi_2 = L_M h. \tag{3.4.2}$$

This equation must be solved with the following boundary condition:

$$h(x, z, \boldsymbol{\xi}) = 0 \qquad ((x, z) \in \partial\Sigma;\ \xi_1 n_1 + \xi_3 n_3 > 0). \tag{3.4.3}$$

Equation (3.4.2) holds for the Poiseuille flow in a tube of any cross section Σ. In the plane case the problem reduces to

$$\xi_1 \partial_x h + k\xi_2 = L_M h \tag{3.4.4}$$

with the boundary condition

$$h(x, \boldsymbol{\xi}) = 0 \qquad (x = 0, \xi_1 > 0 \text{ and } x = L, \xi_1 < 0). \tag{3.4.5}$$

The equation to be solved is an inhomogeneous linearized Boltzmann, with a source term proportional to ξ_2. If we find a particular solution, then we can subtract it and eliminate the inhomogeneity. To this end, we remark that, since the source does not depend on x, the derivative of any particular solution with respect to x will satisfy the homogeneous equation. Since the exponential solutions reproduce themselves by differentiation, the derivative of the simplest particular solution must be a first degree polynomial in x. Hence we look for a particular solution as a second degree polynomial in x. Since the collision invariant multiplying k is x_2, we look for a solution of the form $h_0 = -C\xi_2 x^2 + A(\boldsymbol{\xi})x + B(\boldsymbol{\xi})$ and we expect A and B to be proportional to ξ_2. Here C, and, likewise, D and E (to be soon introduced), are constants. If we insert the last expression into (3.4.4) and equate the coefficients of first and zeroth degree in x in the left- and right-hand side, we obtain

$$-2C\xi_1\xi_2 x = L_M A, \qquad A\xi_1 + k\xi_2 = L_M B. \tag{3.4.6}$$

Solving for the first equation, we have $A = -2CL_M^{-1}(\xi_1\xi_2) + D\xi_2$, and the equation for B becomes

$$L_M B = -2C\xi_1 L_M^{-1}(\xi_1\xi_2) + D\xi_2\xi_1 + k\xi_2. \tag{3.4.7}$$

This equation can be solved if and only if the right-hand side is orthogonal to the collision invariants. This condition is trivially satisfied, by symmetry (Problem 3.4.1), for all invariants except ξ_2. If we impose that the source in Eq. (3.4.7) is orthogonal to ξ_2 as well, we obtain

$$-2C\left(\xi_2\xi_1, \xi_2 L_M^{-1}(\xi_1\xi_2)\right) + k(\xi_2, \xi_2) = 0, \tag{3.4.8}$$

where we have used the notation (g, h) introduced in Section 3.1. This relation determines C in terms of k. Since $(\xi_2, \xi_2) = RT_0$ and, according to (3.3.13),

$$\mu_0 = -(RT_0)^{-1}\left(\xi_1\xi_2 L_M^{-1}(\xi_1\xi_2)\right), \tag{3.4.9}$$

we have

$$C = -k(RT_0)^2\mu_0. \tag{3.4.10}$$

Then we obtain from Eq. (3.4.7)

$$B = kL_M^{-1}\left\{\xi_1 L_M^{-1}[(2RT_0)^2\mu_0(\xi_1\xi_2)] + \xi_2\right\} + DL_M^{-1}(\xi_1\xi_2) + E\xi_2. \tag{3.4.11}$$

Hence we have obtained a particular solution that contains two arbitrary constants, D and E. These will be chosen in such a way as to comply with the symmetry with respect to $x = L/2$ and to simplify the boundary conditions for $h_1 = h - h_0$ as much as possible. This leads to $D = k(RT_0)^2\mu_0$ and $E = 0$. Thus we have

$$\begin{aligned} h_0 = {} & k(RT_0)^2\mu_0\xi_2 x(x - L) + k(RT_0)^2\mu_0 L_M^{-1}(\xi_1\xi_2)(2x - L) \\ & + kL_M^{-1}\left\{\xi_1 L_M^{-1}[(2RT_0)^2\mu_0(\xi_1\xi_2) + \xi_2]\right\}. \end{aligned} \tag{3.4.12}$$

Thus h_1 satisfies the homogeneous linearized Boltzmann equation and the boundary conditions

$$h_1(x,\boldsymbol{\xi}) = -2k(RT_0)^2\mu_0 L_M^{-1}(|\xi_1|\xi_2) + kL_M^{-1}\{\xi_1 L_M^{-1}[(2RT_0)^2\mu_0(\xi_1\xi_2)] + \xi_2\}$$
$$(x=0,\ \xi_1>0 \text{ and } x=L,\ \xi_1<0). \quad (3.4.13)$$

We remark that for small values of the Knudsen number, the part of the solution arising from h_0 will dominate and we shall obtain the familiar parabolic profile. The flow rate will be, for fixed values of L and the pressure difference, inversely proportional to the Knudsen number Kn and hence will increase for an increasing average pressure.

However, this behavior will not be always correct, because of the increasing role played by the Knudsen layers. In order to see what happens at low pressures, we remark that for free-molecular conditions, Eq. (3.4.4) formally reduces to

$$\xi_1 \partial_x h + k\xi_2 = 0, \quad (3.4.14)$$

or, if we take into account the boundary conditions,

$$h = -k\xi_2\left(\frac{x - L/2}{\xi_1} + \frac{L/2}{|\xi_1|}\right). \quad (3.4.15)$$

A singularity at $\xi_1 = 0$ is apparent. We know, however, that molecules traveling almost parallel to the walls or with extremely low speeds are not well described by this equation, because, no matter how few molecules there are, they will collide more frequently with other molecules than with the wall. Thus Eq. (3.4.2) can be assumed to hold for $|\xi_1| \geq (RT_0)^{1/2}/\mathrm{Kn}$. Otherwise we have to compute the effect of the molecular collisions, which will change the solution at abnormally low values of ξ_1. Then we can compute the bulk velocity by integrating over the molecules having a value of ξ_1 satisfying the above restriction, since for the others the solution is not good. We find (Problem 3.4.2)

$$\rho v = \int_{\xi_1 \leq (RT_0)^{1/2}/\mathrm{Kn}} Mh\,d\boldsymbol{\xi} = -\frac{\partial \rho}{\partial y}\frac{LRT_0}{2}\int_{|\xi_1| \geq (RT_0)^{1/2}/\mathrm{Kn}} \frac{\exp(-\xi_1^2/2RT_0)}{|\xi_1|}d\xi_1$$
$$= -\frac{\partial \rho}{\partial y} LRT_0 \log(\mathrm{Kn}) + O(1), \quad (3.4.16)$$

where $O(1)$ denotes a term that remains finite when the Knudsen number goes to ∞.

The first-order effects of collisions will be to remove the molecules that would make v infinity, and the above estimate is correct. A detailed study has been carried out for the BGK model using either the method of collision iteration[20] or a method based on the exact solution,[21] and these results confirm the estimate in Eq. (3.4.16). Thus the flow rate diverges for $\mathrm{Kn} \to \infty$. This behavior confirms that the flow rate must have at least one minimum, since it decreases for small average pressure and increases for large average pressures.

Actually, for a slab, the minimum is rather more marked than in a cylindrical pipe and even rough calculations[20] indicate that it must occur for $\mathrm{Kn} \cong 1$. More accurate results can be obtained with the finite ordinate schemes of the kind mentioned in Section 3.2. The second scheme can be shown to approximate the solution from above while the first appears to approximate the solution from below. The minimum appears to be located at approximately $\mathrm{Kn} = 1.1$. Calculations with the variational method have also been made,[13] since the flow rate is naturally related to the functional to be minimized (Problem 3.4.3).

It is not possible, of course, to make experiments in a real slab, but experimental data for a channel with a rather thin rectangular cross section are available.[22] Provided we keep the Knudsen number not too large, so as to avoid the effects of the finite width of the channel, it is possible to make a comparison, which shows a good agreement between theory and experiment (see Fig. 3.3).

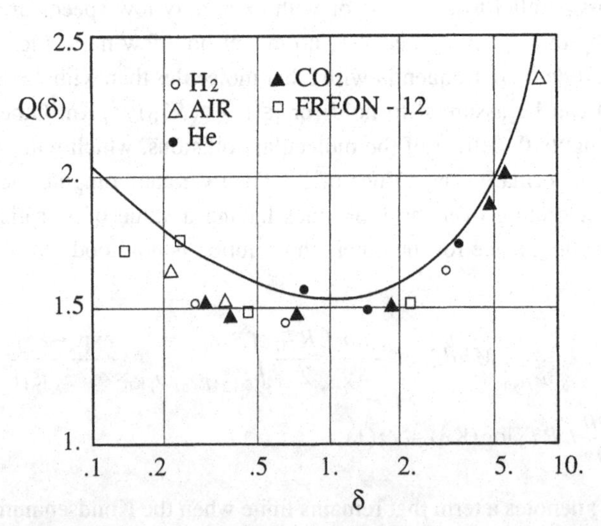

Figure 3.3. Normalized flow rate in plane Poiseuille flow. Comparison between the variational solution and Dong's data. The presence of a minimum is clear.

Problems

3.4.1 Prove that the condition of orthogonality to the collision invariants of the right-hand side of (3.4.7) is trivially satisfied, by symmetry, for all invariants except ξ_2.

3.4.2 Prove that the flow rate ρv is given by (3.4.16) when the perturbation h is given by (3.4.15) for $|\xi_1| \geq (RT_0)^{1/2}/\text{Kn}$.

3.4.3 Prove that the flow rate ρv is related to the functional to be minimized to obtain the solution of plane Poiseuille flow (see Ref. 13).

3.5. Half-Space Problems

In the previous section we mentioned Knudsen layers several times. They provide an important concept when the gas is not so rarefied, because then the Navier–Stokes equations can be applied, except in thin kinetic layers, the most common example being the Knudsen layers near the boundaries. Their study is required to obtain the boundary conditions for the Navier–Stokes equations. We shall comment on this aspect of the matter at the end of this section.

To study the Knudsen layers in detail, at least in the case of a flat or almost flat boundary (where almost flat means that the radius of curvature is much larger than the mean free path), it is convenient to think of a plane Couette flow in which the distance of the plates tends to infinity. We thus obtain a problem in a half-space. To avoid trivialities, however, we must also let the speed V of the upper plate go to infinity with L in such a way that the gradient $k = V/L$ remains finite.

The simplest case occurs when we are interested in the effects of the velocity gradient and disregard the effects of the temperature gradient. Then far from the plate at $x = 0$ the solution provided by the Hilbert expansion (Hilbert solution) applies, whereas near the boundary a correction occurs; the latter, however, can be considered to be small if the nondimensional gradient defined in Eq. (3.2.1) is small. Thus, in this case we can proceed to linearizing the Boltzmann equation about a Maxwellian distribution with bulk velocity v in the y direction. Then we obtain the equation

$$k\xi_1 c_2 (2RT_0)^{-1} + \xi_1 \partial_x h = L_M h, \tag{3.5.1}$$

where $\mathbf{c} = \boldsymbol{\xi} - kx\mathbf{j}$. Then

$$h = k(2RT_0)^{-1} L_M^{-1} \xi_1 c_2 + h_1, \tag{3.5.2}$$

where h_1 satisfies the homogeneous linearized Boltzmann equation. This problem was first proposed by Kramers in 1949[23] and hence is called *Kramers' problem*. Since, however, a similar problem was met by Milne in radiative transfer, the more generic name *Milne problem* is sometimes used.

The term h_1 in Eq. (3.5.2) must be of the form (2.7.7) (with $\mathbf{c}$ in place of $\boldsymbol{\xi}$). However, some simplifications occur. First, since far away from the plate we expect the solution to be of the Hilbert type, the growing exponentials must be discarded, and the term growing linearly with x should also be omitted because we have already included the velocity growth in the basic Maxwellian M. Second, the only collision invariant that should contribute is $\psi_2 = c_2$. Then we are left with

$$h_1 = Ac_2 + \int_{\bar{\gamma}}^{\infty} C(\gamma) g_\gamma(\mathbf{c}) \exp(-\gamma x) d\gamma, \qquad (3.5.3)$$

where, for convenience, γ is here the opposite of what it was in Eq. (2.7.7). It is also clear that not all the eigensolutions will contribute, but, as a consequence of the boundary conditions to be specified below, only those that have the right symmetry to produce a contribution to v. In particular, the eigensolutions present in Eq. (3.5.3) will not contribute to the density. The bulk velocity is then given by

$$v = kx + A(RT_0)^{1/2} + \int_{\bar{\gamma}}^{\infty} C(\gamma) \exp(-\gamma x) d\gamma, \qquad (3.5.4)$$

where the first term comes from the Maxwellian M and the eigensolutions $g_\gamma(\mathbf{c})$ have been normalized in such a way that

$$\int c_2 g_\gamma(\mathbf{c}) M d\mathbf{c} = 1. \qquad (3.5.5)$$

The boundary conditions at $x = 0$ will be of the form (Problem 3.5.1)

$$h^+ = Kh^-. \qquad (3.5.6)$$

In terms of h_1 the boundary conditions take on the form

$$h_1^+ = -k(2RT_0)^{-1} L_M^{-1} \xi_1 c_2 + Kh^- \quad (\xi_1 > 0). \qquad (3.5.6)$$

The simplest kind of boundary conditions is diffuse reflection according to a Maxwellian distribution, which turns out to be M (because we chose the bulk velocity in M to be kx and the temperature is constant). Then the operator K disappears from Eq. (3.5.6) (since it produces M times the mass density corresponding to the function upon which K acts) and we have

$$h_1^+ = -k(2RT_0)^{-1} L_M^{-1} \xi_1 c_2 \quad (\xi_1 > 0). \qquad (3.5.7)$$

This is the simplest kind of boundary condition because it fully assigns the distribution function for the molecules entering the half-space. The expression of h_1 given by Eq. (3.5.3) for $x = 0$ leads to

$$Ac_2 + \int_{\overline{\gamma}}^{\infty} C(\gamma) g_\gamma(\mathbf{c}) d\gamma = -k(2RT_0)^{-1} L_M^{-1} \xi_1 c_2 \quad (\xi_1 > 0). \qquad (3.5.8)$$

In other words we must find a constant A and a function of $C(\gamma)$ in such a way that Eq. (3.5.8) is satisfied. Further, it should be impossible to find more than one solution to this problem; otherwise the solution of Kramers' problem would not be unique. This remark leads to one of the deepest properties of the eigensolutions g_γ, which goes under the name of *half-range completeness*. When we take the general solution (2.7.7) for some fixed value of x (which, without loss of generality, can be taken to be $x = 0$), omit the terms diverging with x, and equate it to some given function for $\xi_1 > 0$, then there must be uniquely determined values of the coefficients A_α and $C(\gamma)$ $(\gamma > 0)$ such that this equation is satisfied. This property has been proved for some models[1–5] in an explicit fashion and by operator techniques in the general case.[24]

Once A and $C(\gamma)$ are determined, the solution of the problem is completely known. In particular, the velocity profile is provided by Eq. (3.5.4). Even without computing the solution, we can sketch a qualitative profile of v as a function of x (see Fig. 3.4). $v(0)$ will be some finite value, which is called the *microscopic slip velocity*. The value of v will start from there and increase (because of the term kx) approaching the asymptotic behavior $kx + A(RT_0)^{1/2}$. The constant $A(RT_0)^{1/2}$ is called the *macroscopic slip velocity* and is far more important than $v(0)$. In fact, by knowing it, we can reconstruct the correct behavior of the bulk velocity outside the Knudsen layer.

Of course, the solution is proportional to k and, in particular, we find that $A(RT_0)^{1/2} = \zeta k$, where the coefficient ζ is called the *slip coefficient*.

We shall add a few details about the solution of the BGK model for Kramers' problem. The solution can be computed in closed form[25,1,2] and in particular the slip coefficient ζ can be expressed in terms of an integral that is easy to compute numerically,[25] with the following result:

$$\zeta = (1.01615)\lambda = (1.1466)\ell, \qquad (3.5.9)$$

where ℓ and $\lambda = 2\pi^{-1/2}\lambda$ are the two mean free paths we used before.

The microscopic slip velocity can also be evaluated, with the following result:

$$v(0) = 2^{-1/2}\lambda k = \left(\frac{2}{\pi}\right)^{1/2} \ell k. \qquad (3.5.10)$$

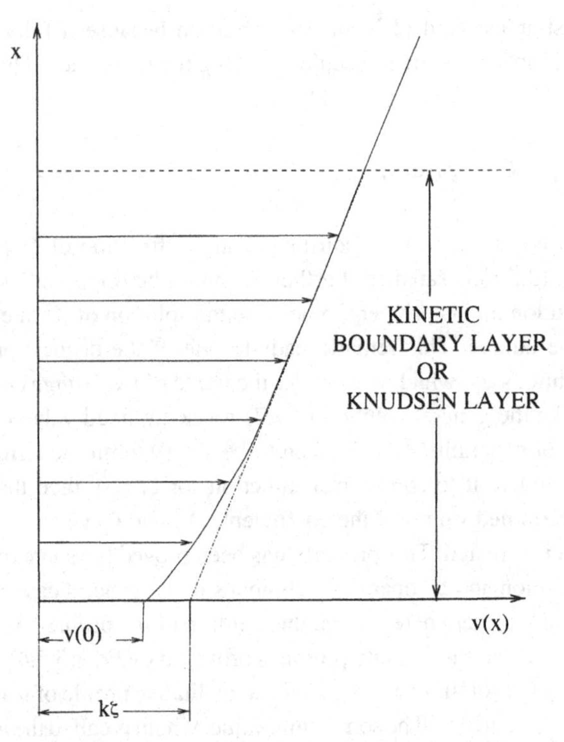

Figure 3.4. Sketch of the velocity profile in the Knudsen layer. The microscopic and macroscopic slip are indicated.

The velocity profile is given by

$$v(x) = k\left[x + \zeta - \frac{\pi^{-1/2}\lambda}{2} I\left(\frac{x}{\lambda}\right)\right], \tag{3.5.11}$$

where the function $I(x/\lambda)$ is virtually zero outside the Knudsen layer and can be also easily computed.[26] A plot is given in Fig. 3.5.

It is also possible to evaluate the distribution function of the molecules arriving at the plate. It has the form

$$h(\mathbf{c}) = \lambda \frac{kc_2}{RT_0} P(|\xi_1|(2RT_0)^{-1/2}) \qquad (\xi_1 < 0), \tag{3.5.12}$$

where the function P varies smoothly and satisfies the relation

$$|\xi_1|(2RT_0)^{-1/2} + 0.7071 < P(|\xi_1|(2RT_0)^{-1/2}) < |\xi_1|(2RT_0)^{-1/2} + 1.01615.$$

We remark that the distribution function is discontinuous at $\xi_1 = 0$ at the wall;

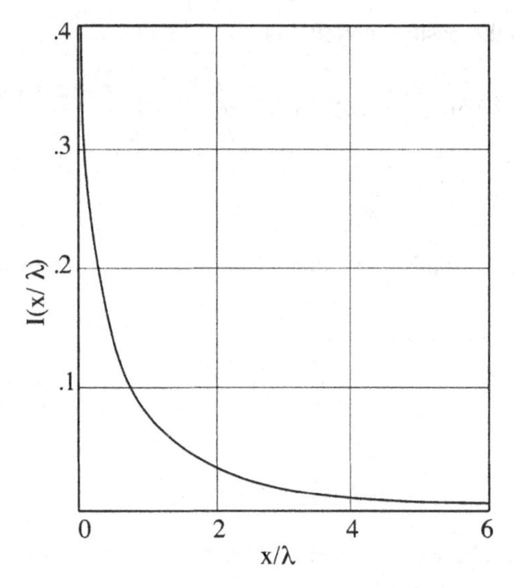

Figure 3.5. The velocity defect in the Knudsen layer as a function of the nondimensional coordinate normal to the wall.

in fact, h is zero if $\xi_1 > 0$, whereas $P(0)$ is different from zero as shown by (3.5.13). The same inequality implies that the distribution function for the molecules arriving at the wall is rather close to a Hilbert solution (for which the function P should be replaced by a linear function). This is not surprising: In fact, each molecule has the velocity acquired after its last collision, which, on the average, occurred a mean free path away from the wall, nearly outside the Knudsen layer and hence in a region where the distribution function is close to that of a Hilbert solution. It is interesting to recall that Maxwell[27] assumed that the distribution function of the arriving molecules was exactly of the form prevailing far from the wall; then, using momentum conservation (see also Chapter 2, Section 2.6), he was able to evaluate the slip coefficient, without solving Kramers' problem. He found $\zeta = \ell$ (with an error of 15%) and

$$h(\mathbf{c}) = \lambda \frac{kc_2}{RT_0}\left(|\xi_1|(2RT_0)^{-1/2} + 0.8863\right) \qquad (\xi_1 > 0), \tag{3.5.13}$$

which is a good approximation to the correct result given by Eqs. (3.5.12).

The Kramers problem can be also easily attacked with variational methods. If we use the BGK model, we can use the integral equation. It is convenient to write an equation for $\phi = (v - kx)/(k\lambda)$, which (Problem 3.5.2) turns out to be

$$\phi(x) = \pi^{-1/2} T_1\left(\frac{x}{\lambda}\right) + (\lambda)^{-1}\pi^{-1/2}\int_0^\infty T_{-1}\left(\frac{|x - x_*|}{\lambda}\right)\phi(x_*)dx_* \quad (x \geq 0). \tag{3.5.14}$$

For this problem the basic functional is

$$J(\tilde{\phi}) = \int_0^\infty dx \left\{ [\tilde{\phi}(x)]^2 - (\lambda)^{-1}\pi^{-1/2} \int_0^\infty T_{-1}\left(\frac{|x-x_*|}{\lambda}\right) \tilde{\phi}(x)\tilde{\phi}(x_*)dx_* \right.$$
$$\left. - 2\tilde{\phi}(x)\pi^{-1/2}T_1\left(\frac{x}{\lambda}\right) \right\} \tag{3.5.15}$$

In order to relate the minimum value of the functional to the slip coefficient ζ, let us put

$$\sigma = \zeta/\lambda = \lim_{x\to\infty} \phi. \tag{3.5.16}$$

In this way

$$\psi(x) = \phi(x) - \sigma \tag{3.5.17}$$

has a finite integral between 0 and ∞. If we rewrite Eq. (3.5.14) in terms of ψ, we obtain (Problem 3.5.3)

$$\psi(x) = \pi^{-1/2}\left[T_1\left(\frac{x}{\lambda}\right) - \sigma T_0\left(\frac{x}{\lambda}\right)\right]$$
$$+ (\lambda)^{-1}\pi^{-1/2} \int_0^\infty T_{-1}\left(\frac{|x-x_*|}{\lambda}\right) \psi(x_*)dx_* \quad (x \geq 0), \tag{3.5.18}$$

where we used the basic property of the derivatives of the Abramowitz functions. Using again this property we can now integrate Eq. (3.5.18) twice between x and ∞ to obtain (Problem 3.5.4)

$$\sigma T_2\left(\frac{x}{\lambda}\right) - T_3\left(\frac{x}{\lambda}\right) = (\lambda)^{-1} \int_0^\infty T_1\left(\frac{|x-x_*|}{\lambda}\right) \psi(x_*)dx_* \quad (x \geq 0). \tag{3.5.19}$$

Letting $x = 0$, we obtain

$$\sigma\frac{\pi^{1/2}}{4} - \frac{1}{2} = (\lambda)^{-1} \int_0^\infty T_1\left(\frac{|x_*|}{\lambda}\right) \psi(x_*)dx_*, \tag{3.5.20}$$

or, in terms of ϕ (Problem 3.5.5),

$$\sigma\frac{\pi^{1/2}}{2} - \frac{1}{2} = (\lambda)^{-1} \int_0^\infty T_1\left(\frac{|x_*|}{\lambda}\right) \phi(x_*)dx_*. \tag{3.5.21}$$

Now, for $\tilde{\phi} = \phi$, we have

$$J(\phi) = -\int_0^\infty \phi(x)\pi^{-1/2}T_1\left(\frac{x}{\lambda}\right)dx \tag{3.5.22}$$

and Eq. (3.5.21) becomes

$$\sigma = \pi^{-1/2} - 2\lambda^{-1} \min J(\tilde{\phi}). \tag{3.5.23}$$

Let us take the simplest possible trial function, a constant c. A straightforward computation gives (Problem 3.5.6)

$$J(c) = \left(\frac{1}{2\pi^{1/2}} c^2 - \frac{1}{2} c \right) \lambda, \tag{3.5.24}$$

which attains the minimum value for $c = \pi^{1/2}/2$, corresponding to $J(c) = -\pi^{1/2}/8$. Equation (3.5.21) then gives the following approximate value for σ:

$$\sigma = \pi^{-1/2} + \frac{\pi^{1/2}}{4}. \tag{3.5.25}$$

Hence the approximate value of the slip coefficient is

$$\zeta = \left(\pi^{-1/2} + \frac{\pi^{1/2}}{4} \right) \lambda = \left(\frac{2}{\pi} + \frac{1}{2} \right) \ell \cong (1.1366)\ell. \tag{3.5.26}$$

This is to be compared with the value $\zeta = (1.1466)\ell$, obtained by a numerically accurate evaluation of (5.9). We see that even a very simple choice for the trial function, certainly inadequate to describe the Knudsen layer, yields a rather accurate estimate of ζ (the error is less than 1%).

The same procedure can be applied to models with a velocity-dependent collision frequency.[28,29] It is also possible to apply the procedure with the variational principle for the linearized Boltzmann equation in an integro-differential form and extend it to arbitrarily assigned kernels for gas-surface interaction.[30] In the latter case the result depends on a small number of accommodation coefficients, which can be computed once the kernel is given (Problem 3.5.7).

In the mid 1970s experimental methods were developed to measure velocity profiles in the Knudsen layer on a flat wall.[31,32] In particular, Reynolds et al.[31] reported that in the Knudsen layer the deviation of the velocity defect (i.e., the deviation of the actual velocity from the continuum velocity profile) shows a behavior quantitatively different from the results obtained by the BGK model, which we have just described (Fig. 3.6).

Loyalka[33] pointed out that such a discrepancy could be due to a basic deficiency of the BGK model, in that it does not allow for the dependence of the collision frequency upon the molecular velocity. With this in view, he carried out a detailed numerical study of the model with a velocity-dependent collision frequency, which had been used to find a closed-form solution for the structure of the kinetic layer as early as 1966.[34,4] The latter solution was considered by

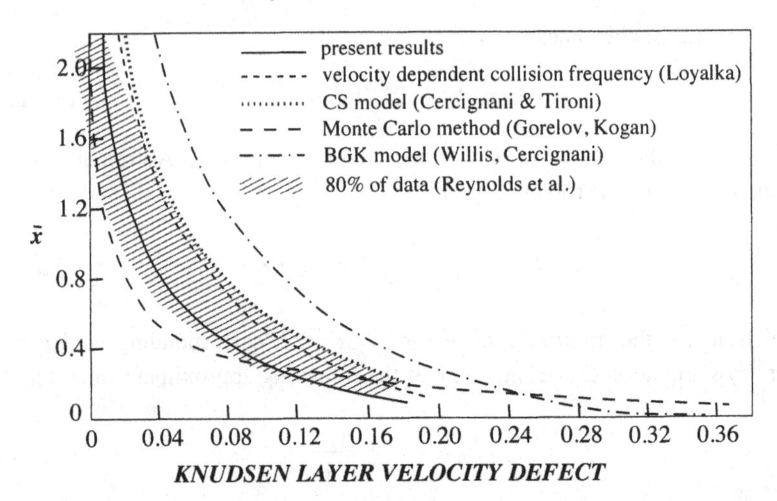

Figure 3.6. Comparison between the velocity defect in the Knudsen layer according to different models, Monte Carlo simulation, and experimental data. The data fall in the shadowed region. The results by Bird[39] are not shown but they fall between the two lower theoretical curves. $\bar{x}$ denotes the ratio between the coordinate normal to the wall and the mean free path $\ell = \pi^{1/2}\lambda/2$. The picture is taken from Ref. 40, where the results shown by the full line are obtained.

Loyalka not to be useful for the purpose of a numerical evaluation, for which he preferred to use a direct numerical technique. His results show that the velocity dependence of the collision frequency appropriate for hard spheres does indeed have an important effect on the velocity defect. In fact, the numerical solution practically coincides with the upper boundary of the region containing 80% of the results of Reynolds et al..

Another well-known deficiency of the BGK model (see Section 3.1) is that it yields a Prandtl number not appropriate to a monatomic gas (Pr = 1 rather than Pr = 2/3). This unsatisfactory aspect motivated the use of the ES model (see Chapter 1, Section 1.9) to investigate the structure of the velocity profile in Knudsen layers as early as 1966.[35] Unfortunately this solution contained a trivial misprint, in that a numerator and a denominator were erroneously interchanged. When the mistake is corrected the results are in a reasonably good agreement with the experiments; in fact, they are almost undistinguishable from the results presented by Loyalka.[33] This fact was pointed out in 1976.[36] At about the same time, Abe and Oguchi[37] examined the Knudsen layers with a model equation of the kind discussed in Section 3.1 involving thirteen moments in the collision term. This collision model contains two nondimensional parameters – the Prandtl number and a further parameter, which takes the value unity when the model reduces to the linearized ES model. Abe and Oguchi conclude that the solution indicates a "weak dependence" on this parameter.

As early as 1968, Gorelov and Kogan[38] reported the results of a Monte Carlo simulation, which appear to fall on the lower boundary of the aforementioned region containing 80% of the experimental data. Their results were confirmed by Bird.[39] Then it was pointed out[40] that one can concoct a new model having the desirable properties of both a correct Prandtl number and variable collision frequency, while amenable to a simple solution. The velocity profile computed by means of this model turns out to be in exceptionally good agreement[40] with the experiments of Reynolds et. al..[31]

If we consider the Couette flow with bounce-back boundary conditions (see Section 2.2 of Chapter 2) we discover that the Knudsen layers are absent. This is, however, only true for Couette flow, since the second derivatives of the bulk velocity vanish. This is not true in the case of Poiseuille flow, where a second-order slip (proportional to the square of the mean free path) is present even for bounce-back boundary conditions. For a discussion of this point we refer to a paper of the author.[41] In general, second-order effects had been previously treated in the literature,[42] and a systematic treatment is available;[43–45] it is true that, in general, the effects of second-order slip show up when the Knudsen layers are so thick that they begin to merge with the main core of the flow.

So far we have dealt with the problem of the velocity profile in the presence of a velocity gradient outside the kinetic layer. Analogous problems arise in connection with a temperature gradient either normal or tangential to the wall. The first problem, the so-called temperature jump problem, leads to a mathematically more complex situation than the velocity slip problem, since, even for the BGK model, it is governed by a system of two equations rather than a single equation (see Eq. (3.1.4)). There is no difficulty in obtaining a numerical solution, as was first done in the pioneering paper by Welander,[8] but obtaining a closed-form solution is much harder. A brilliant but unsuccessful attempt to overcome this difficulty was made by Darrozès;[46] he suggested that the Riemann–Hilbert problem that one meets can be solved by diagonalizing two-by-two matrices. This is, of course, possible but introduces new singularities in the complex plane, leading to a difficult problem, which Darrozès was not able to master. In fact the problem was solved later[47] by using concepts from the theory of the integrals of algebraic functions. The problem was worked out in more detail by Siewert and Kelley in 1980.[48] The problem was brought to completion in 1982, when a procedure to compute analytically the partial indices of the Riemann–Hilbert problem was indicated.[49] For details we refer to the original papers[47–49] and Ref. 4.

Cassel and Williams[50] discovered that if the collision frequency is taken exactly proportional to the molecular speed $|\boldsymbol{\xi}|$, it becomes possible to solve the corresponding model equation in closed form. For details see Refs. 47 and 4.

The value of the temperature jump coefficient τ for the BGK model turns out to be $\tau = (2.1904)\ell$;[15] for the model used by Cassel and Williams $\tau = (2.247)\ell$.[50]

When there is a temperature gradient in a direction tangential to the wall, a new phenomenon – thermal creep – arises. First discovered by Maxwell,[27] in thermal creep the gas slips on the wall with a bulk velocity proportional to the temperature gradient. The corresponding coefficient ω has been computed by Sone,[51] Williams,[52] and Loyalka;[53] the results agree in giving $\omega = (0.648)\ell$.

A comparison of these results with accurate numerical solutions of the linearized Boltzmann equation will be made later (Section 3.9).

Finally, we discuss the application of the results of this section to the problem of boundary conditions for the Navier–Stokes equations. We have found that both the bulk velocity and the temperature of the gas differ from the corresponding quantities of the wall, by an amount proportional to gradients of macroscopic quantities, the coefficient being of the order of the mean free path. Hence when the Knudsen number becomes very small, the classical boundary conditions of no slip and no temperature jump apply. When the Knudsen number increases, the effects of the presence of the Knudsen layers become noticeable, and, accordingly, one must adopt modified boundary conditions.

As an example, we take up again Couette flow, which was treated in detail in Section 3.3. Using the auxiliary variable s, we found (Eq. (3.3.20)$_1$)

$$v = \tau_w \int_0^x \frac{dx}{\mu(T(x))} + v_w. \tag{3.5.27}$$

If we impose the slip boundary condition at $x = s = 0$ and at $x + L$, we have

$$v_w = \frac{\zeta_0}{\mu_0}\tau_w, \tag{3.5.28$_1$}$$

$$v(L) = \tau_w \int_0^L \frac{dx}{\mu(T(x))} + \frac{\zeta_w}{\mu_w}\tau_w = V - \frac{\bar{\zeta}}{\bar{\mu}}\tau_w, \tag{3.5.28$_2$}$$

where, in agreement with the notation of Section 3.3, the subscript w denotes a quantity evaluated at $x = 0$ and an overline a quantity evaluated at $x = L$. Hence we obtain the stress

$$\tau_w = V\left[\int_0^L \frac{dx}{\mu(T(x))} + \frac{\zeta_w}{\mu_w} + \frac{\bar{\zeta}}{\bar{\mu}}\right]^{-1}. \tag{3.5.29}$$

In the linearized case we can assume isothermal conditions and let ζ and μ be constant. In this case we have

$$\tau_w = \mu\frac{V}{L + 2\zeta}. \tag{3.5.30}$$

Thus the stress is reduced because of the presence of the velocity slip. If we could assume the formula to be correct throughout the transition regime, we would find in the free-molecular limit

$$\tau_{w\, f.m.} = \frac{1}{4}\rho(2RT)^{1/2}V\frac{\lambda}{\zeta}. \tag{3.5.31}$$

If we compare this approximate result with the correct free-molecular value $(\rho V 2RT/\pi^{1/2}/2)$, we see that Eq. (3.5.31) would be correct for $\zeta = \pi^{1/2}\lambda/2$, a value fortuitously equal to the value computed by Maxwell! This accidental coincidence explains why even in the early 1960s there remained competent aerodynamicists who believed that they could dispense with kinetic theory and compute rarefied flows with the Navier–Stokes equations. The Knudsen minimum effect in Poiseuille flow and other examples convinced everybody that kinetic theory was required.[54]

Problems

3.5.1 Check that the boundary condition for h has the form shown in (3.5.6).

3.5.2 Prove that Eq. (3.5.14) holds.

3.5.3 Insert (3.5.17) into (3.5.14) and check that (3.5.18) holds.

3.5.4 Obtain (3.5.19) from (3.5.18).

3.5.5 Check that (3.5.21) holds.

3.5.6 Check that (3.5.15) is a functional that attains an extremum if and only if (3.5.15) holds and that it reduces to (3.5.24) when $\tilde{\phi}$ is a constant c.

3.5.7 Extend the variational calculation of the slip coefficient to a model with velocity-dependent collision frequency (see Refs. 28 and 29).

3.5.8 Extend the variational calculation of the slip coefficient to arbitrary models for the gas–surface interaction (see Ref. 30).

3.6. Numerical Methods

The linearized approach discussed in the previous sections lends itself to an elegant treatment of many problems. Its usefulness is however restricted to small speeds and small temperature differences.

For problems in which conditions greatly removed from equilibrium occur, the situation is not so simple. The BGK and other models have been used for some years and have provided useful insight into nonequilibrium problems. Although they have been very useful in obtaining approximate solutions[4,5] and forming qualitative ideas on the solutions of practical problems, they in general do not provide us with detailed and precise answers to the sort of questions that are posed by the space engineer. For real physical situations, in fact, such as the flow pattern around an object that moves inside a rarefied gas, we need methods to actually calculate or approximate solutions of the Boltzmann equation. For most situations, it is hopeless to look for explicit solutions of the latter equation or even of the BGK model. The five-dimensional integral in the collision term makes numerical approximations a difficult task as well.

Suppose we want to approximate the collision integral $Q(f, f)$ by a quadrature formula that requires the evaluation of the integrand at a number of points. Obviously, the integrand must decay fast enough at infinity to give us reasonable accuracy with a finite number of evaluation points.

For the sake of our argument, let us assume that the formula then requires 30 function evaluations to approximate a one-dimensional integral. Then $30^2 = 900$ evaluations would be needed to achieve the same accuracy for a comparable two-dimensional integral, 27,000 for a three-dimensional integral, 810,000 for a four-dimensional one, and 243,000,000 for a five-dimensional integral as in $Q(f, f)$ above. Clearly, the numerical effort for such a procedure would be unreasonable, in particular because the evaluation would have to be repeated at each space point and each velocity on the grid, and, for time-dependent problems, for each time step.

This explains why a successful method for practical calculations is the technique of Hicks, Yen, and Nordsiek,[55,56] which is based on a Monte Carlo quadrature method to evaluate the collision integral and was further developed by Aristov and Tcheremissine.[57,58] This tecnique was used with some success in a few two-dimensional flows.[59,60]

As we shall see in Chapter 6, polyatomic gases and chemically reacting and thermally radiating flows are hard to describe with theoretical models having the same degree of accuracy as the Boltzmann equation for monatomic nonreacting and nonradiating gases. Thus, as we shall discusss in Chapter 7, simulations schemes, which started with the work of Bird on the *Direct Simulation Monte Carlo* (DSMC) method[61,62] have become the only practical tool for solving the Boltzmann equation.

The key idea of particle simulation is rather simple. Recall that the underlying physical reality is a gas with, say, 10^{20} molecules. As it is impossible to keep a record of all of them, one introduces the Boltzmann equation as a way to keep track of the distribution function (i.e., a function containing information about the average density, energy, temperature, etc. of the particle system). It is useful to recall at this time that the Boltzmann equation is only a mathematical approximation to the physical reality (which becomes exact for classical hard spheres in the Boltzmann–Grad limit). The basic idea of particle simulation is really just to return to the particle level, but to restrict the number of molecules to a tractable figure (which depends not only on the situation to be modeled, but also on the available computer). If, say, 10^5 molecules are to be used, interaction rules must be given that will reflect the influence of the collisions on the behavior of the gas; it is clear that for any physically reasonable number of molecules it is hopeless to follow the time evolution of a system of, say, N hard spheres.

Such an approach would not only be unreasonable from a mathematical point of view, but also not physically meaningful. After all, our target is the simulation of rarefied gas dynamics, not the solution of N-body problems with large N. Since the former is believed to evolve by averaging over the latter, some averaging should be part of the procedure.

There are two options to arrive at a simulation method. First, one can design procedures based on the fundamental properties of a rarefied gas alone, such as free-molecular flow, the mean free path, and the collision frequency. Such schemes need not have an a priori relationship to the Boltzmann equation, but they will reflect many of the ideas and concepts employed in the derivation of the latter; in the best case, they will turn out to be consistent with and converge to solutions of the Boltzmann equation.

The Bird scheme, which has been successfully employed for the simulation of rarefied gases for decades, belongs to this category. The main advantage of this method is its practicability and success in applications such as reentry calculations for spacecraft. There are obvious relations to the derivation of the Boltzmann equation, but it has only recently been shown that the Bird simulation is actually convergent to solutions of the Boltzmann equation in the correct limit.[63] The proof, based on a reinterpretation of the simulation scheme as a measure-valued stochastic process, is based on compactness arguments and therefore not constructive. Still more recently, a constructive proof of the convergence was given[65] based on the strategy used in the validation proof of O. Lanford[64] (i.e., by a direct control of the s-particle distribution function for each s).

There appear to be very few limitations to the complexity of the flow fields that the Bird approach can deal with. Chemically reacting and ionized flows can and have been analyzed by these methods. In the DSMC method, the molecular collisions are considered on a probabilistic rather than a deterministic basis. Furthermore, the gases of real life are modeled by some hundreds of thousands of simulated molecules on a computer. For each of them the space coordinates and velocity components (as well as the variables describing the internal state, if we deal with a polyatomic molecule) are stored in the memory and are modified with time as the molecules are simultaneously followed through representative collision and gas–surface interaction events in the simulated region of space. The calculation is unsteady and the steady solutions are obtained as asymptotic limits of unsteady ones. The flow field is subdivided into cells, which are taken to be small enough for the solution to be approximately constant through the cell. The time variable is advanced in discrete steps of size Δt, small with respect to the mean free time (i.e., the time between two subsequent collisions of a molecule). This permits a separation of the inertial motion of the molecules from the collision process: One first moves the molecules according to collision-free dynamics and subsequently the velocities are modified according to the collisions occurring in each cell.

Some variations of Bird's method have appeared, due to Koura,[66] Belotserkorvskii and Yanitskii,[67] and Deshpande.[68] They differ from Bird's method because of the method used to sample the time interval between two subsequent collisions.

The second option for arriving at simulation schemes is to actually start from the Boltzmann equation, as first suggested by Nanbu,[69] and derive simulation schemes that model the Boltzmann collision term as accurately as possible. Consistency and convergence of such methods should be much easier to verify; the big question is whether the result will be practicable. Indeed, many authors have suggested particle simulations derived from the Boltzmann equation. Some of these procedures have been largely of theoretical interest, while others have been put to practical use.

The computing task of a simulation method varies with the molecular model. Typically, it is proportional to N for Bird's method, while it is proportional to N^2 for Nanbu's method. Babovsky,[70–72] however, found a procedure to reduce the computing task of Nanbu's method and make it proportional to N (see Chapter 7).

Bird's scheme was designed in the 1960s and has been applied consistently ever since, with good success. The derivation of the method is a priori independent of the Boltzmann equation (one could call the procedure "pre-Boltzmann"). Nevertheless, many of the steps closely resemble those taken by Boltzmann in his classical derivation of the equation bearing his name.

We mention that there are modifications of both the original Bird scheme and the low discrepancy scheme that Babovsky[70,71] derived from Nanbu's,[69] which "interpolate" between the two in the sense that features of one method are transferred to the other. For example, there is a "no–time–counter" version of the Bird simulation that avoids the advancement of the "clock" typical of the original Bird method.[61] The idea is simply to allow "fake" collisions between molecules[66] (see Chapter 7).

In spite of this similarity between the Bird scheme and the low discrepancy methods, there is a profound conceptual difference. In the Nanbu–Babovsky method, one aims at a direct approximation of the Boltzmann equation. Here, the molecules are just computational elements. However, if we ignore the fact that the molecules have no position in a given cell, the Bird method is a genuine particle method; the correlations between molecules generated by the dynamics are accounted for, and the state of the system is described by a probability density that does not factorize. In other words, the simulation works on a level prior to the factorization of the N-particle distribution function typical of the Boltzmann equation.

So which is "the best" method? It is not easy to give a conclusive answer to this question. All the methods discussed above are used with good success, for example for the simulation of the space shuttle reentry. Their efficiency and accuracy depend on many more aspects than the ones we discussed here, for example, how best to create and use the spatial grid, how to produce the random collision pairs, how to handle boundaries, and so on.

The theoretical analysis of particle simulation methods[63,65,73] has two significant objectives (and, indeed, results): It clarifies the relationships between the Boltzmann equation and particle simulation, and it leads to improvements of the simulation schemes. A particular challenge is to address situations where the mean free path between collisions becomes so small that we come close to the realm of ordinary gas dynamics.

3.7. The Direct Simulation Monte Carlo Method

A Monte Carlo simulation of the Boltzmann equation can be devised rather easily. Here we shall keep in mind the problem of Couette flow, which has been the main theme of this chapter. In this case the solution depends on just one space coordinate and one can make use of this fact to simplify the program from the viewpoint of both occupation of memory and computation time.

In order to check the stability of the solution against two-dimensional perturbations, and in preparation for dealing with more complicated problems, it may be useful to consider both a one-dimensional and a two-dimensional simulation. In the two-dimensional simulation we must introduce artificial boundaries in the y direction and consider a rectangle with sides L and $\hat{L}$. At the artificial boundaries we introduce periodicity conditions. Sometimes specular reflection may be used, but this is not appropriate for the case of Couette flow. In the present description of the method we consider the two-dimensional case; the modifications for the one-dimensional one are obvious.

The basic steps of the simulation are as follows:

a) The time interval $[0, T]$, over which the solution is sought for, is subdivided into subintervals with step Δt.
b) The space domain is subdivided into cells with sides Δx, Δy.
c) The gas molecules are simulated in the gap G with a stochastic system of N points having positions $x_i(t)$, $y_i(t)$ and velocities $\xi_i(t)$.

d) At each time there are N_m molecules in the mth cell; this number is varied by computing its evolution in the following two stages:

   Stage 1. The binary collisions in each cell are calculated without moving the molecules.

   Stage 2. The molecules are moved, with the new initial velocities acquired after collision (no collisions in this stage).

e) Stages 1 and 2 are repeated until $t = T$.

f) The important moments of the distribution functions are calculated by averaging over the particle in a cell. An additional space averaging is introduced for moments that are space-homogeneous.

Let us describe now the two stages of the calculation in some detail:

Stage 1. We describe the "no time counter" scheme,[1] which envisages the following two steps:

Step 1. The maximum number of binary collisions in a box, N_{cmax}, is computed. For Maxwell molecules this is independent of the relative speed and thus it represents the average number of collisions that must occur in a time interval Δt for the probability of a collision to become certainty. For other molecular models, the relative speed must also be taken into account.

Step 2. N_{cmax} pairs (i, j) of molecules are randomly chosen. Each of these pairs is "collided" with probability $|\boldsymbol{\xi}_i - \boldsymbol{\xi}_j| / < |\boldsymbol{\xi}_i - \boldsymbol{\xi}_j| >_{\max}$. If it turns out that the collisional event must occur, the velocities after collisions are calculated in the following way:

$$\boldsymbol{\xi}_i^+ = \frac{1}{2}(\boldsymbol{\xi}_i + \boldsymbol{\xi}_j + \boldsymbol{\omega}|\boldsymbol{\xi}_i - \boldsymbol{\xi}_j|), \tag{3.7.1}$$

$$\boldsymbol{\xi}_j^+ = \frac{1}{2}(\boldsymbol{\xi}_i + \boldsymbol{\xi}_j - \boldsymbol{\omega}|\boldsymbol{\xi}_i - \boldsymbol{\xi}_j|), \tag{3.7.2}$$

where $\boldsymbol{\omega}$ is a vector randomly distributed on the unit sphere. Otherwise the velocities are left unchanged.

Stage 2. The new positions and velocities of the molecules are computed through the equations:

$$\begin{aligned} x_i^+ &= x_i + \xi_{1i}\Delta t, \\ y_i^+ &= y_i + \xi_{2i}\Delta t, \\ \boldsymbol{\xi}_i^+ &= \boldsymbol{\xi}_i. \end{aligned} \tag{3.7.3}$$

The molecules with $x_i^+ \leq 0$ or $x_i^+ \geq L$ are replaced by others with a random choice of their velocity dictated by the boundary conditions (the velocity is not random but determined in the case of deterministic boundary conditions, such as specular and reverse reflection). The molecules with $y_i^+ \leq 0$ or $y_i^+ \geq \hat{L}$ are reinjected at $y_i^+ + \hat{L}$ or $y_i^+ - \hat{L}$, respectively, with their velocities.

3.8. A Test Case: Couette Flow with Reverse Reflection

In this section we consider the one-dimensional simulation for Couette flow with reverse reflection boundary conditions. This case was treated in detail in Section 2.2 of the previous chapter and provides a test case, because it is possible to compute the exact solution at least at the level of moments, and thus gives a test of the accuracy of the Direct Simulation Monte Carlo method. We follow Ref. 73 where a one-dimensional simulation of 100 cells with 10,000 molecules per cell is first presented. The data were collected only at the end of certain chosen intervals of length $T \gg \Delta t$. The time unit chosen was the mean free time and the length unit was the mean free path.

The results for the moments p_{12} and p_{11} and their ratio are given in Figs. 3.7–3.9 for $\mathcal{T} = 10.04$.

From these pictures it is easy to verify the following expectations, based on the discussion in Chapter 2, Section 2.2:

a) There is an unsteady solution with constant density, while the stresses are uniform and the velocity profile is steady. The latter statement is verified by linear interpolation at a given time instant.
b) Interchanging the values of the Mach and Knudsen numbers has no consequences thanks to the dynamical self-similarity of the flow, which implies that the solution only depends on the product of these two numbers.
c) The time evolution of second-order moments corresponds to the solution given by Truesdell and summarized in Section 2.2.

Fluctuations are large, but in the case of the bulk velocity they always remain within the standard deviation bounds, even if the pictures refer to a given time instant and not to a time average. In the case of temperature fluctuations, the standard bounds are exceeded at $t = 0$, Ma $= .5$, and Kn $= 0.2$ and, marginally, for Ma $= 31$ at $pt/\mu = 0.96$. In the first case the effect reduces considerably at the next time step and is probably due to the way the initial distribution is implemented. In the second case probably the time step is too large for the average speed of the molecules. In fact the large scatter of data in this case appears to be due to the fact that the temperature increases by a factor 10^{14} during the computation time and thus the sample turns out to be poor. In spite of this, in the long run, the linear behavior of the bulk velocity and the

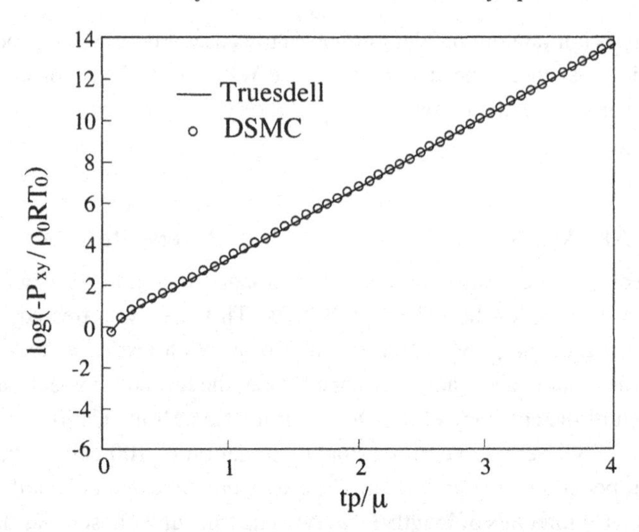

Figure 3.7. Time evolution of the tangential stress in a homonenergetic shear flow for Kn = 0.2, M = 31. The results of a one-dimensional Monte Carlo simulation are compared with the exact Truesdell solution.

homogeneity of the other moments is well confirmed by the calculations. The comparisons in Figs. 3.7–3.9 show indeed a perfect agreement.

The distribution function in the one-dimensional simulation was also computed and discussed in Ref. 73. Since the density is constant, f is normalized as a probability density in velocity space and compared with an anisotropic normal distribution f_N. The latter, when referred to suitable axes, can be written as follows:

$$f_N = \frac{1}{2\pi(T_1T_2)^{1/2}} \exp\left(-\frac{(c_1')^2}{2T_1} - \frac{(c_2')^2}{2T_2}\right), \tag{3.8.1}$$

where c_1' and c_2' are the molecular velocity components with respect to a Cartesian coordinate system rotated by an angle θ with respect to (x, y). Thus

$$c_1' = c_1' \cos\theta - c_2' \sin\theta, \qquad c_2' = c_1' \sin\theta + c_2' \cos\theta. \tag{3.8.2}$$

Equation (3.8.1) is a good representation for the part of the distribution close to the bulk velocity, but the tails of the distribution are more populated in the actual distribution than in (3.8.1). As time passes by, the less populated levels differ more and more widely in the approximate analytical representation (3.8.1) and in the results of the calculations. The angle $\theta = 1.080$ computed by best-fitting is in reasonably good agreement with the value 1.097 for the ellipse of

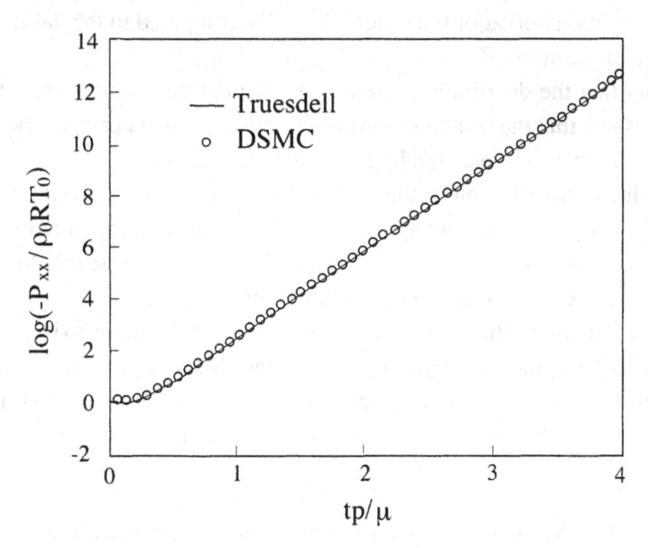

Figure 3.8. Time evolution of the stress normal to the flow in a homonenergetic shear flow for Kn = 0.2, M = 31. The results of a one-dimensional Monte Carlo simulation are compared with the exact Truesdell solution.

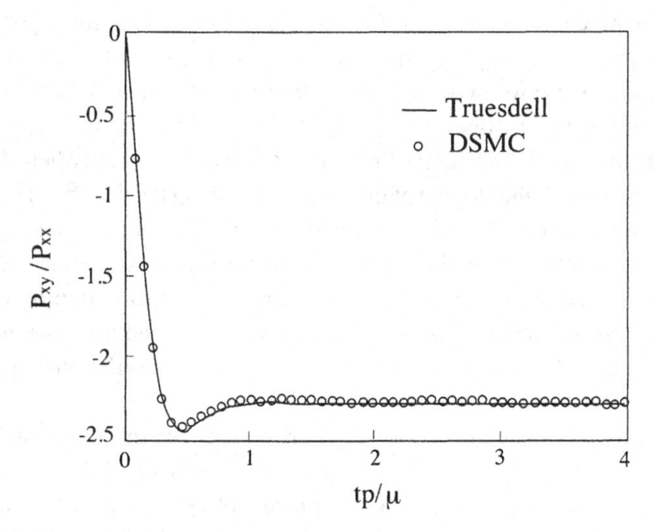

Figure 3.9. Time evolution of the ratio between the tangential and the normal stress in a homonenergetic shear flow for Kn = 0.2, M = 31. The results of a one-dimensional Monte Carlo simulation are compared with the exact Truesdell solution. The results of a two-dimensional simulation are also presented in the same reference.[73]

the stresses. For the higher Mach number Ma = 31, the angle varies from 1.32 to 1.42 in a time interval of the order $3.5\mu/p$ as compared to the value 1.36 of the ellipse of the stresses.

The fact that the distributions are always symmetrical with respect to their centers ensures that the centered third-order moments, and hence the heat flow, vanish at all times, as predicted by the analytical solution.

The velocity distribution in the (x, z) plane is perfectly isotropic, ensuring the equality of p_{11} and p_{33}, in agreement with the closed-form solution.

The validation presented in Ref. 73 and discussed above seems to guarantee that Monte Carlo simulation provides results in good agreement with the Boltzmann equation. This is important not only for applications to high altitude flight[1] but also because because this type of simulation has proved to be a useful tool for investigating the development of instabilities and possibly the transition of a rarefied gas to turbulence, as we shall see in Chapter 7.

3.9. Accurate Numerical Solutions of the Linearized Boltzmann Equation

As indicated above, accurate numerical solutions of the nonlinear Boltzmann equation are usually carried out by Monte Carlo methods, although other methods are being developed that might turn out to provide feasible approaches if further progress in high speed computers can be achieved in the next few years. Among these we quote the recent progress in the approximation of the Boltzmann equation by discrete velocity models[74–76] (see Chapter 7).

The situation is different for the linearized Boltzmann equation where discrete methods can rely on the storage of the matrix describing the linearized collision operator in a discrete form. This method has been especially investigated in Japan by Sone and his coworkers in the case of hard-sphere molecules[77–81] and produces very accurate results for the velocity and temperature profiles. The results for global quantities, such as drag, heat transfer, and flow rate, are also very accurate and can provide a test of accuracy for the variational approach discussed in Section 3.3.

For linearized Couette flow[77] the comparison between the variational result for the drag on the plates π_{12} given in Section 3.3 (Eqs. (3.3.30) and (3.3.31)) and the results of Sone et al.[77] for hard-sphere molecules is quite favorable, as shown in Table 3.3 (where $k = \delta/\gamma_1$, with $\gamma_1 = 1.270042427$, is the Knudsen number used in Ref. 77). As one can see, the maximum disagreement is of the order of 1%. The numerical values of the heat flow between two parallel plates for hard spheres are compared in Ref. 79 with the variational results[15] and

Table 3.3. *Comparison between the variational*[14] *and numerical*[77] *results for the nondimensional drag on the plates in a linearized Couette flow of a gas of hard spheres.*

k	δ	$\pi_{12}{}^{14}$	$\pi_{12}{}^{77}$
20	0.039369	0.9609	0.9663
15	0.052491	0.9498	0.9562
10	0.078738	0.9296	0.9375
8	0.098422	0.9158	0.9243
6	0.13123	0.8948	0.9037
4	0.19684	0.8583	0.8666
3	0.26246	0.8267	0.8339
2	0.39369	0.7731	0.7777
1.5	0.52492	0.7280	0.7303
1	0.78738	0.6540	0.6535
0.8	0.98422	0.6085	0.6069
0.6	1.31229	0.5458	0.5434
0.4	1.96844	0.4532	0.4509
0.3	2.62459	0.3878	0.3859
0.2	3.93688	0.3011	0.2999
0.15	5.24917	0.2461	0.2453
0.1	7.87376	0.1804	0.1799

numerical results from the BGK model.[15] The mass flow rate for the linearized Poiseuille flow is compared with the BGK results[13] and Dong's experimental data[22] in Ref. 78.

The numerical method has also been applied to the computation of the slip,[80] temperature jump, and thermal creep coefficients.[81] The slip coefficient turns out to be

$$\zeta = (0.9874)\lambda, \tag{3.9.1}$$

which is about 2.5% less than the value provided by the BGK model (see Eq. (3.5.9)). For the temperature jump the difference is more marked (about 4% less than that of the BGK model).

In Ref. 80 the velocity profile for hard spheres is compared with the experimental results by Reynolds et al..[31]

Before ending this chapter we should also remark that an early accurate analytical method (the so-called half-range polynomial expansion) was developed for problems between parallel plates by Gross, Jackson, and Ziering[82] and applied to several problems.[83,84] We note that in the original formulation the method becomes exact in both the free-molecular and continuum limits for Maxwell molecules (and Maxwell's boundary conditions), but it is slightly inaccurate in the continuum limit for other molecular models.[85]

References

[1] C. Cercignani, "Elementary solutions of the linearized gas dynamics Boltzmann equation and their application to the slip flow problem," *Annals of Physics* **20**, 219–233 (1962).

[2] C. Cercignani, "Elementary solutions of linearized kinetic models and boundary value problems in the kinetic theory of gases," Brown University Report (1965).

[3] C. Cercignani, "Analytical solution of the temperature jump problem for the BGK model," *Transp. Theory Stat. Phys.* **6**, 29–56 (1977).

[4] C. Cercignani, *The Boltzmann Equation and Its Applications*, Springer-Verlag, New York (1988).

[5] C. Cercignani, *Mathematical Methods in Kinetic Theory,* Plenum Press, New York (1990).

[6] C. Cercignani, "Plane Couette flow according to the method of elementary solutions," *J. Math. Anal. Appl.* **11**, 93–101 (1965).

[7] C. Cercignani and A. Daneri, "Flow of a rarefied gas between two parallel plates," *J. Appl. Phys.* **34**, 3509–3513 (1963).

[8] P. Welander, "On the temperature jump in a rarefied gas," *Arkiv för Fysik*, **7**, 507–553 (1954).

[9] D. R. Willis, "Comparison of kinetic theory analyses of linearized Couette flow," *Phys. Fluids* **5**, 127–135 (1962).

[10] D. R. Willis, "Heat transfer in a rarefied gas between parallel plates at large temperature ratios," in *Rarefied Gas Dynamics*, J. A. Laurman, ed., vol. I, 209–225, Academic Press, New York (1963).

[11] D. Anderson, "Numerical solution of the Krook kinetic equation," *J. Fluid Mech.* **25**, 271–298 (1966).

[12] C. Cercignani and G. Tironi, "Nonlinear heat transfer between two parallel plates according to a model with correct Prandtl number," in *Rarefied Gas Dynamics*, C. L. Brundin, ed., Vol. I, 441–453, Academic Press, New York (1967).

[13] C. Cercignani and C. D. Pagani, "Variational approach to boundary-value problems in kinetic theory," *Phys. Fluids* **9**, 1167–1173 (1966).

[14] C. Cercignani, "A variational principle for boundary value problems in kinetic theory," *J. Stat. Phys.* **1**, 297–311 (1969).

[15] P. Bassanini, C. Cercignani, and C. D. Pagani, "Comparison of kinetic theory analyses of linearized heat transfer between parallel plates," *Int. J. Heat Mass Transfer* **10**, 447–460 (1967).

[16] W. P. Teagan and G. S. Springer, "Heat transfer and density-distribution measurements between parallel plates in the transition regime," *Phys. Fluids* **11**, 497–506 (1968).

[17] J. W. Cipolla and C. Cercignani, "Effect of molecular model and boundary conditions on linearized heat transfer," in *Rarefied Gas Dynamics*, D. Dini et al., ed., Vol. II, 767–777, Editrice Tecnico Scientifica, Pisa (1971).

[18] G. S. Springer, private communication (1970).

[19] M. Knudsen, "Die Gesetze der molekular strömung und der inneren reibungströmung der gase durch röhren," *Ann. der Physik* **28**, 75–130 (1909).

[20] C. Cercignani, "Plane Poiseuille flow and Knudsen minimum effect," in *Rarefied Gas Dynamics*, J. A. Laurman, ed., Vol. II, 92–101, Academic Press, New York (1963).

[21] C. Cercignani, "Plane Poiseuille flow according to the method of elementary solutions," *J. Math. Anal. Appl.* **12**, 254–262 (1965).
[22] W. Dong, University of California Report UCRL 3353 (1956).
[23] H. Kramers, "On the behavior of a gas near a wall," *Suppl. Nuovo Cimento* **6**, 297–304 (1949).
[24] W. Greenberg, C. van der Mee, and V. Protopopescu, *Boundary Value Problems in Abstract Kinetic Theory*, Operator Theory: Advances and Applications, vol. 23, Birkhäuser, Basel (1987).
[25] S. Albertoni, C. Cercignani, and L. Gotusso, "Numerical evaluation of the slip coefficient," *Phys. Fluids* **6**, 993–996 (1963).
[26] C. Cercignani and F. Sernagiotto, "Rayleigh's problem at low Mach number according to kinetic theory," in *Rarefied Gas Dynamics*, J. H. de Leeuw, ed., Vol I, 332–353, Academic Press, New York (1965).
[27] J. C. Maxwell "On stresses in rarified gases arising from inequalities of temperature," *Phil. Trans. Roy. Society* **170**, 231–256, Appendix, 1879.
[28] C. Cercignani, P. Foresti, and F. Sernagiotto, "Dependence of the slip coefficient on the form of the collision frequency," *Nuovo Cimento* **57**B, 297–306 (1968).
[29] S. K. Loyalka and J. Ferziger, "Model dependence of the slip coefficient," *Phys. Fluids* **10**, 1833–1839 (1967).
[30] C. Cercignani and M. Lampis, "Variational calculation of the slip coefficient and the temperature jump for arbitrary gas surface interactions," in *Rarefied Gas Dynamics: Space Related Studies*, E. P. Muntz, D. P. Weaver, and D. H. Campbell, eds., 553–561, AIAA, Washington (1989).
[31] M. A. Reynolds, J. J. Smolderen, and J. F. Wendt, "Velocity profile measurements in the Knudsen layer for the Kramers problem," in *Rarefied Gas Dynamics*, M. Becker and M. Fiebig, eds., Vol. I, A.21-1-14, DFVLR-Press, Porz-Wahn (1974).
[32] W. Rixen and F.Adomeit, "Simple moments of the molecular velocity distribution function in plane Poiseuille flow," in *Rarefied Gas Dynamics*, M. Becker and M. Fiebig, eds., Vol. I, B.18-1-12, DFVLR-Press, Porz-Wahn (1974).
[33] S. K. Loyalka, "Velocity profile in the Knudsen layer for the Kramers's problem," *Phys. Fluids* **18**, 1666–1669 (1975).
[34] C. Cercignani, "The method of elementary solutions for kinetic models with velocity-dependent collision frequency," *Annals of Physics* **40**, 469–481 (1967).
[35] C. Cercignani and G. Tironi, "Some applications of a linearized kinetic model with correct Prandtl number," *Nuovo Cimento*, **43**, 64–78 (1966).
[36] C. Cercignani, "Knudsen layers; some problems and a solution technique," in *Rarefied Gas Dynamics*, J. L. Potter, ed., Part II, 795–807, AIAA, New York (1977).
[37] T. Abe and H. Oguchi, ISAS Rep. No. 553, Vol. 42, no. 8, Tokyo (1977).
[38] S. L. Gorelov and M. N. Kogan, "Solutions of linear problems of rarefied gas dynamics by the Monte Carlo method," *Izv. Akad. Nauk SSR, Mekh. Zhidk. Gaza* **3**, 136–138 (1968); translated in: *Fluid Dynamics* **3**, 96–98 (1968).
[39] G. A. Bird, "Direct simulation of the incompressible Kramers problem," in *Rarefied Gas Dynamics*, J. L. Potter, ed., Part I, 323–333, AIAA, New York (1977).
[40] C. Cercignani, "Knudsen layers; theory and experiment," in *Recent Developments in Theoretical and Experimental Fluid Mechanics*, U. Müller, K. G. Rösner, and B. Schmidt, eds., 187–195, Springer-Verlag, Berlin (1979).
[41] C. Cercignani, "Kinetic theory with 'bounce-back' boundary conditions," *Transp. Theory Stat. Phys.* **18**, 125–131 (1989).
[42] C. Cercignani, "Higher order slip according to the linearized Boltzmann equation," University of California Report No. AS-64-18 (1964).

[43] Y. Sone, "Asymptotic theory of flow of rarefied gas over a smooth boundary. I," in *Rarefied Gas Dynamics*, L. Trilling and H. Y. Wachman, eds., Vol. I, 243–253, Academic Press, New York (1969).
[44] Y. Sone, "Asymptotic theory of flow of rarefied gas over a smooth boundary. II," in *Rarefied Gas Dynamics*, D. Dini, ed., Vol. II, 737–749, Editrice Tecnico Scientifica, Pisa (1971).
[45] Y. Sone, "Asymptotic theory of a steady flow of a rarefied gas past bodies for small Knudsen numbers," in *Advances in Kinetic Theory and Continuum Mechanics*, R. Gatignol and Soubbaramayer, eds., 19–31, Springer-Verlag, Berlin–Heidelberg (1991).
[46] J. S. Darrozès, "Le glissements d'un gaz parfait sur les parois dans un ecoulement de Couette," *Rech. Aérospat.* **119**, 13–21 (1967).
[47] C. Cercignani, "Analytical solution of the temperature jump problem for the BGK model," *Transp. Theory Stat. Phys.* **6**, 29–56 (1977).
[48] C. E. Siewert and C. T. Kelley, "An analytical solution to a matrix Riemann–Hilbert problem," *J. Appl. Math. Phys. (ZAMP)* **31**, 344–351 (1980).
[49] C. Cercignani and C. E. Siewert, "On partial indices for a matrix Riemann–Hilbert problem," *J. Appl. Math. Phys. (ZAMP)* **33**, 297–300 (1981).
[50] J. S. Cassel and M. M. R. Williams, "An exact solution of the temperature slip problem in a rarefied gas," *Transp. Theory Stat. Phys.* **2**, 81–90 (1972).
[51] Y. Sone, "Thermal creep in a rarefied gas," *J. Phys. Japan* **21**, 1836–1837 (1966).
[52] M. M. R. Williams, "Boundary value problems in the kinetic theory of gases. Part 2. Thermal creep," *J. Fluid Mech.* **45**, 759–768 (1971).
[53] S. K. Loyalka, "Thermal transpiration in a cylindrical tube," *Phys. Fluids* **12**, 2301–2305 (1969).
[54] F. S. Sherman, "A survey of experimental results and methods for the transition regime of rarefied gas dynamics," in *Rarefied Gas Dynamics*, J. A. Laurman, ed., Vol. II, 228–260, Academic Press, New York (1963).
[55] A. Nordsiek and B. Hicks, "Monte Carlo evaluation of the Boltzmann collision integral," *Rarefied Gas Dynamics*, C. L. Brundin, ed., Vol. I, 695–710, Academic Press, New York (1967).
[56] S. M. Yen, B. Hicks, and R. M. Osteen, "Further development of a Monte Carlo method for the evaluation of Boltzmann collision integral," in *Rarefied Gas Dynamics*, M. Becker and M. Fiebig, eds., Vol. I, A.12-1–A.12-10, DFVLR-Press, Porz-Wahn (1974).
[57] V. Aristov, F. G. Tcheremissine, "The conservative splitting method for solving the Boltzmann equation," *U.S.S.R. Computational Mathematics and Mathematical Physics* **20**, 208–225 (1980).
[58] F. G. Tcheremissine, "Numerical methods for the direct solution of the kinetic Boltzmann equation," *U.S.S.R. Computational Mathematics and Mathematical Physics* **25**, 156–166 (1985).
[59] F. G. Tcheremissine, "Advancement of the method of direct numerical solving of the Boltzmann equation," in *Rarefied Gas Dynamics: Theoretical and Computational Techniques*, E. P. Muntz, D. P. Weaver, and D. H. Campbell, eds., 343–358, AIAA, Washington (1989).
[60] C. Cercignani and A. Frezzotti, "Numerical simulation of supersonic rarefied gas flows past a flat plate: effects of the gas–surface interaction model on the flow-field," in *Rarefied Gas Dynamics: Theoretical and Computational Techniques*, E. P. Muntz, D. P. Weaver, and D. H. Campbell, eds., 552–566, AIAA, Washington (1989).

[61] G. A. Bird, "Direct simulation and the Boltzmann equation," *Phys. Fluids* **13**, 2676–2687 (1970).

[62] G. A. Bird, "Monte Carlo simulation in an engineering context," in *Rarefied Gas Dynamics*, S. S. Fischer, ed., Vol. I, 239–255, AIAA, New York (1981).

[63] W. Wagner, "A convergence proof for Bird's direct simulation Monte Carlo method for the Boltzmann equation," *J. Stat. Phys.* **66**, 1011–1044 (1992).

[64] O. Lanford III, "Time evolution of large classical systems," in *Dynamic Systems, Theory and Applications*, E. J. Moser, ed., Lecture Notes in Physics **38**, 1–111, Springer-Verlag (1975).

[65] M. Pulvirenti, W. Wagner, and M. B. Zavelani, "Convergence of particle schemes for the Boltzmann equation," *Euro. J. Mech. B: Fluids* **7**, 339–351 (1994).

[66] K. Koura: "Transient Couette flow of rarefied binary gas mixtures," *Phys. Fluids* **13**, 1457–1466 (1970).

[67] O. M. Belotserkovskii and V. Yanitskii, "Statistical particle-in-cell method for solving rarefied gas dynamics problems," *Zhurnal Vychitelnik Matematiki i Matematicheskii Fiziki* **15**, 1195–1203 (1975) (in Russian).

[68] S. M. Deshpande, Department of Aeronautical Engineering of the Indian Institute for Science Report No. 78, FM4 (1978).

[69] K. Nanbu, "Direct simulation scheme derived from the Boltzmann equation," *J. Phys. Soci. Japan* **49**, 2042–2049 (1980).

[70] H. Babovsky, "A convergence proof for Nanbu's Boltzmann simulation scheme," *Euro. J. Mech. B: Fluids* **8(1)**, 41–55 (1989).

[71] H. Babovsky, and R. Illner: "A convergence proof for Nanbu's simulation method for the full Boltzmann equation," *SIAM J. Numerical Analysis*, **26**, 45–65, (1989).

[72] C. Cercignani, R. Illner, and M. Pulvirenti, *The Mathematical Theory of Dilute Gases*, Springer-Verlag, New York (1994).

[73] C. Cercignani and S. Cortese, "Validation of a Monte Carlo simulation of the plane Couette flow of a rarefied gas," *J. Stat. Phys.* **75**, 817–838 (1994).

[74] D. Goldstein, B. Sturtevant, and J. E. Broadwell, "Investigations of the motion of discrete-velocity gases," in *Rarefied Gas Dynamics: Theoretical and Computational Techniques*, E. P. Muntz, D. P. Weaver, and D. H. Campbell, eds., 110–117, AIAA, Washington (1989).

[75] T. Inamuro and B. Sturtevant, "Numerical study of discrete velocity gases," *Phys. Fluids A* **2**, 2196–2203 (1990).

[76] F. Rogier and J. Scheider, "A direct method for solving the Boltzmann equation," *Transp. Theory Stat. Phys.* **23**, 313–338 (1994).

[77] Y. Sone, S. Takata, and T. Ohwada, "Numerical analysis of the plane Couette flow of a rarefied gas on the basis of the linearized Boltzmann equation for hard-sphere molecules," *Euro. J. Mech. B: Fluids* **9**, 273–288 (1990).

[78] T. Ohwada, Y. Sone, and K. Aoki, "Numerical analysis of the Poiseuille and thermal transpiration flows between two parallel plates on the basis of the linearized Boltzmann equation for hard-sphere molecules," *Phys. Fluids A* **1**, 2042–2049 (1989).

[79] T. Ohwada, K. Aoki, and Y. Sone, "Heat transfer and temperature distribution in a rarefied gas between two parallel plates with different temperatures: Numerical analysis of the Boltzmann equation for a hard-sphere molecule," in *Rarefied Gas Dynamics: Theoretical and Computational Techniques*, E. P. Muntz, D. P. Weaver, and D. H. Campbell, eds., 70–81, AIAA, Washington (1989).

[80] T. Ohwada, Y. Sone, and K. Aoki, "Numerical analysis of the shear and thermal creep flows of a rarefied gas over a plane wall on the basis of the linearized

Boltzmann equation for hard-sphere molecules," *Phys. Fluids A* **1**, 1588–1599 (1989).
[81] Y. Sone, T. Ohwada, and K. Aoki, "Temperature jump and Knudsen layer in a rarefied gas over a plane wall: Numerical analysis of the linearized Boltzmann equation for hard-sphere molecules," *Phys. Fluids A* **1**, 363–370 (1989).
[82] E. P. Gross, E. A. Jackson, and S. Ziering, "Boundary value problems in kinetic theory of gases," *Ann. Phys.* **1**, 141–167 (1957).
[83] E. P. Gross and S. Ziering, "Kinetic theory of linear shear flow," *Phys. Fluids* **1**, 215–224 (1958).
[84] E. P. Gross and S. Ziering, "Heat flow between parallel plates," *Phys. Fluids* **2**, 701–712 (1959).
[85] C. Cercignani, "Shear flow for gas molecules interacting with an arbitrary central force," *Nuovo Cimento* **27**, 1240–1248 (1963).

4

Propagation Phenomena and Shock Waves in Rarefied Gases

4.1. Introduction

As is the case for other media, the propagation of disturbances plays an important role in a rarefied gas. The variety of regimes and the unusual form of the basic equation make their study rather different from the corresponding one in continuum mechanics. Yet, the classification of these phenomena is not so different, and the term *wave* is applied indifferently to completely different situations. The common feature seems to be the propagation of a peculiar aspect, such as a sharp change or an oscillating behavior, which travels between different parts of the medium. The phase speed and the possible attenuation rate of the waves are typical objects of study, since they are general features that can help us in understanding many qualitative features of more complicated situations.

We shall classify the propagation phenomena in a rarefied gas into three classes:

1. The propagation of discontinuities in the distribution function or its derivatives. This phenomenon does not occur spontaneously in the kinetic model, at variance with other models, such as the Euler equations for an ideal fluid, but can be a consequence of discontinuous initial data or be induced by the presence of boundaries.
2. The propagation of small oscillations (such as sound waves or shear waves) against an equilibrium background. This phenomenon has the interesting feature of changing smoothly with the frequency or wavelength. Also, one can distinguish between free waves, in which the wavenumber is fixed, and forced waves, in which the frequency is fixed.
3. Traveling waves, typically shock waves. As explained before, the shock waves cannot be classified among the discontinuities, as is the case for a Euler fluid. In kinetic theory the situation is, in a certain sense, similar to that occurring for the Navier–Stokes equations. The similarity is related to

the circumstance that the equations are linear in the higher order derivatives. The fact, however, that these derivatives are the only ones occurring in the Boltzmann equation produces significant differences with respect to the case of Navier–Stokes equations. Nonetheless, in both cases we find that the solution is smooth but changes very rapidly through a thin layer. The name traveling waves comes from the fact that we deal with solutions that travel at a constant speed without changing their shape.

In this last classification we should have mentioned a fourth type of wave, intermediate between type 2 and type 3. There might be waves that have an oscillatory character but cannot be strictly classified under 2, either because the background is not an equilibrium state or because their amplitude is large. Virtually nothing is known on this aspect of the solutions of the Boltzmann equation. In the same category we should also put solutions, if any, that have the aspect of a solitary wave (i.e., solutions dominated by dispersion and nonlinearity, rather than by dissipation). Since dissipation and nonlinearity are very deeply tied with each other in the Boltzmann equation, it seems reasonable to expect that these phenomena might occur only for very small mean free paths, when a sort of decoupling between the left- and right-hand sides of the Boltzmann equation occurs, as we have seen when studying the case of Couette flow in the previous chapter.

In the next few sections we shall proceed to a study of the propagation phenomena, in the order suggested by the above classification.

4.2. Propagation of Discontinuities

In this section we shall deal with solutions of the Boltzmann equation exhibiting discontinuities in the distribution function f or its derivatives with respect to time and space coordinates.

Let us start from the problem of studying solutions with discontinuities in the distribution function itself. We shall assume, for simplicity, that, although discontinuous, f remains finite and the collision term $Q(f, f)$ also remains finite.

Since the left-hand side of the Boltzmann equation contains the derivatives of f with respect to time and space coordinates, a solution discontinuous in these variables cannot be an ordinary solution of the Boltzmann equation but must be what is called a *weak* solution. Given the simple structure of the left-hand side of the equation, we do not need to enter into mathematical subtleties, but we need only remark that that side of the equation is a directional derivative, a derivative of f taken along the trajectories that the molecules would have in the absence of collisions (to be called simply the *trajectories* henceforth). This directional derivative must be finite since it equals the collision term, which has been assumed to have this property. Hence the distribution function cannot have discontinuities along the trajectories. Thus the latter are the only possible *loci* of discontinuities, or to adopt a standard terminology, are possible *wave fronts*.

We have reached the conclusion that the possible wave fronts are given by the following equation:

$$\mathbf{x} = \mathbf{x}_0 + \boldsymbol{\xi} t, \tag{4.2.1}$$

where $\mathbf{x}_0$ may depend on $\boldsymbol{\xi}$. Taken literally, these fronts reduce to a point at any given instant of time. This is satisfactory if the solution depends on just one space coordinate. The fronts expected in a two-dimensional case are typically curves. Thus in the plane case we can expect an arbitrary curve at time $t = 0$; the points of this curve can propagate with an arbitrary speed $|\boldsymbol{\xi}|$ in an arbitrary direction and produce a moving curve. Thus if there is a parabola of discontinuity $y_0 = ax_0^2$ at $t = 0$, it propagates as the parabola $y = a(x - \xi_1 t)^2 + \xi_2 t$ at subsequent times. We remark that the curve at any given time instant can depend on $\boldsymbol{\xi}$. In the three-dimensional case, the wave fronts are usually surfaces (a line would usually not be very significant as a *locus* of discontinuity, just as a point in the plane case); again if we assign the form of the surface at $t = 0$, Eq. (4.2.1) gives us a rule to construct the motion of the points of this surface. Another way of obtaining possible discontinuity curves in the two-dimensional case and discontinuity surfaces in the three-dimensional case is to keep $\mathbf{x}_0$ fixed and let $\boldsymbol{\xi}$ vary under a given geometric condition (such as tangency to a given boundary). In the three-dimensional case the surfaces cannot be arbitrarily shaped but must be ruled surfaces, such as cylinders, cones, one-sheet hyperboloids, etc. or parts of these surfaces, which contain families of straight lines of the form (4.2.1). As in the plane case these wave fronts can propagate with an arbitrary speed in an arbitrary direction. Then we have a weaker singularity, because for each $\boldsymbol{\xi}$ we have just a point for any fixed value of t.

It is also interesting to discuss discontinuous solutions in the steady case. In this case, it is even more important to distinguish the different situations according to the number of space coordinates occurring in the solution. In the case that f does not depend on y and z, then the only possible discontinuity occurs at $\xi_1 = 0$; this occurs usually at the boundary of a problem in a slab or a half-space as we have seen in the previous chapter (Section 3.5). It disappears at interior points, except in the case of free molecular flows, as we shall see later, when discussing the attenuation of the discontinuities. In the two-dimensional case, the discontinuities may be located on straight lines of the equation

$$\xi_2 x - \xi_1 y = C(\boldsymbol{\xi}). \tag{4.2.2}$$

A limiting case is $\boldsymbol{\xi} = \mathbf{0}$. Similarly, in three dimensions, we have the relation

$$\boldsymbol{\xi} \wedge \mathbf{x} = \mathbf{C}(\boldsymbol{\xi}), \tag{4.2.3}$$

where $\mathbf{C}(\boldsymbol{\xi})$ is an arbitrary vector orthogonal to $\boldsymbol{\xi}$. As above, Eqs. (4.2.3) essentially give the equation of a straight line in a three-dimensional space, whereas we expect a discontinuity surface. If we rewrite it (assuming $\boldsymbol{\xi} \neq \mathbf{0}$) in the equivalent form

$$\mathbf{x} = \mathbf{x}_0 + s\boldsymbol{\xi}, \tag{4.2.4}$$

where s is a real-valued parameter, we can proceed as above. We can assign a surface in space for $s = 0$ and compute a possible discontinuity surface by elimination, or we can take $\mathbf{x}_0$ fixed and let $\boldsymbol{\xi}$ vary.

Let us comment on two aspects of the discontinuous solutions that we are investigating. The propagation takes place at a definite speed in a definite direction, if we fix a value of $\boldsymbol{\xi}$, but all kinds of speeds and directions arise if we consider the full distribution function.

This circumstance shows up in the fact that the moments of the distribution function are usually continuous, even when the distribution function is not. An exception occurs when the point, line, or surface is independent of $\boldsymbol{\xi}$ for some value t, or when the range of integration of some variable is restricted by the geometry of the problem. To show this, let us first consider the case in which the solution depends on just one space coordinate x and time t. At time $t = 0$ we have just one discontinuity point. Without loss of generality, we can assume it to be the origin. We can disregard the components of $\boldsymbol{\xi}$ different from ξ_1 (which we simply denote by ξ) and consider, as a typical moment, the density. The discontinuity line in a space–time diagram is $x = \xi t$. We have for $t > 0$:

$$\rho(x,t) = \int f(x,\xi,t)d\xi = \int_{-\infty}^{x/t} f(x,\xi,t)\,d\xi + \int_{x/t}^{\infty} f(x,\xi,t)d\xi. \tag{4.2.5}$$

Since the function f is continuous at each point of the (x, t) plane except those of the straight line $x = \xi t$, the above formula shows that the discontinuity disappears by integration. To give an explicit example, let us assume that $f = A\exp(-\xi)H(\xi)$ for $x < \xi t$, and $f = B\exp(\xi)H(-\xi)$ for $x > \xi t$ $(A \neq B,$

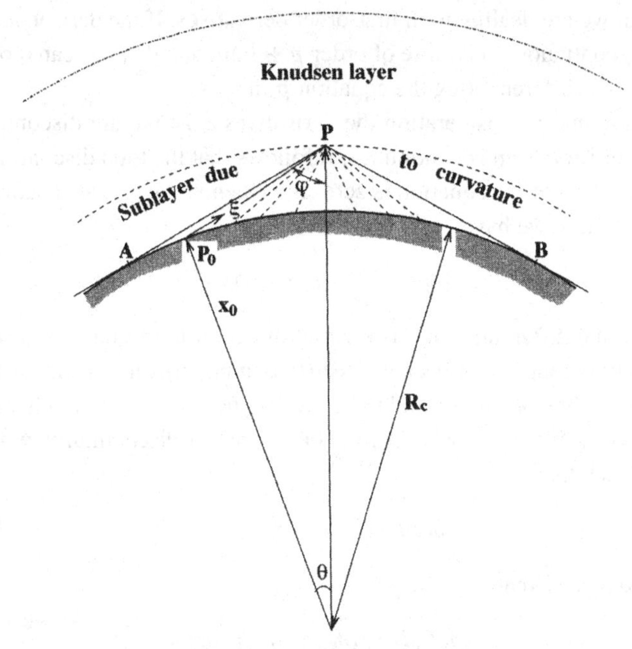

Figure 4.1. The Sone sublayer due to curvature in the case of a convex body.

and H denotes the Heaviside step function). Then

$$\begin{aligned}\rho(x,t) &= B\int_{-\infty}^{\min(x/t,0)} \exp(\xi)d\xi + A\int_{\max(x/t,0)}^{\infty} \exp(-\xi)d\xi \\ &= B\exp(H(-x)x/t) + A\exp(-H(x)x/t)\end{aligned} \tag{4.2.6}$$

and the discontinuity is smoothed out.

The occurrence of discontinuities in steady problems was known for a long time in free molecular flow; when the mean free path is finite, the discontinuities are damped out by collisions (see below) but still exist. When the mean free path is very small, they are only felt in the Knudsen layers. This fact was discovered by Sone,[1] who worked with a model equation with the BGK model. In particular, he showed that a sublayer, due to this discontinuity, is present; the Sone sublayer (Fig. 4.1) has a thickness of the order of the square of the mean free path divided by the radius of curvature and is characterized by important changes due to the discontinuity just discussed.

Let us consider now the case of a discontinuity in the derivatives, when we assume the function to be continuous. Without any loss of generality we can assume that we are dealing with first-order derivatives. If the derivatives up to order p are continuous and those of order $p+1$ are not, then we can repeat the argument after differentiating the equation p times.

In the case under consideration the derivatives exist but are discontinuous. Since the collision term is continuous, it follows that the total discontinuity of the left-hand side of the equation is zero. If we denote by $[g]$ the discontinuity of any function g, we have

$$[\partial_t f] + \boldsymbol{\xi} \cdot [\partial_{\mathbf{x}} f] = 0. \tag{4.2.7}$$

In addition, the differential in a tangential direction to the line, or surface of discontinuity, is continuous, because it can be computed using only the values on one side of the line or surface and these values are continuous. In other words, if $\tau(\mathbf{x}, t) = 0$ is the equation of a line, or surface of discontinuity, whenever dt, $d\mathbf{x}$ are such that

$$\partial_t \tau dt + \partial_{\mathbf{x}} \tau \cdot d\mathbf{x} = 0 \tag{4.2.8}$$

it must also be true that

$$[\partial_t f] dt + [\partial_{\mathbf{x}} f] \cdot d\mathbf{x} = 0. \tag{4.2.9}$$

Then it follows that the coefficients of the various differentials in the above relations must be proportional or, in other words, that a function $\lambda(\mathbf{x}, t)$ must exist, such that

$$[\partial_t f] = \lambda \partial_t \tau, \quad [\partial_{\mathbf{x}} f] = \lambda \partial_{\mathbf{x}} \tau. \tag{4.2.10}$$

Hence, assuming $\lambda \neq 0$ (otherwise there is no discontinuity) and inserting Eqs. (4.2.10) into Eq. (4.2.7), we obtain

$$\partial_t \tau + \boldsymbol{\xi} \cdot \partial_{\mathbf{x}} \tau = 0. \tag{4.2.11}$$

Therefore $\tau = \tau(\mathbf{x} - \boldsymbol{\xi} t, \boldsymbol{\xi})$ and we find again that the discontinuities can occur only on sets built out of points that move according to Eq. (4.2.1). The discussion is similar to the case of the discontinuities of f. The considerations on the steady case also remain true. (The fact that the sets of possible discontinuities for the function and the derivatives are the same is a general statement holding for any equation or system in which the higher order derivatives appear only in linear terms.)

The speed of propagation is thus arbitrary because the discontinuities can travel with any value taken by the molecular speed.

A word must be added about the damping of the discontinuities. We shall restrict ourselves to the case of discontinuities of the function f. Because the gain term tends to smooth out the irregularities of f (when they are $\boldsymbol{\xi}$-dependent, which is usually the case; see the above discussion on moments), we can assume that in most cases the discontinuities of f will evolve according to

$$\frac{d}{dt}[f] = -\nu(f)[f], \tag{4.2.12}$$

where d/dt indicates the derivative along the trajectory of a molecule and $\nu(f)$ (the collision frequency) is the factor multiplying f in the collision term (which can be assumed to be continuous as we have done for the gain term). Then

$$[f] = [f_0] \exp\left(-\int_0^t \nu(\mathbf{x} + \boldsymbol{\xi} s) ds\right). \tag{4.2.13}$$

Thus we see that the discontinuities are damped on the scale of the mean free time (inverse of the collision frequency ν). In the steady case, the previous formula must be slightly modified: s becomes a convenient parameter, $\mathbf{x}$ is replaced by $\mathbf{x}_0$, a point where $[f]_0$ is evaluated, and $t = |\mathbf{x} - \mathbf{x}_0|/|\boldsymbol{\xi}|$ is the value of s corresponding to another point $\mathbf{x}$ on the straight line in the direction of $\boldsymbol{\xi}$, where $[f]$ is evaluated. We remark that the discontinuities are thus damped on a distance of the order of a mean free path. The discontinuities at $\boldsymbol{\xi} = \mathbf{0}$ disappear suddenly; this applies, in particular, to the discontinuities at $\xi_{\mathbf{n}} = 0$ on flat boundaries.

Problems

4.2.1 Verify that (4.2.10) follows if (4.2.9) holds whenever (4.2.8) does.
4.2.2 Check that (4.2.13) follows from (4.2.12).

4.3. Shear, Thermal, and Sound Waves

Let us consider now the waves produced by mechanical or thermal oscillations, assuming that the amplitude of the wave is so small that the distribution function is close to a Maxwellian distribution M (with zero bulk velocity). The linearized Boltzmann equation reads as follows:

$$\partial_t h + \boldsymbol{\xi} \cdot \partial_{\mathbf{x}} h = L_M h, \tag{4.3.1}$$

and we look for solutions of the form

$$h(\mathbf{x}, \xi, t) = h_0(\omega, \mathbf{k}) \exp(i\omega t - i\mathbf{k} \cdot \mathbf{x}). \tag{4.3.2}$$

As mentioned in the introduction, we distinguish between two kinds of waves, according to whether we fix a vector $\mathbf{k}$ with real-valued components and look for a (generally complex) ω for which a solution exists or we fix a real ω

(positive, without loss of generality) and look for a vector **k** (generally with complex components) for which a solution exists. In the second case we speak of forced waves, because the oscillation is maintained with a fixed frequency by a suitable boundary condition, whereas in the first case we speak of free waves, because we can think of a periodic state in space, created in some way at $t = 0$ and evolving at subsequent times (with damping). We shall concentrate on the second case, because it is the easiest to be investigated experimentally. In addition, we simplify the discussion by assuming that the wave has a fixed direction of propagation, that we take as the x axis. Then we have

$$i\omega h_0 - ik\xi_1 h_0 = L_M h_0. \tag{4.3.3}$$

Let us multiply by $\overline{h}_0$ (where the overline denotes complex conjugation) and integrate with respect to $\boldsymbol{\xi}$. Before writing the result, we adopt the notation introduced in Chapter 1 (Section 1.10) and let (g, h) denote $\int \overline{g} h M d\boldsymbol{\xi}$. Then we have

$$i\omega(h_0, h_0) - ik(h_0, \xi_1 h_0) = (h_0, L_M h_0). \tag{4.3.4}$$

All the expressions $(\cdot, \cdot)$ are real (this is trivial for the expressions appearing in the left-hand side; for $(h_0, L_M h_0)$ see Problem 4.3.1). Hence, taking the real and imaginary parts of Eq. (4.3.4) and letting k_r and k_i denote the real and imaginary parts of k, we obtain

$$k_i(h_0, \xi_1 h_0) = (h_0, L_M h_0), \quad \omega(h_0, h_0) - k_r(h_0, \xi_1 h_0) = 0. \tag{4.3.5}$$

The second relation gives for the phase speed $v_{\text{ph}} = \omega / k_r$:

$$v_{\text{ph}} = \frac{(h_0, \xi_1 h_0)}{(h_0, h_0)}. \tag{4.3.6}$$

This relation immediately shows that the phase speed can be arbitrarily large if we concentrate enough molecules on the tail with, say, $\xi_1 > 0$ and sufficiently large. If, on the contrary, we produce a disturbance by letting the basic Maxwellian distribution have slightly oscillating parameters, we can expect a speed of the order of $(RT_0)^{1/2}$ (i.e., of the order of the ordinary speed of sound for a monatomic gas $c_s = (5RT_0/3)^{1/2}$).

The first equation of the system (4.3.5) also gives an important piece of information; the imaginary part of k is negative in the direction of propagation. In fact, according to Eq. (4.3.6), the wave propagates toward increasing or decreasing values of x according to whether $(h_0, \xi_1 h_0)$ is positive or negative. Then the statement follows, because $(h_0, L_M h_0) \leq 0$. This means that the wave decreases in amplitude in the direction of propagation (Problem 4.3.2).

There are two typical problems for which this theory is of interest: the propagation of sound waves and of shear waves. One may think of a plate oscillating in the direction of its normal or in its own plane, with an assigned frequency ω; a periodic disturbance propagates through a gas filling the region at one side of the plate.

The computation of the relation between ω and k (the *dispersion relation*) is not a completely straightforward matter. If the frequency is low, then one can apply the Navier–Stokes equations, but if the frequency becomes sufficiently high this approach fails even at ordinary densities, because ω^{-1} can become of the order of the mean free time.

Traditional moment methods also fail at high frequencies. Methods akin to Grad's procedure, discussed in Chapter 2, were used by Wang Chang and Uhlenbeck[2] and Pekeris et al.[3] for sound propagation, but the results for the attenuation rate are in complete disagreement with experiments at high frequencies. Since Pekeris et al. used 483 moments, their solution, if convergent, does not certainly converge to the correct solutions for large values of ω.

Another moment approach is due to Kahn and Mintzer[4] and is more akin to the Liu and Lees moment method (see Chapter 2, Section 2.6) and hence is suitable for high frequencies. In fact, they expanded the solution into a series of orthogonal polynomials, taking as a weight function not a Maxwellian distribution, but the free molecular solution. Unexpectedly, their results turned out to have the correct behavior not only at large but also at low frequencies and attracted favorable comments.[5–12] Subsequent work showed, however, that this exceptional accuracy is due to some mistakes; in fact Toba[13] pointed out that there was a mistake in the boundary conditions and Hanson and Morse[14] found a mistake in the asymptotic evaluation of certain integrals. Hanson and Morse recomputed the asymptotics correctly and found that the agreement did not improve; on the contrary, the behavior for low frequencies was completely wrong and even physically nonsensical (growing rather than damped modes). The agreement for very high frequencies is reasonably good, as was to be expected.

The use of models proves useful in this problem as well. For shear waves, a solution was provided by Cercignani and Sernagiotto[15] in 1964 (see also Refs. 12 and 16–19), by extending the procedure used in the steady case[20] to deal with perturbations tangential to the boundary by means of the BGK model, discussed in the previous chapter (Sections 3.1 and 3.5). At variance with the latter case all the solutions are of an exponential form. However, there is a limiting frequency ω_0, such that, for $\omega \leq \omega_0$, in addition to the values of γ in the continuous spectrum, producing an integral similar to that appearing in (2.7.7), there are also two discrete terms $\pm\gamma_0(\omega)$. In the case of propagation in a half-space, just one discrete term enters in the solution, because we discard exponentials growing with the distance from the oscillating plate. It is possible, however, to exhibit a discrete term for $\omega > \omega_0$ as well. This can be done by moving the integration path over the continuous spectrum in the complex plane of the variable γ. In so doing, it is possible[17–19] to uncover a pole γ_0 that is the analytic continuation of the discrete term present before, at least for frequencies larger than, but still close to, ω_0, which, as a consequence, loses, in a sense, its character of a critical frequency.

Intuitively the discrete term should represent the dominating mode far away from the boundary. Yet, the situation is more complicated, because if we go very far away from the plate, all the terms are small, but the contribution from the continuous spectrum dominates over the discrete term. This is a feature of the BGK model and would not be present if the collision frequency increased linearly for large values of the molecular speed (as in the case of hard spheres). The experimental verification of these predictions appears to lie beyond the available techniques. The discrete term (either genuine or arising from the analytic continuation) appears to dominate in the physically relevant region (between 10^{-1} and 10 mean free paths), according to estimates made by Dorning and Thurber[21] for a similar problem concerning neutron waves. We have to exclude the regions very close to and very far from the boundary. In the first region free molecular conditions prevail; in the second high speed molecules dominate (unless the collision frequency grows sufficiently fast with the molecular speed) and the distribution function is influenced by the high speed tail of the wall Maxwellian.[22,23,17,11]

The case of shear and thermal waves was also analyzed by Cercignani and Majorana in 1985.[24]

The case of sound waves is more complicated, even for the BGK model, because here we meet a system of integral equations rather than just one equation. Although it is easy to write down the general solution, the discussion of the problem of sound propagation in a half-space cannot rely on a simple exact solution. Simplifications leading to just one equation have been considered: One can drop the conservation of energy for the purpose of studying the so-called isothermal waves,[22,23,25] or conserve energy but allow only one-dimensional collisions,[22] or, finally, decouple one of the three degrees of freedom from the remaining two in a somewhat artificial way.[17] The case of models with velocity-dependent collision frequency leads to very interesting mathematical problems[19,26–28] but offers no new ways to deal with an explicit solution of the problem of sound propagation.

The first attempt to deal with an explicit solution of the BGK model for the problem of sound propagation in a half-space is due to Mason,[29] who assumed specular reflection at the plate. The agreement with experiments is not good, but it is not clear whether this is due to the special form of the boundary condition (which would be an interesting result) or, more likely, to some approximations introduced by Mason in the numerical evaluation of his analytic results.

The next attempt was due to Buckner and Ferziger.[5] Since the solution of a problem in the full space is relatively easy, they proposed to replace the half-space problem by a related one in the entire space. For this purpose they inserted a source distribution at $x = 0$ containing adjustable parameters and thereafter chose these parameters in such a way as to minimize the solution for $x < 0$ (where it should be zero). They extended their method to a more refined model containing five moments rather than three (the latter is the case for the sound wave problem in a half-space with the BGK model). The results of Buckner and Ferziger are in a good agreement with experiments for low and very high frequencies (continuum and free molecular limits). In the transition regime, the agreement is good as far as the phase speed is concerned, but the attenuation coefficient is in error by about 30% with respect to the experimental data. Due to an oversight, however, most of the calculations of Buckner and Ferziger were not performed with the approximate source term that would appear appropriate on physical grounds.

The exact closed-form solution for the problem of sound propagation in a half-space according to the BGK model was produced in 1984[30,31] by extending the analytical technique used for the steady case.[32] The numerical evaluation of this solution turned out to agree completely with a previous numerical solution by Thomas and Siewert[33] (except for a value that turned out to be a misprint in Ref. 33). There is a problem when comparing theoretical results with experimental data, in that for high frequencies the solution remains wavelike but is no longer a classical plane wave, with the consequence that the phase speed and the attenuation parameter are no longer clearly defined. The solutions[31,5] produce nearly linear plots for the real and imaginary parts of the logarithm of the pressure, but the average slopes change if we change the range of the space coordinate within which the calculations are performed. Buckner and Ferziger[5] correctly chose the range where the experimental data by Meyer and Sessler[33] were taken. This work seems not to have been done for the results of Thomas and Siewert.[33]

A simple and successful early approach is due to Sirovich and Thurber,[35] who used the idea that it is not necessary to compute the entire solution of the half-space problem. It is enough to use the discrete modes or, when the latter are absent, their analytic continuation. This concept and the validity of the corresponding dispersion relation has been explained above in connection with the shear waves. It consists in assuming that the discrete spectrum is largely dominant.

According to the aforementioned estimates of Dorning and Thurber,[21] this should be the case[36] for the experiments of Greenspan[37] and Meyer and Sessler.[34] We remark that the emitter and the receiver in a typical experiment on high frequency sound waves are separated by a fraction of the mean free path and this seems to put the method of Sirovich and Thurber at jeopardy. However, according to an estimate of Sirovich and Thurber,[38] even if the emitter and the receiver are a distance apart of $1/10$ of the mean free path, 25% of the molecules leaving the emitter collide before reaching the receiver. This high percentage is due to the fact that the molecules have velocity components parallel to the planes of the emitter and the receiver.

The advantage of the method used by Sirovich and Thurber[35] is that it bypasses the problem of solving the boundary value problem and requires just a study of the dispersion relation (and its analytic continuation). Sirovich and Thurber[35] computed the sound speed and the attenuation rate for the Gross and Jackson models,[39] Eq. (3.1.6), with 3, 5, 8, and 11 moments (with coefficients ν_α appropriate for Maxwell molecules) and the generalized models suggested by Sirovich,[40] Eq. (3.1.7), with the same number of moments and coefficients appropriate for hard spheres. The results for a gas of Maxwell molecules (with 11 moments) are in qualitative agreement with experimental data; in particular, the behavior of the phase speed is affected by an error of order 15%, whereas the errors in the attenuation rate are of order 25% for $\omega\theta > 5$ (here $\theta = \mu/p$ is a suitable mean free time; μ and p denote the viscosity coefficient and the pressure, respectively, in the unperturbed gas). The results for the hard sphere gas (11 moments) are in good agreement with experiments in the high frequency region ($\omega\theta > 5$) but the attenuation is in error by 20% in the transition regime. A word of caution is required here; in fact the results for the 8-moment model of hard spheres are closer to experiments than those for the corresponding 11-moment method.

For a graphical representation of the results discussed here, we refer to the aforementioned paper by Thurber[36] and to a previous book of the author.[19]

The fact that today we could compute the same results with arbitrarily high accuracy has not stimulated any further work, especially because no new experimental data, after those mentioned above (going back to the 1950s), appear to be available.

Problems

4.3.1 Check that $(h_0, L_M h_0)$ is real and negative.

4.3.2 Check that the linearized waves are damped in the direction of propagation.

4.4. Shock Waves

So far the nonlinear nature of the Boltzmann equation has played little role in the solutions discussed in this chapter. We are now going to discuss the simplest problem dominated by nonlinearity, the problem of the shock wave structure. We look for traveling waves of the form $f = f(x - ct, \boldsymbol{\xi})$, where c is a constant with the condition that f tends to two different Maxwellians $M_{\pm}$ when $x \to \pm\infty$. Then

$$(\xi_1 - c)f' = Q(f, f), \tag{4.4.1}$$

where f' exceptionally denotes the derivative with respect to the first argument of f. These waves are called traveling waves because they involve solutions that travel at a constant speed c without changing their shape.

It is clear that the constant c plays no significant role in the solution; sometimes it is used to impose some extra condition (e.g., that the bulk velocity goes to zero when $x \to \infty$). Although c has the role of the speed of propagation of the wave, we can let $c = 0$ by a convenient choice of the observer. In this case we can replace f' by $\partial f/\partial x$ and write

$$\xi_1 \frac{\partial f}{\partial x} = Q(f, f); \; f \to M_{\pm} \quad \text{when } x \to \pm\infty, \tag{4.4.2}$$

where

$$M_{\pm} = \rho_{\pm}(2\pi R T_{\pm})^{-3/2} \exp\left(-\frac{|\boldsymbol{\xi} - u_{\pm}\mathbf{i}|^2}{2RT_{\pm}}\right). \tag{4.4.3}$$

It is conventional to take $u_+ > 0$ (and hence also $u_- > 0$; see Eq. (4.4.4) below). Thus the fluid flows from $-\infty$ (upstream) to $+\infty$ (downstream).

The parameters in the two Maxwellians cannot be chosen at will. In fact, conservation of mass, momentum, and energy imply that the flows of these quantities must be constant and, in particular, they must be the same at $\pm\infty$. Hence

$$\rho_+ u_+ = \rho_- u_-,$$

$$\rho_+(u_+^2 + RT_+) = \rho_-(u_-^2 + RT_-),$$

$$\rho_+ u_+\left(\frac{1}{2}u_+^2 + \frac{5}{2}RT_+\right) = \rho_- u_-\left(\frac{1}{2}u_-^2 + \frac{5}{2}RT_-\right). \tag{4.4.4}$$

These conditions are just the well-known Rankine–Hugoniot conditions, which relate the upstream and downstream values of density, bulk velocity, and temperature in an ideal compressible fluid.

The problem that we have just defined is the problem of the shock wave structure. In kinetic theory, a shock wave cannot be classified among the discontinuities, as is the case for an Euler fluid. The same situation occurs for the Navier–Stokes equations. In both cases we find that the solution is smooth but changes very rapidly through a thin layer. The solution for the Navier–Stokes equations is not hard to find,[41] but it is in a sense contradictory, unless the shock is very weak (the strength of a shock can be defined in several ways, e.g., by the ratio r between the downstream und upstream pressures). In fact, unless r is close to unity, the shock layer turns out to be just a few mean free paths thick, and we expect the Navier–Stokes equations to be invalid on this scale.

Before studying the problem of the shock wave structure we remark that, since we can arbitrarily shift the origin along the x axis without changing the Boltzmann equation, the solution is defined up to a constant. The latter can be fixed by putting the origin at a special point, which could be the point where the density takes on the arithmetic mean of the downstream and upstream values, or where its derivative takes on the maximum value.

We start our study from the case of weak shock waves. In this case the parameter $s = z - 1$ is small and shall be denoted by ϵ, which thus measures the difference between upstream and downstream values, and we can expect the transition between these values to be not steep and the space derivative to be of the order of ϵ. This latter point is not a priori evident but can be justified by starting with some undefined expansion parameter and then identifying it later with ϵ on the basis of the results.

Let us then apply an expansion similar to the Hilbert expansion discussed in the Chapter 2 (Section 2.3), with one (simplifying) exception: The zeroth order term, a Maxwellian distribution, can be assumed to be uniform, since the upstream and downstream parameters are assumed to differ by terms of order ϵ. Thus we are led to writing the Boltzmann equation with a parameter ϵ in front of the space derivative (Eq. (4.3.1)) and looking for a solution of our problem in the form[42]

$$f = \sum_{n=0}^{\infty} \epsilon^n f_n. \tag{4.4.5}$$

By inserting this formal series into Eq. (4.3.1) and matching the various orders in ϵ, we obtain equations that one can hope to solve recursively:

$$Q(f_0, f_0) = 0, \qquad (4.4.6)_0$$

$$2Q(f_1, f_0) = \xi_1 \partial_x f_0 \equiv S_0, \qquad (4.4.6)_1$$

$$\dots,$$

$$2Q(f_j, f_0) = \xi_1 \partial_x f_{j-1} - \sum_{i=1}^{j-1} Q(f_i, f_{j-i}) \equiv S_{j-1}, \qquad (4.4.6)_j$$

$$\dots,$$

where $Q(f, g)$ denotes the symmetrized collision operator and the sum is empty for $j = 1$. The first equation, namely Eq. $(4.4.6)_0$, gives

$$f_0 = M \qquad (4.4.7)$$

with the parameters $(\rho_0, \mathbf{v}_0 = u_0\mathbf{i}, T_0)$ assumed to be constant. As a consequence $S_0 = 0$.

Hence Eq. $(4.4.6)_1$ can be written as

$$L_M h_1 = 0, \qquad (4.4.8)$$

where L_M is the linearized collision operator around M and $h_j = f_j/M$. Thus h_1 is a linear combination of the collision invariants (with x dependent coefficients), which may be written as follows:

$$h_1 = \frac{\rho_1}{\rho_0} + \frac{\xi_1 - u_0}{RT_0} u_1 + \left(\frac{|\boldsymbol{\xi} - u_0\mathbf{i}|^2}{2RT_0} - \frac{3}{2} \right) \frac{T_1}{T_0}, \qquad (4.4.9)$$

where ρ_1, u_1, and T_1 are the first-order contributions to ρ, u, and T.

As in the Hilbert method, Eq. $(4.4.6)_2$ can be solved only if S_2 is orthogonal to the collision invariants. This leads to

$$u_0 \frac{d\rho_1}{dx} + \rho_0 \frac{du_1}{dx} = 0,$$

$$u_0^2 \frac{d\rho_1}{dx} + 2\rho_0 u_0 \frac{du_1}{dx} + RT_0 \frac{d\rho_1}{dx} + R\rho_0 \frac{dT_1}{dx} = 0,$$

$$\frac{5}{2} RT_0^2 u_0 \frac{d\rho_1}{dx} + \frac{5}{2}\rho_0 u_0^2 \frac{du_1}{dx} + \frac{5}{2}\rho_0 u_0 \frac{dT_1}{dx} + \frac{1}{2} u_0^3 \frac{d\rho_1}{dx} + \rho_0 RT_0 \frac{du_1}{dx} = 0. \qquad (4.4.10)$$

These equations form a homogeneous algebraic system for $d\rho_1/dx$, du_1/dx, and dT_1/dx; in order to have a nonvanishing solution (i.e., a truly inhomogeneous solution of the Boltzmann equation), we must impose that the determinant of the system

$$D = \rho_0^2 u_0 \left(\frac{5}{3} R T_0 - u_0^2\right) \tag{4.4.11}$$

vanishes. If we exclude $u_0 = 0$, we are led to

$$u_0 = \left(\frac{5}{3} R T_0\right)^{1/2} = c_0, \tag{4.4.12}$$

where c_0 is the sound speed (at temperature T_0) for an ideal gas. This result means that a weak shock can be a small perturbation of a uniform flow only if the latter has a Mach number unity; this is quite clear from the inviscid theory of shock waves. If we proceed further[40] we can compute (see Problem 4.4.1) the expression of f_2. To find f_3 we must satisfy another orthogonality condition, which leads to an inhomogeneous algebraic system for $d\rho_2/dx$, du_2/dx, and dT_2/dx that has determinant $D = 0$. Then the source term of the system must satisfy a condition, which, after a lengthy calculation, turns out to be (see Problem 4.4.2)

$$\frac{d}{dx}\left(\frac{\rho_1}{\rho_1^+}\right) = -L \frac{d^2}{dx^2}\left(\frac{\rho_1}{\rho_1^+}\right), \tag{4.4.13}$$

where ρ_1^+ is the downstream value of ρ_1 and

$$L = \frac{4}{5\rho_1^+ c_0}\left(\frac{\kappa_0}{R} + 5\mu_0\right). \tag{4.4.14}$$

Here κ_0 and μ_0 denote the heat conductivity and the viscosity coefficient at temperature T_0 (given by Eqs. (2.3.13)). Integration of Eq. (4.4.13) yields

$$\frac{\rho_1}{\rho_1^+} = \mathrm{Th}\left(\frac{x - x_0}{L}\right). \tag{4.4.15}$$

This result indicates that the constant L given by Eq. (4.4.14) is a measure of the shock thickness. The latter then turns out to be of the order of the ratio of the mean free path to the strength of the shock $s \cong \rho_1^+/\rho_0$. The same result also holds for the Navier–Stokes equations,[42,43] which thus turn out to be correct to the lowest order in the shock strength.

Nicolaenko[44] put the above analysis on a more rigorous basis. In particular he showed that the formal expansions based on stretched variables are not uniformly valid beyond the Navier–Stokes level.

The other extreme case is the infinitely strong shock profile. H. Grad[45] suggested that the limit of the shock profile for $r \to \infty$ exists (at least for collision operators with a finite collision frequency) and is given by a multiple of the delta function (centered at the upstream bulk velocity) plus a comparatively smooth function for which it is not hard to derive an equation. The latter seems more complicated than the Boltzmann equation itself, but the presumed smoothness of its solution should allow a simple approximate solution to be obtained. The simplest choice for the smooth remainder is a Maxwellian distribution,[45] the parameters of which are determined by the conservation equations.

Based on this remark, we write

$$f = \hat{f} + \tilde{f}, \tag{4.4.16}$$

where

$$\hat{f} = \hat{\rho}\delta(\boldsymbol{\xi} - u_{-}\mathbf{i}). \tag{4.4.17}$$

Then

$$u_{-}\frac{d\hat{\rho}}{dx} = -[\nu(\tilde{f})]_{\boldsymbol{\xi}=u_{-}\mathbf{i}}\hat{\rho},$$

$$\partial_t \tilde{f} + \boldsymbol{\xi} \cdot \partial_{\mathbf{x}} \tilde{f} = Q(\tilde{f}, \tilde{f}) + Q_{+}(\hat{f}, \tilde{f}) - \nu(\hat{f})\tilde{f}, \tag{4.4.18}$$

where $\nu(\tilde{f})$ and $\nu(\hat{f})$ denote the collision frequency evaluated at $f = \tilde{f}$ and $f = \hat{f}$, respectively. $Q_{+}(\cdot, \cdot)$ denotes, of course, the gain part of the collision term.

So far we have not made the simplifying assumption that $\tilde{f}$ is a Maxwellian distribution with parameters determined by the conservation equations. If we now use this assumption and denote by $\tilde{\rho}$, $\tilde{u}$, and $\tilde{p}$ the density, bulk velocity, and pressure of $\tilde{f}$ (in general x dependent), we obtain (see Problem 4.4.3)

$$\tilde{\rho}\tilde{u} = \sigma u_{-}, \quad \tilde{\rho}\tilde{u}^2 + \tilde{p} = \sigma u_{-}^2, \quad \tilde{\rho}\tilde{u}^3 + 5\tilde{u}\tilde{p} = \sigma u_{-}^3, \tag{4.4.19}$$

where $\sigma = \rho_{-} - \hat{\rho}$.

If we take the first relation into account, the second yields

$$\tilde{p} = \sigma u_{-}(u_{-} - \tilde{u}), \tag{4.4.20}$$

which inserted in the last equation yields (after division by σu_{-})

$$4\tilde{u}^2 - 5u_{-}\tilde{u} + u_{-}^2 = 0. \tag{4.4.21}$$

The important fact is that this is an equation for $\tilde{u}$ containing just u_-, a constant, in the coefficients. Thus $\tilde{u}$ itself is constant. A glance at the first equation in (4.4.19) and at (4.4.20) shows that the temperature in the Maxwellian distribution, $\tilde{T}$, is also constant. Then the collision frequency in Eq. (4.4.18) is proportional to $\tilde{\rho}$ for any molecular model. The solution of Eq. (4.4.21) is more than elementary, since we can discard the trivial root $\tilde{u} = u_-$ and notice that the other must be $\tilde{u} = u_-/4$ (by the relation between the roots and the coefficients of a second-degree equation). Then $\tilde{\rho} = 4\sigma$ and we can let

$$[\nu(\tilde{f})]_{\xi=u_-\mathrm{i}} = A\tilde{\rho} = 4A(\rho_- - \hat{\rho}), \tag{4.4.22}$$

where A is a constant depending parametrically on u_-. Then Eq. (4.4.18) gives

$$u_- \frac{d\hat{\rho}}{dx} = -4A(\rho_- - \hat{\rho})\hat{\rho}. \tag{4.4.23}$$

This equation is easily integrated to yield

$$\hat{\rho} = \rho_- \frac{1}{\exp(Bx) + 1}, \tag{4.4.24}$$

where $B = 4A/u_-$.

This concludes the computation. The simple details necessary to reconstruct how the various quantities vary through the shock can be left to the reader (see Problem 4.4.4). Here we restrict ourselves to a couple of remarks. First, the deviation of the various quantities from their values at $\pm\infty$ is exponential, of the order of $\exp(-B|x|)$. An exception is offered by the temperature, which has a flatter profile downstream, with a deviation of the order of $\exp(-2B|x|)$.

We also remark that the density profile turns out to be given by a hyperbolic tangent, exactly as in the case of a weak shock. Also, if we express the shock thickness in terms of the downstream mean free path and compare the two expressions for weak and strong shocks, we find that a strong shock has a thickness that is about twice the weak shock thickness, suitably extrapolated.

Grad's approach is one of the most elegant analytical methods for dealing with the problem of shock wave structure. As we have described it, it appears to be strictly related to, and perhaps inspired by, the first approach to the same problem used by H. M. Mott-Smith in 1954[46] (a similar approach had been suggested by Tamm[47] in the Soviet Union). This method consists in approximating the distribution by a linear combination of the upstream and downstream

Maxwellians with x-dependent coefficients $a_\pm(x)$. Then the conservation equations yield

$$a_+\rho_+u_+ + a_-\rho_-u_- = \rho_\pm u_\pm,$$

$$a_+\rho_+(u_+^2 + RT_+) + a_-\rho_-(u_-^2 + RT_-) = \rho_\pm(u_\pm^2 + RT_\pm),$$

$$a_+\rho_+u_+\left(\frac{1}{2}u_+^2 + \frac{5}{2}RT_+\right) + a_-\rho_-u_-\left(\frac{1}{2}u_-^2 + \frac{5}{2}RT_-\right) = \rho_\pm u_\pm\left(\frac{1}{2}u_\pm^2 + \frac{5}{2}RT_\pm\right). \tag{4.4.25}$$

It is clear that these equations are satisfied if, and only if, $a_+(x) + a_-(x) = 1$ (provided the Rankine–Hugoniot conditions (4.4.4) are satisfied) (Problem 4.4.5). Hence we can write $a_+(x) = a(x)$ and $a_-(x) = 1 - a(x)$. An additional equation is needed to determine $a = a(x)$. This was not required in Grad's approach, since an additional relation was automatically provided by identifying the singular and nonsingular terms in the Boltzmann equation. The necessity of a further condition, to be chosen in an essentially arbitrary way, is the main weakness of the Mott-Smith approach.

A possible choice for the extra relation is offered by the moment equation obtained by multiplying the Boltzmann equation by ξ_1^2 and integrating. The result is

$$\rho_+u_+(u_+^2 + 3RT_+) - \rho_-u_-(u_-^2 - 3RT_-)\frac{da}{dx} = -\alpha a(1-a), \tag{4.4.26}$$

where

$$\alpha = \int \left(\xi_1^2 + \xi_{*1}^2 - \xi_1'^2 - \xi_{*1}'^2\right) M_+(\boldsymbol{\xi})M_-(\boldsymbol{\xi}_*)d\boldsymbol{\xi}d\boldsymbol{\xi}_*d\theta d\epsilon. \tag{4.4.27}$$

By using the third Rankine–Hugoniot relation, Eq. (4.4.26) can be simplified to

$$2\rho_+u_+(RT_+ - RT_-)\frac{da}{dx} = \alpha a(1-a). \tag{4.4.28}$$

This equation easily integrates to

$$a(x) = \frac{e^{\beta x}}{e^{\beta x} + 1}, \tag{4.4.29}$$

where

$$\beta = \frac{\alpha}{2\rho_+u_+(RT_+ - RT_-)} \tag{4.4.30}$$

and an arbitrary constant has been set equal to zero, since this amounts to fixing the position of the center of the shock layer (which, as we know, is arbitrary). Hence

$$\rho = a\rho_+ + (1-a)\rho_- = \frac{\rho_+ e^{\beta x} + \rho_-}{e^{\beta x}+1} = \rho_- \frac{\left(\frac{4\mathrm{M}_-^2}{3+\mathrm{M}_-^2}\right) e^{\beta x} + 1}{e^{\beta x}+1}, \tag{4.4.31}$$

where

$$\mathrm{M}_- = u_- \left(\frac{5RT_-}{3}\right)^{1/2} \tag{4.4.32}$$

is the upstream Mach number. The other physical quantities can be easily computed (Problem 4.4.6).

The calculation of the constant β is complicated, except for Maxwell molecules; in the latter case (see Problem 4.4.7)

$$\beta = \left(\frac{15}{2\pi}\right)^{1/2} \left[\frac{90\mathrm{M}_-^2}{3+\mathrm{M}_-^2} \frac{(\mathrm{M}_-^2-1)^2}{15\mathrm{M}_-^4 - 6\mathrm{M}_-^2 - 9}\right] \ell_-, \tag{4.4.33}$$

where ℓ_- is the upstream mean free path:

$$\ell_- = \frac{\mu_-}{p_-}\left(\frac{\pi RT_-}{2}\right)^{1/2}. \tag{4.4.34}$$

When $M_- \to \infty$, the ratio of the upstream mean free path to the shock thickness, essentially given by $\beta\ell_- \cong 6(15)^{1/2}/[(2\pi)^{1/2}\mathrm{M}_-]$, tends to zero. This occurs for any power law of molecular interaction; for hard spheres, however, the same ratio tends to a finite limit.

We have arbitrarily chosen the moment equation to be that obtained by multiplication by $\varphi = \xi_1^2$. If we modify this choice, we are led to the same result, except for a change in the value of the constant β. This change is, however, very important since β is essentially the inverse of the thickness of the shock wave. Choosing the moment corresponding to multiplication by $\varphi = \xi_1^3$ we obtain changes of the order of 25% in the shock wave thickness. This indicates that the Mott-Smith method is a good recipe for a qualitative description of the low-order moments but requires some extra information to fix the shock thickness. In addition, the Mott-Smith method does not agree with the Navier–Stokes equations for weak shocks.

Because of all these circumstances, several authors have presented modifications of the Mott-Smith method. The most interesting ones appear to be those of Salwen et al.[48] and Holway.[49] According to the first of these two methods, the linear combination of the downstream and upstream Maxwellians $M_\pm$ is replaced by a trimodal ansatz:

$$f(x, \boldsymbol{\xi}) = a_+(x)M_+(\boldsymbol{\xi}) + a_-(x)M_-(\boldsymbol{\xi}) + a_0(x) f_0(\boldsymbol{\xi}), \tag{4.4.35}$$

where f_0 is given by

$$f_0(\boldsymbol{\xi}) = (\xi_1 - w_0)\left(\frac{h_0}{\pi}\right)^2 \exp\left(-h_0|\boldsymbol{\xi} - w_0\mathbf{i}|^2\right). \tag{4.4.36}$$

Here h_0 and w_0 are constants to be determined by certain compatibility conditions that arise when discussing the solution of the conservation equations, which also give a relation among the three functions $a_0(x)$ and $a_\pm(x)$ (see Problem 4.4.8). Accordingly, two further moment equations are necessary. Salwen et al.[48] considered two choices $\varphi_{1,2} = (\xi_1^2, \xi_1^3)$ and $\varphi_{1,2} = (\xi_1^2, \xi_1|\boldsymbol{\xi}|^2)$. The resulting differential equations were then solved numerically. The discrepancy between the solutions corresponding to the two choices is considerably less than for the Mott-Smith bimodal distribution. The most remarkable improvement is, however, the removal of the disagreement with the weak shock theory discussed above.

Holway's method amounts to replacing the downstream Maxwellian distribution in Mott-Smith's ansatz by an ellipsoidal distribution:

$$f_+ = \left(\frac{h_1}{\pi}\right)^{1/2}\left(\frac{h_2}{\pi}\right) \exp\left[-h_1(\xi_1 - w_0)^2 - h_2\xi_2^2 - h_2\xi_2^2\right], \tag{4.4.37}$$

where h_1, h_2, and w_0 are three functions to be determined along with $a_\pm$.

Both modifications of the Mott-Smith method produce an overshoot of the temperature downstream. This might be an unphysical byproduct of the approximations but is confirmed by accurate Monte Carlo simulations (see next section). In fact, as we remarked above when discussing Grad's approach to the strong shock structure, the temperature profile is flatter at the downstream end than that of any other variable and may be very sensitive to the specific approximation being used. In any case, the heat flow does not change sign when the temperature derivative does.

The different results for the shock thickness defined by

$$L = (\rho_+ - \rho_-) \Big/ \left(\frac{d\rho}{dx}\right)_{\max} \tag{4.4.38}$$

are compared in Fig. 4.2. A comparison of the Mott-Smith bimodal distribution for a Lennard–Jones intermolecular potential[50] with the experimental data is given in Fig. 4.3.[51]

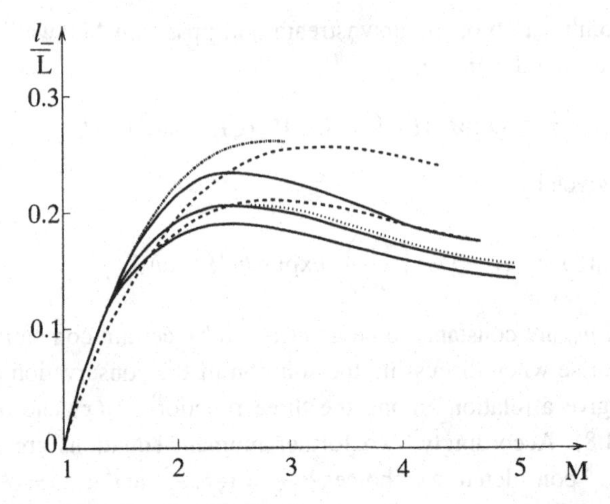

Figure 4.2. Different results for the ratio between the upstream mean free path and the shock thickness defined in (4.38) The solid curves correspond to the results of Salwen et al.[48] for different choices of the moment equations (from above: (ξ_1^3; $\xi_1|\boldsymbol{\xi}|^2$), ($\xi_1^2$; $\xi_1|\boldsymbol{\xi}|^2$); ($\xi_1^2$; $\xi_1^2\xi_1^3$)). The dashed curves correspond to Mott-Smith's results for two choices of moments (from above, ξ_1^2 and ξ_1^3). The dot and dash curve is the Navier–Stokes result.

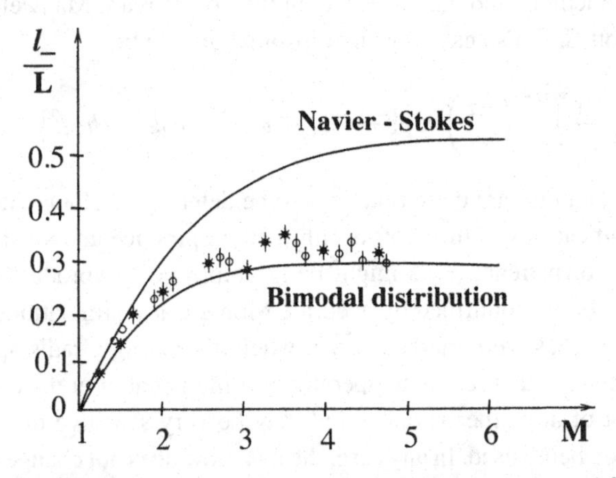

Figure 4.3. Comparison between the inverse shock thickness computed for a bimodal distribution with a Lennard–Jones potential[50] and experimental data from different authors.[51]

We conclude the discussion on the variations on the Mott-Smith approach by reporting on three methods that share a certain commonality. In 1977, Lampis[52] published a short note, in which she suggested using entropy balance in place of a moment equation. This, however, leads to a complicated equation. The same idea occurred later[53] in a paper using the Navier–Stokes equations to deal with the problem of shock structure. In this paper, the profile is assumed to be given by a hyperbolic tangent and the constant related to the shock thickness is determined by a global entropy balance. The agreement with the experiments by Alsemeyer[54] is rather good. Hosokawa and Inage[55] used a similar idea; they assumed the hyperbolic profile and used the equation proposed by Lampis[52] at just one point to determine the thickness. The agreement with experiments is reasonably good.

It is clear that the Navier–Stokes equations disagree with experiments for $M_- > 2$. The situation is not improved by going to higher orders in the Hilbert or related expansions (see however Refs. 56 and 57). As for Grad's thirteen moment equations, they do not possess a solution for the shock structure profile if $M_- > 1.65$. This is related to the fact that the Navier–Stokes and Fourier constitutive relations for the stress tensor and the heat flow are not valid at the upstream and downstream points, even though $f - M_\pm$ becomes vanishingly small there, a fact discussed in detail by Elliot et al.[58,59] These authors proposed new closure relations for the thirteen-moment theory and succeeded in modeling the upstream flow quite well. The picture downstream remains incomplete although superior to anything previously available. In addition, since the upstream point is more critical, the theory of Elliot et al. removes the difficulties encountered in computing the shock wave structure at the thirteen-moment level.

As for model equations, an early numerical solution of the BGK model for the problem of shock structure was obtained by Liepmann et al.,[60] who used the integral form of the BGK model; thus they solved a closed system of three integral equations for the three macroscopic quantities ρ, $u(= v_1)$, and T. They used an iteration method with the Navier–Stokes solution as zeroth iterate. The BGK model solution does not exhibit the temperature overshoot discussed above, and the density and velocity profiles appear to be be less antisymmetrical with respect to the appropriate center of the profile than are those obtained by the bimodal and trimodal approximations discussed above.

The BGK solution was recalculated with greater accuracy by Anderson.[61] Later Segal and Ferziger[62] studied the problem using several models of the Boltzmann equation: the BGK and ES models, a nonlinear extension of the polynomials described in Chapter 3 (Section 3.1), and a new model, the so-called trimodal gain function model. The latter was developed supposedly for dealing with the problem of the shock structure; it is essentially based upon replacing the local Maxwellian distribution Φ in the BGK model by a function of the form (4.4.35), where the coefficients and the parameters are suitably expressed in terms of ρ, u, and T and their upstream and downstream values. Thus the qualitative features of the Mott-Smith method are embodied in the model, but no commitment is made to a detailed validity of the multimodal approximation for the solution. The numerical solution shows no evidence of the temperature overshoot.

Problems

4.4.1 Compute the second-order approximation to a weak shock wave (see Ref. 42).
4.4.2 Check that ρ_1 satisfies Eq. (4.4.13) (see Ref. 42).
4.4.3 Prove Eqs. (4.4.19) (see Ref. 45).
4.4.4 Compute the density, bulk velocity, pressure, and temperature in an infinitely strong shock by Grad's method (see Ref. 45).
4.4.5 Show that the relations (4.4.25) are satisfied if, and only if, $a_+(x) + a_-(x) = 1$ (provided the Rankine-Hugoniot conditions (4.4.4) are satisfied).
4.4.6 Compute the bulk velocity, pressure, and temperature in a shock wave by Mott-Smith's method when the moment equation is (4.4.28).
4.4.7 Check that (4.4.33) holds for Maxwell's molecules.
4.4.8 Develop the details of the method proposed by Salwen et al. (see Ref. 48).

4.5. Monte Carlo Simulation and the Problem of Shock Wave Structure

The problem of shock wave structure is in a sense ideal to test the accuracy of numerical methods, in particular Monte Carlo simulation. In fact there are no solid boundaries, the upstream and downstream boundary conditions are clearly defined, and only one space coordinate is involved; in addition, the nonlinear collision term is essential for a sensible solution to exist and methods incapable of simulating that term accurately should fail.

This circumstance did not escape the attention of the authors who first attempted to handle the Boltzmann equation on a computer. In the same year when Liepmann et al.[60] attacked the problem by means of the BGK model, Haviland and Lavin presented the first results of a Monte Carlo simulation for the shock structure. The first simulation by Bird appeared in 1965.[63] The flow was modeled by between one and two thousand simulated molecules and a time-averaged result with a sample of about one thousand in each cell required a run of thirty minutes on a computer that was claimed at the time to be the fastest in the world. A similar calculation on a contemporary personal computer requires fifteen seconds. The next simulation at $M_- = 10$ is due to Perlmutter;[64] the results are in much better agreement with the Mott-Smith solution than with the Navier–Stokes equations, except for the density profile upstream. Similar conclusions were drawn from simulations between $M_- = 1.05$ and $M_- = 10$ based on a Monte Carlo technique to evaluate the collision term.[65] Further simulations with the same method appeared in 1972. A more detailed study by Bird was made in 1970.[66] Most of the theoretical results available at that time seemed to agree in giving no evidence of the temperature overshoot downstream. In fact, it turns out that the actual theoretical overshoot is less than 1%, as predicted by both the numerical solution with a Monte Carlo quadrature formula for the collision term[67] and the Monte Carlo simulation.[68] The standard deviation of the statistical scatter for strong shocks is as low as 0.1%. For a Mach 8 shock wave in a monatomic gas with viscosity proportional to the power 0.68 of the absolute temperature, the overshoot is barely noticeable.

We remark that the temperature overshoot is the consequence of a sizable temperature overshoot in the longitudinal temperature, defined by

$$RT_{\|} = \frac{1}{\rho}\int (\xi_1 - u)^2 f \, d\boldsymbol{\xi} = \frac{p_{11}}{\rho} - u^2. \tag{4.5.1}$$

Hence, since p_{11} and ρu are constant and the constants can be expressed in terms of the upstream Mach number, we obtain the following identity, due to Yen:[69]

$$T_{\|} = \frac{T_{\|-}}{3}\left[(5\mathrm{M}_-^2 + 3)\frac{\rho_1}{\rho} - 5\mathrm{M}_-^2\left(\frac{\rho_1}{\rho}\right)^2\right]. \tag{4.5.2}$$

If we differentiate $T_{\|}$ with respect to ρ, and evaluate it downstream ($\rho = \rho_+$), we obtain (apart from positive constant factors) $9 - 5\mathrm{M}_-^2$. We thus see that the parallel temperature has always an overshoot (since the density is monotonic) for $\mathrm{M}_- > (9/5)^{1/2}$. This overshoot is rather marked and this explains the overshoot in the temperature (the transversal temperature being always monotonic).

The long time span between the first simulations and simulations that could be regarded with some confidence is understandable: The method is very demanding of computer resources. In 1964, even with the fastest computers, the restriction on the number of molecules that could be used was such that large random fluctuations had to be expected in the results, and it was difficult to arrive at definite conclusions. Thus the number of simulated molecules and the sample sizes in the computations that could be performed in those years were extremely small in comparison with those that have become common in more recent years.

Although it has been claimed[70] that "the Monte Carlo techniques are not suited to very low values of the Mach number because the solution is near equilibrium, and small deviations from equilibrium would be lost in the statistical fluctuations," this is not correct. What is true is that more care must be exercised with details, such as sample size and boundary conditions.[68] In particular it seemed convenient to introduce a "jumping" piston. At the upstream and downstream ends one inserts specularly reflecting plane pistons with velocities determined by the Rankine-Hugoniot conditions, but the piston must "jump" back at its original position at the end of each time step with additional equilibrium molecules being inserted at the upstream end and removed at the downstream end. This jump is required if the boundaries are to retain a fixed distance from the wave. These boundary conditions react instantaneously to the scatter so that weak disturbances arise, due to a mismatch between the fluctuating exiting flux and and the entering flux. These disturbances do not affect strong shocks, but care must be exercised for weak shocks.

Concerning the comparison of calculations with experiments, it was not clear, at the time, whether the experimental data (including those on the properties of the gases used in the experiments) were known to a sufficient accuracy to discriminate between different results. In fact, as remarked by Bird,[66] a comparison based on either the shock wave thickness or even the complete density profile did not provide a verdict on which of the methods provides the best description of the shock structure, and the prospects for a sufficiently accurate experimental determination of the higher moments were not hopeful.

Today, as we said above, the results of Monte Carlo simulation are so accurate that we can confidently state that the Boltzmann equation with realistic collision models predicts a very flat overshoot of the temperature downstream.[71]

The firmest conclusion that one can draw from the Monte Carlo simulations of the shock wave structure is that the Mott-Smith method is reasonably accurate as far as the shock thickness is concerned. A detailed examination of the distribution function, however, shows that the bimodal ansatz is grossly inaccurate.

The refinements we have discussed in the previous section may succeed in capturing some features of the profiles of physical quantities, such as the temperature overshoot or the anisotropy of the distribution function, but only the Direct Simulation Monte Carlo method is capable of capturing the details of the distribution function in an accurate fashion.[72] The distribution of the first component of velocity has a peculiar shape for large values of the Mach number. If one subtracts the part of the profile due to the upstream Maxwellian distribution, the remainder does not look like a Maxwellian but seems to have a high curvature part near the value of the upstream bulk velocity. One can actually perform an accurate Monte Carlo simulation of an infinitely strong

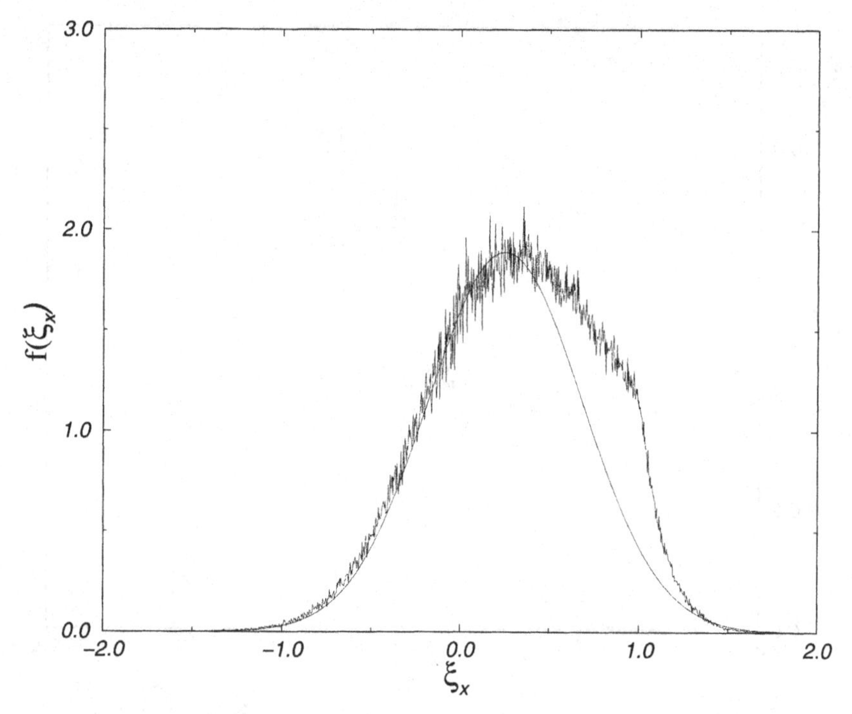

Figure 4.4. The velocity distribution, integrated with respect to the square of the transversal speed $\xi_\perp$, for an infinite Mach number shock wave. Comparison between the Monte Carlo results and the downstream Maxwellian assumed in Grad's and Mott–Smith's methods. The distributions are normalized to the total density. (From Ref. 73.)

shock wave[73] and show that a corner point shows up in the distribution function, exactly at the value of the upstream bulk velocity (Figs. 4.4 and 4.5).

The isolines in the plane of ξ_1 and $\xi_\perp = (\xi_2^2 + \xi_3^2)^{1/2}$ show a common corner point at $\xi_1 = u_+$ and $\xi_\perp = 0$ (Fig. 4.6).

Thus the problem of the shock wave structure has continued to be an important test case for numerical simulations. Important studies have included comparisons of measured and computed velocity distribution functions within strong shock waves in helium.[74]

Before ending this section, we remark that in the experiments by Garen et al.[75] a small but systematic deviation of the Navier–Stokes solution even at small values of the Mach number was found. This inaccuracy might be compatible with the theoretical analysis given in the previous section. The agreement with the extrapolation of the numerical results obtained by Hicks et al.[76] for hard spheres is much better. Yet the results obtained by both Hicks et al.[77] and

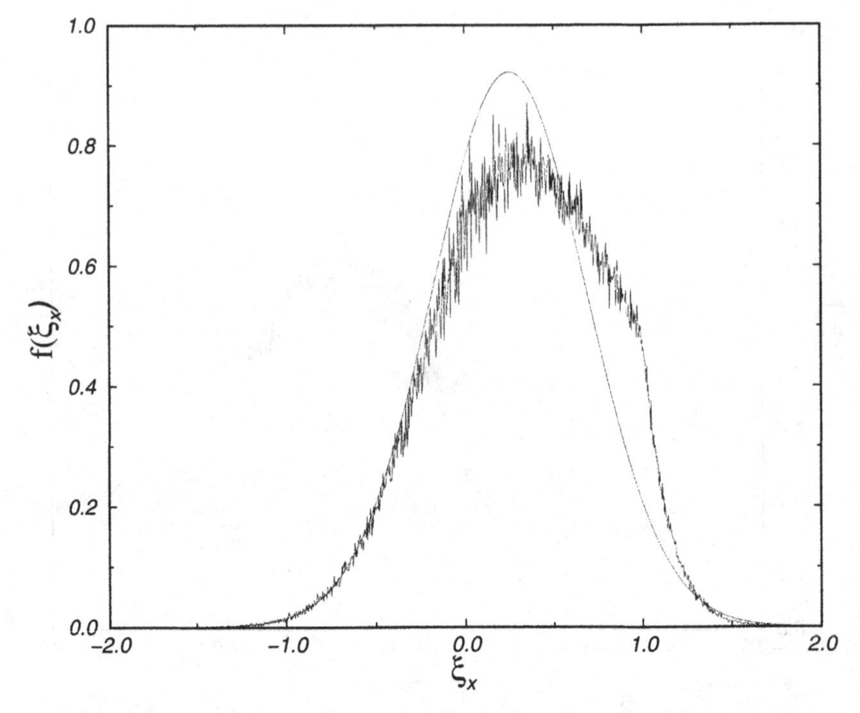

Figure 4.5. The velocity distribution, integrated with respect to the square of transversal speed $\xi_\perp$, for an infinite Mach number shock wave. Comparison between the Monte Carlo results and the downstream Maxwellian assumed in Grad's and Mott–Smith's methods. The distributions are normalized to unity. (From Ref. 73.)

Bird[68] for $M = 1.2$ are in agreement with the Navier–Stokes equations. Bird[68] offers an explanation for this apparent contradiction: The shock was not fully formed in the shock tube experiments and the early calculations by Hicks et al. were not accurate.

4.6. Concluding Remarks

By its own nature, a Monte Carlo simulation describes the problem of the shock wave structure as a time-dependent flow and thus the structure is formed during the calculation. If, however, the interest of the investigator lies just in the wave structure itself, the initial conditions are chosen in an artificial way, in order to reach a steady solution as fast as possible. Initial and boundary data leading to a description of the formation of a shock wave must be different and must be based, for example, on modeling the motion of a piston that compresses the gas.

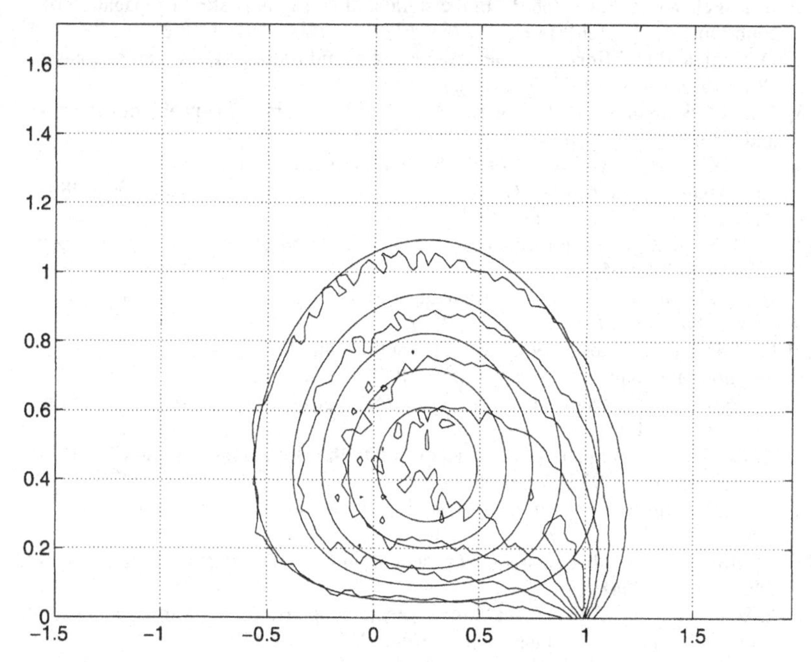

Figure 4.6. The isolines of the velocity distribution in the plane of ξ_1 and of the transversal speed $\xi_\perp$, for an infinite Mach number shock wave. The circles (slightly deformed because of the factor $\xi_\perp$ in front of the Maxwellian) indicate the isolines for a Maxwellian distribution. (From Ref. 73.)

A similar problem arises if the piston withdraws and lets the gas expand. Then, as is well known,[78] continuum mechanics predicts an expansion wave and this can also be found by Monte Carlo simulation.[79]

The other wave phenomena described in this chapter may also be described by Monte Carlo simulation. But, whereas the discontinuities in the distribution function are clearly seen in several simulations, not much attention seems to have been paid to the simulation of periodic solutions.

References

[1] Y. Sone, "On a new kind of boundary layer over a convex solid boundary in rarefied gas," Kyoto University Research Report No. 24 (1972).

[2] C. S. Wang Chang and G. E. Uhlenbeck, "On the propagation of sound in monatomic gases," Univ. Michigan Engineering Research Institute Project M29, Ann Harbor (1952). Reprinted in: C. S. Wang Chang and G. E. Uhlenbeck, *The Kinetic Theory of Gases*, Studies in Statistical Mechanics, Vol. V, North-Holland, Amsterdam.

[3] C. L. Pekeris, Z. Alterman, L. Finkelstein, and K. Frankowski, "Propagation of sound in a gas of rigid spheres," *Phys. Fluids* **5**, 1608–1616 (1962).
[4] D. Kahn and D. Mintzer, "Kinetic theory of sound propagation in rarefied gases," *Phys. Fluids* **8**, 1090–1102 (1965).
[5] J. K. Buckner and J. H. Ferziger, "Linearized boundary value problem for a gas and sound propagation," *Phys. Fluids* **9**, 2315–2322 (1966).
[6] S. S. Abarbanel, "Uni-directional perturbation methods in rarefied gas-dynamics," in *Rarefied Gas Dynamics*, C. L. Brundin, ed., Vol. I, 369–380, Academic Press, New York (1967).
[7] G. H. Sessler, "Free-molecule propagation in rarefied gases," *J. Acoust. Soc. Am.* **38**, 974–977 (1965).
[8] G. Maidanik and H. L. Fox, "Propagation and reflection of sound in rarefied gases," *Phys. Fluids* **8**, 259–272 (1965).
[9] L. H. Holway, "Time-varying weight functions and the convergence of polynomial expansions," *Phys. Fluids* **10**, 35–48 (1967).
[10] L. Lees, "Kinetic theory description of rarefied gas flows," *SIAM J. Appl. Math.* **13**, 278–311 (1965).
[11] H. Grad, "High frequency sound according to the Boltzmann equation," *SIAM J. Appl. Math.* **14**, 935–955 (1966).
[12] C. Cercignani, *Mathematical Methods in Kinetic Theory*, Plenum Press, New York (1969).
[13] K. Toba, "Kinetic theory of sound propagation in a rarefied gas," *Phys. Fluids* **11**, 2495–2497 (1968).
[14] F. B. Hanson and T. F. Morse, "Free-molecule expansion polynomials and sound propagation in a rarefied gas," *Phys. Fluids* **12**, 1564–1572 (1969).
[15] C. Cercignani and F. Sernagiotto, "Some results about a kinetic model with velocity-dependent collision frequency," in *Rarefied Gas Dynamics*, C. L. Brundin, ed., Vol I, 381–394, Academic Press, New York (1967).
[16] C. Cercignani and F. Sernagiotto, "The method of elementary solutions for time-dependent problems in linearized kinetic theory," *Annals of Physics* **30**, 154–167 (1964).
[17] C. Cercignani, "Elementary solutions of linearized kinetic models and boundary value problems in the kinetic theory of gases," Brown University Report (1965).
[18] C. Cercignani, *Mathematical Methods in Kinetic Theory*, 2nd ed., Plenum Press, New York (1990).
[19] C. Cercignani, *The Boltzmann Equation and Its Applications*, Springer-Verlag, New York (1988).
[20] C. Cercignani, "Elementary solutions of the linearized gas dynamics Boltzmann equation and their application to the slip flow problem," *Annals of Physics* **20**, 219–233 (1962).
[21] J. J. Dorning and J. K. Thurber, in *Transport Theory Conference*, AEC Report ORO-3588–1, Blacksburgh, VA. (1969).
[22] H. Weitzner, "Steady-state oscillations in a gas," in *Rarefied Gas Dynamics*, J. H. de Leeuw, ed., Vol. I, 1–20 (1965).
[23] H. S. Ostrowski and D. J. Kleitman, "Steady state oscillations in gases," *Nuovo Cimento* **44**, 49–67 (1966).
[24] C. Cercignani and A. Majorana, "Analysis of thermal and shear waves according to BGK kinetic model," *J. Appl. Math. Phys. (ZAMP)* **36**, 699–711 (1985)
[25] R. J. Mason, "Forced sound propagation induced by a fractionally accommodating piston," in *Rarefied Gas Dynamics*, C. L. Brundin, ed., Vol. I, 395–409, Academic Press, New York (1967).

[26] C. Cercignani, "The method of elementary solutions for kinetic models with velocity-dependent collision frequency," *Annals of Physics* **40**, 469–481 (1967).
[27] C. Cercignani, "Unsteady solutions of kinetic models with velocity-dependent collision frequency," *Annals of Physics* **40**, 454–468 (1967).
[28] T. Klinç and I. Kuščer, "Effect of lateral leakage in neutron waves," *Transp. Theory Stat. Phys.* **1**, 41–58 (1971).
[29] R. J. Mason, Jr., "Forced sound propagation in gases of arbitrary density," in *Rarefied Gas Dynamics*, J. H. de Leeuw, ed., Vol. I, 48–70, Academic Press, New York (1965).
[30] K. Aoki and C. Cercignani, "A technique for time-dependent boundary value problems in the kinetic theory of gases. Part I. Basic analysis," *J. Appl. Math. Phys. (ZAMP)* **35**, 127–143 (1984).
[31] K. Aoki and C. Cercignani, "A technique for time-dependent boundary value problems in the kinetic theory of gases. Part II. Application to sound propagation," *J. Appl. Math. Phys. (ZAMP)* **35**, 345–362 (1984).
[32] C. Cercignani, "Analytical solution of the temperature jump problem for the BGK model," *Transp. Theory Stat. Phys.* **6**, 29–56 (1977).
[33] J. R. Thomas, Jr. and C. E. Siewert, "Sound-wave propagation in a rarefied gas," *Transp. Theory Stat. Phys.* **8**, 219–240 (1979).
[34] E. Meyer and G. Sessler, "Schallausbreitung in Gasen hohen Frequenzen," *Z. Physik* **149**, 15–39 (1957).
[35] L. Sirovich and J. K. Thurber, "Propagation of forced sound waves in rarefied gas-dynamics," *J. Acoust. Soc. Am.* **37**, 329–339 (1965).
[36] J. K. Thurber, "Spectral concentration and high frequency sound propagation," in *The Boltzmann Equation*, F.A. Grünbaum, ed., 211–262, New York University, New York (1972).
[37] M. Greenspan, "Propagation of sound in five monatomic gases," *J. Acoust. Soc. Am.* **28**, 644–648 (1956).
[38] L. Sirovich and J. K. Thurber, "Comparison of theory and experiment for forced sound-wave propagation in rarefied gasdynamics: Reply to comments on 'Propagation of forced sound-waves in rarefied gasdynamics'," *J. Acoust. Soc. Am.* **38**, 478–480 (1965).
[39] E. P. Gross and E. A. Jackson, "Kinetic models and the linearized Boltzmann equation," *Phys. Fluids* **2**, 432–441 (1959).
[40] L. Sirovich, "Kinetic modeling of gas mixtures," *Phys. Fluids* **5**, 908–918 (1962).
[41] D. Gilbarg and D. Paolucci, "The structure of shock waves in the continuum theory of fluids," *J. Rational Mech. Anal.* **2**, 617–642 (1953).
[42] C. Cercignani, "Bifurcation problems in fluid mechanics," *Meccanica* **5**, 7–16 (1970).
[43] G. I. Taylor, "The conditions necessary for a discontinuous motion in gases," *Proc. Roy. Soc. Ser. A* **84**, 371–377 (1910).
[44] B. Nicolaenko, "Shock wave solutions of the Boltzmann equation as a nonlinear bifurcation problem from the essential spectrum," in *Théories Cinétiques Classiques et Relativistes*, G. Pichon, ed., 127–150, CNRS, Paris (1975).
[45] H. Grad, "Singular and non-uniform limits of solutions of the Boltzmann equation," in *Transport Theory*, R. Bellman, G. Birkhoff, and I. Abu-Shumays, eds., 269–308, American Mathematical Society, Providence, RI (1969).
[46] H. Mott-Smith, "The solution of the Boltzmann equation for a shock wave," *Phys. Rev.* **82**, 885–892 (1951).

[47] N. E. Tamm, "O shirine udarnich voln bolshoi intensivnosti" ("On the thickness of shock waves of large intensity"), *Tr. Fiz. Inst. imeni P.N. Lebedeva Akad. Nauk SSSR* **29**, 239–249 (1965) (in Russian).
[48] H. Salwen, C. Grosch, and S. Ziering, "Extension of the Mott–Smith method for a one-dimensional shock wave," *Phys. Fluids* **12**, 180–189 (1964).
[49] L. H. Holway, Jr., "Kinetic theory of shock structure using an ellipsoidal distribution function," in *Rarefied Gas Dynamics*, J. H. de Leeuw, ed., Vol. I, 193–215 (1965).
[50] C. Muckenfuss, "Some aspects of shock structure of strong shock waves," *Phys. Fluids* **5**, 1325–1336 (1962).
[51] M. Linzer and D. F. Hornig, "Structure in shock fronts in argon and nitrogen," *Phys. Fluids* **6**, 1661–1668 (1963).
[52] M. Lampis, "New approach to the Mott–Smith method for shock waves," *Meccanica* **12**, 171–173 (1977).
[53] P. A. Thompson, T. W. Strock, and D. S. Lim, "Estimate of shock thickness based on entropy production," *Phys. Fluids* **26**, 48–49 (1983).
[54] P. A. Alsmeyer, "Density profiles in argon and nitrogen shock waves measured by the absorption of an electric beam," *J. Fluid Mech.* **74**, 497–513 (1976).
[55] I. Hosokawa and S. Inage, "Local entropy balance through the shock wave," *J. Phys. Soc. Japan* **55** 3402–3409 (1986).
[56] C. E. Simon and J. D. Foch, "Numerical integration of the Burnett equations for shock structure in a Maxwell gas," in *Rarefied Gas Dynamics*, J. L. Potter, ed., Part I, 493–500, AIAA, New York (1977).
[57] K. A. Fiscko and D. R. Chapman, "Comparison of Burnett, super-Burnett and Monte Carlo solutions for hypersonic shock structure," in *Rarefied Gas Dynamics: Theoretical and Computational Techniques*, E. P. Muntz, D. P. Weaver, and D. H. Campbell, eds., 374–395, AIAA, Washington (1989).
[58] J. P. Elliot and D. Baganoff, "Solution of the Boltzmann equation at the upstream and downstream singular points in a shock wave," *J. Fluid Mech.* **65**, 603–624 (1974).
[59] J. P. Elliot, D. Baganoff, and R. D. McGregor, "Closure relations based on straining and shifting in velocity space," in *Rarefied Gas Dynamics*, J. L. Potter, ed., Part II, 703–714, AIAA, New York (1977).
[60] H. W. Liepmann, R. Narasimha, and M. T. Chahine, "Structure of a plane shock layer," *Phys. Fluids* **5**, 1313–1324 (1962).
[61] D. Anderson, "Numerical solution of the Krook kinetic equation," *J. Fluid Mech.* **25**, 271–287 (1966).
[62] B. M. Segal and J. H. Ferziger, "Shock-wave structure using nonlinear model Boltzmann equations," *Phys. Fluids* **15**, 1233–1247 (1972).
[63] G. A. Bird, "Shock wave structure in a rigid sphere gas," in *Rarefied Gas Dynamics*, J. H. de Leeuw, ed., Vol. I, 216–222 (1965).
[64] M. Perlmutter, "Model sampling applied to the normal shock problem," in *Rarefied Gas Dynamics*, L. Trilling and H.Y. Wachman, eds., Vol. I, 327–330, Academic Press, New York (1969).
[65] B. L. Hicks and S. M. Yen, "Solution of the non-linear Boltzmann equation for plane shock waves," in *Rarefied Gas Dynamics*, L. Trilling and H. Y. Wachman, eds., Vol. I, 313–318, Academic Press, New York (1969).
[66] G. A. Bird, "Aspects of the structure of strong shock waves," *Phys. Fluids* **15**, 1172–1177 (1970).
[67] S. M. Yen, "Numerical solution of the nonlinear Boltzmann equation for non-equilibrium gas flow problems," *Annu. Revi. Fluid Mech.* **16**, 67–97 (1984).

[68] G. A. Bird, "Perception of numerical methods in rarefied gasdynamics," in *Rarefied Gas Dynamics: Theoretical and Computational Techniques*, E. P. Muntz, D. P. Weaver, and D. H. Campbell, eds., 211–226, AIAA, Washington (1989).

[69] S. M. Yen, "Temperature overshoot in shock waves," *Phys. Fluids* **9**, 1417–1418 (1966).

[70] G. A. Sod, "A numerical solution of the Boltzmann equation," *Comm. Pure Appl. Math.* **30**, 391–419 (1977).

[71] G. A. Bird, "The search for solutions in rarefied gas dynamics," in *Rarefied Gas Dynamics*, J. Harvey and G. Lord, eds., Vol. 2, 753–762, Oxford University Press, Oxford (1995).

[72] E. P. Muntz, D. A. Erwin, and G. C. Pham-Van-Diep, "A review of the kinetic detail required for accurate predictions of normal shock waves," in *Rarefied Gas Dynamics*, A. E. Beylich, ed., 198–206, VCH, Weinheim (1991).

[73] C. Cercignani, A. Frezzotti, and P. Grosfils, "Structure of an infinitely strong shock wave," *Phys. Fluids* **11**, 2757–2764 (1999).

[74] G. C. Pham-Van-Diep, D. A. Erwin, and E. P. Muntz, "Nonequilibrium molecular motion in a hypersonic shock wave," *Science* **245**, 624–626 (1989).

[75] W. Garen, R. Synofzik, G. Wortberg, and A. Frohn, "Experimental investigation of the structure of weak shock waves in noble gases," in *Rarefied Gas Dynamics*, J. L. Potter, ed., Part I, 519–528, AIAA, New York (1977).

[76] B. L. Hicks, S. M. Yen, and B. L. Reilly, "The internal structure of shock waves," *J. Fluid Mech.* **53**, 85–111 (1972).

[77] S. M. Yen and W. Ng, "Shock wave structure and intermolecular collision laws," *J. Fluid Mech.* **65**, 127–144 (1974).

[78] H. W. Liepmann and A. Roshko, *Elements of Gasdynamics*, Wiley, New York (1957).

[79] F. Seiler, B. Schmidt, and N. Kuscherus, "Direct Monte-Carlo simulation technique applied to some rarefied gas dynamic problems," in *Rarefied Gas Dynamics*, V. Boffi and C. Cercignani, eds., Vol. I, 431–441, B. G. Teubner, Stuttgart (1986).

5

Perturbation Methods in More than One Dimension

5.1. Introduction

In the previous chapters we have learned the basic tools required to deal with problems in rarefied gas dynamics. Some of them can be readily extended to problems in more than one dimension (actually, occasionally we presented these tools in a general form, when this did not reduce the clarity of exposition).

In this chapter we first study the extension of the techniques to handle the linearized Boltzmann equation to the case of geometries different from a slab or a half-space. We shall start from perturbations of a fixed, nondrifting Maxwellian distribution. The results on the general solution in the one-dimensional case, suitably modified will give a general idea of the solution in three dimensions (Section 5.2). Then we shall discuss particular cases of both internal (Section 5.3) and external (Section 5.4) steady problems.

This will naturally lead to a general discussion of the continuum limit of the Boltzmann equation.

Free molecular and nearly free molecular flows are discussed in Section 5.7. Finally, methods to deal with flows in the nearly continuum regime are illustrated (Section 5.8). There are even flows in which all the regimes occur; an important example of this kind will be presented in Section 5.9.

5.2. Linearized Steady Problems

In this section we shall examine the linearized Boltzmann equation and study the general form of its solution in three-dimensional steady problems.

We start from the remark that sometimes it is convenient to consider two solutions of the linearized Boltzmann equation simultaneously. We consider these solutions in the steady case and denote them by h and $h^{\dagger}$. In order to reach more generality we consider the presence of a source (as in the case of Poiseuille flow; see Section 3.4) for each solution and we denote them by S

and $S^\dagger$. Then we write

$$Dh = Lh + S, \tag{5.2.1}$$

$$Dh^\dagger = Lh^\dagger + S^\dagger, \tag{5.2.2}$$

where $D = \boldsymbol{\xi} \cdot \partial_{\mathbf{x}}$ and L is the linearized collision operator about some fixed, non-drifting Maxwellian distribution M as in Section 2.7. We also use the parity operator P and adopt the notations $((\cdot, \cdot))$ and $((\cdot, \cdot))_B$ introduced in that section. Then we have

$$\begin{aligned}((Ph^\dagger, S)) - ((h, PS^\dagger)) &= ((Ph^\dagger, (D - L)h)) - ((h, P(D - L)h^\dagger)) \\ &= ((Ph^\dagger, Dh)) - ((h, PDh^\dagger)), \end{aligned} \tag{5.2.3}$$

where we used the property

$$((Pg, Lh)) = ((LPg, h)) = ((PLg, h)) = ((h, PLg))$$

that we discussed in Chapter 2 to eliminate the terms containing the linearized collision operator L. The terms remaining can be clearly transformed into boundary terms, as we did in Chapter 2. Then we obtain (see Problem 5.2.1)

$$((Ph^\dagger, S)) - ((h, PS^\dagger)) = ((h^{\dagger +}, Ph^-))_B - ((h^+, Ph^{\dagger -}))_B. \tag{5.2.4}$$

This is a general relation between the solutions h, $h^\dagger$ and their sources S, $S^\dagger$.

A particular case of interest is offered by the choice

$$S^\dagger = \delta(\mathbf{x} - \mathbf{x}')\delta(\boldsymbol{\xi} - \boldsymbol{\xi}'), \tag{5.2.5}$$

where, as usual, δ denotes Dirac's delta function. The corresponding solution is denoted by $G(\mathbf{x}, \boldsymbol{\xi}; \mathbf{x}', \boldsymbol{\xi}')$ and is called the Green's function (in the absence of boundaries, if we choose that Eq. (5.2.2) is satisfied in the entire space with suitable conditions at infinity). If h is a solution satisfying Eq. (5.2.1) in some region Ω with boundary conditions of the usual kind on the boundary $\partial\Omega$, Eq. (5.2.4) becomes

$$\begin{aligned}&\int_{\Re^3}\int_{\Omega} G(\mathbf{x}, -\boldsymbol{\xi}; \mathbf{x}', \boldsymbol{\xi}')S(\mathbf{x}, \boldsymbol{\xi})M(|\boldsymbol{\xi}|)d\mathbf{x}d\boldsymbol{\xi} - h(\mathbf{x}', -\boldsymbol{\xi}')M(|\boldsymbol{\xi}'|) \\ &= -\int_{\Re^3}\int_{\partial\Omega} \boldsymbol{\xi}\cdot\mathbf{n}G(\mathbf{x}, -\boldsymbol{\xi}; \mathbf{x}', \boldsymbol{\xi}')h(\mathbf{x}, \boldsymbol{\xi})M(|\boldsymbol{\xi}|)d\sigma d\boldsymbol{\xi}, \end{aligned} \tag{5.2.6}$$

or, rearranging,

$$h(\mathbf{x}, \boldsymbol{\xi})M(|\boldsymbol{\xi}|) = \int_{\Re^3}\int_{\Omega} G(\mathbf{x}', -\boldsymbol{\xi}'; \mathbf{x}, -\boldsymbol{\xi})S(\mathbf{x}', \boldsymbol{\xi}')M(|\boldsymbol{\xi}'|)d\mathbf{x}d\boldsymbol{\xi}$$
$$+ \int_{\Re^3}\int_{\partial\Omega} \boldsymbol{\xi}' \cdot \mathbf{n}'G(\mathbf{x}', -\boldsymbol{\xi}'; \mathbf{x}, -\boldsymbol{\xi})h(\mathbf{x}', \boldsymbol{\xi}')M(|\boldsymbol{\xi}'|)d\sigma d\boldsymbol{\xi}'. \tag{5.2.7}$$

This equation shows that the solution h of Eq. (5.2.1) can be expressed in terms of the source S and the boundary values of h on $\partial\Omega$, once G is known. The problem is, of course, that the boundary values are not known (in the simplest case, they are known for $\boldsymbol{\xi}' \cdot \mathbf{n} > 0$, but not for $\boldsymbol{\xi}' \cdot \mathbf{n} < 0$). It is possible, using the Green's function, to obtain integral equations that determine the boundary values (Problem 5.2.2). Because of the translation invariance of Eq. (5.2.1), G will actually depend just on the combination $\mathbf{x} - \mathbf{x}'$ and not separately on $\mathbf{x}$ and $\mathbf{x}'$. Taking this remark into account and exchanging $\mathbf{x}$ for $\mathbf{x}'$ and $\boldsymbol{\xi}$ for $\boldsymbol{\xi}'$ simultaneously (Problem 5.2.3), we get

$$G(\mathbf{x}, -\boldsymbol{\xi}; \mathbf{x}', -\boldsymbol{\xi}') = G(\mathbf{x}', \boldsymbol{\xi}'; \mathbf{x}, \boldsymbol{\xi}). \tag{5.2.8}$$

Then Eq. (5.2.7) can be slightly simplified to

$$h(\mathbf{x}, \boldsymbol{\xi})M(|\boldsymbol{\xi}|) = \int_{\Re^3}\int_{\Omega} G(\mathbf{x}, \boldsymbol{\xi}; \mathbf{x}', \boldsymbol{\xi}')S(\mathbf{x}', \boldsymbol{\xi}')M(|\boldsymbol{\xi}'|)d\mathbf{x}d\boldsymbol{\xi}'$$
$$+ \int_{\Re^3}\int_{\partial\Omega} \boldsymbol{\xi}' \cdot \mathbf{n}'G(\mathbf{x}, \boldsymbol{\xi}; \mathbf{x}', \boldsymbol{\xi}')h(\mathbf{x}', \boldsymbol{\xi}')M(|\boldsymbol{\xi}'|)d\sigma' d\boldsymbol{\xi}'. \tag{5.2.9}$$

In spite of the fact that Eq. (5.2.9) still contains unknown boundary values, it can give us information on the structure of the solutions of the linearized Boltzmann equation, provided we know G. In particular, if the source S is absent, the solution inside Ω will depend on $\mathbf{x}$ in the same way as G does, because only the boundary integral will survive and $\mathbf{x} \neq \mathbf{x}'$ in Ω.

Let us then look at the structure of the Green's function. Since we know that G depends on just the combination $\mathbf{x} - \mathbf{x}'$, we can let $\mathbf{x}' = 0$ without loss of generality and write $G(\mathbf{x}, \boldsymbol{\xi}; \boldsymbol{\xi}')$ for $G(\mathbf{x}, \boldsymbol{\xi}; 0, \boldsymbol{\xi}')$. We have then to solve the equation

$$\boldsymbol{\xi} \cdot \partial_{\mathbf{x}} G = LG + \delta(\mathbf{x})\delta(\boldsymbol{\xi} - \boldsymbol{\xi}'). \tag{5.2.10}$$

A suitable method is to take the Fourier transform with respect to $\mathbf{x}$:

$$\hat{G}(\mathbf{k}, \boldsymbol{\xi}; \boldsymbol{\xi}') = \int G(\mathbf{x}, \boldsymbol{\xi}; \boldsymbol{\xi}')e^{i\mathbf{k}\cdot\mathbf{x}}d\mathbf{x}, \tag{5.2.11}$$

which can be inverted to give

$$G(\mathbf{x}, \boldsymbol{\xi}; \boldsymbol{\xi}') = \frac{1}{2\pi} \int \hat{G}(\mathbf{k}, \boldsymbol{\xi}; \boldsymbol{\xi}') e^{-i\mathbf{k}\cdot\mathbf{x}} d\mathbf{k}. \tag{5.2.12}$$

Then Eq. (5.2.10) gives

$$-i\boldsymbol{\xi} \cdot \mathbf{k}\hat{G} = L\hat{G} + \delta(\boldsymbol{\xi} - \boldsymbol{\xi}'). \tag{5.2.13}$$

Here $\mathbf{k}$ is just a (vector-valued) parameter. If we let $\xi_1 = \boldsymbol{\xi} \cdot \mathbf{k}/|\mathbf{k}|$ $(\mathbf{k} \neq 0)$, we have

$$-i\xi_1|\mathbf{k}|\hat{G} = L\hat{G} + \delta(\boldsymbol{\xi} - \boldsymbol{\xi}'). \tag{5.2.14}$$

Then we are reduced to finding the Green's function in the one-dimensional case, where the other two components of $\boldsymbol{\xi}$ form a vector in the plane orthogonal to $\mathbf{k}$. Since we know the general solution for one-dimensional problems (Section 2.7), a lengthy but simple calculation leads to the Green's function for this case, G_1. We have (Problem 5.2.3)

$$G_1 = \left\{ \frac{\xi_1' + \xi_1}{2C_1} + \sum_{\alpha=2}^{4} \frac{\psi_\alpha [L^{-1}(\xi_1\psi_\alpha)]' + \psi_\alpha' [L^{-1}(\xi_1\psi_\alpha)]}{2C_\alpha} \right\} \operatorname{sgn} x$$
$$+ \sum_{\alpha=2}^{4} \frac{\psi_\alpha \psi_\alpha' |x|}{2C_\alpha} + \operatorname{sgn} x \int_{\bar{\gamma}}^{\infty} g_\gamma(\boldsymbol{\xi}) g_\gamma(\boldsymbol{\xi}') \exp(-\gamma|x|) d\gamma. \tag{5.2.15}$$

To pass to the three-dimensional case one must be careful, because of the contribution from $\mathbf{k} = \mathbf{0}$. Thus it is better to split the contribution from the discrete terms and from the integral. The latter poses no problems and provides a solution that decays exponentially when going far from the source. The Fourier transforms of the discrete terms are singular; in addition, the transformation from the three-dimensional to the one-dimensional case is singular for $|\mathbf{k}| = 0$. To avoid difficulties and possible mistakes we deal with the contribution not decaying exponentially directly, using the remark that it can be expanded, at least asymptotically, in powers of the mean free path (as in the Hilbert expansion). Thus we have

$$h = h^A + h^B, \tag{5.2.16}$$

where h^B is important near boundaries and h^A survives also far from the boundaries (the asymptotic part) and hence cannot depend exponentially on $\mathbf{x}/\ell$, where ℓ is the mean free path.

To compute h^A we proceed as follows. We write

$$h^A = \sum_{\alpha=1}^{4} A_\alpha(\mathbf{x})\psi_\alpha + \epsilon A_0(\mathbf{x}) + \epsilon L^{-1}\boldsymbol{\xi} \cdot \partial_{\mathbf{x}} h^A, \tag{5.2.17}$$

where, as before (Chapters 2 and 3), $\psi_4 = |\boldsymbol{\xi}|^2 - 5RT_0$ and we have taken into account the fact that the term with $\psi_0 = 1$ must be of order ϵ when we iterate because $\psi_0\boldsymbol{\xi}$ is not orthogonal to the collision invariants. If we now iterate, we obtain

$$h^B = \sum_{\alpha=1}^{4} A_\alpha(\mathbf{x})\psi_\alpha + \epsilon A_0(\mathbf{x}) + \epsilon L^{-1}\sum_{\alpha=1}^{4} \boldsymbol{\xi}\cdot\partial_{\mathbf{x}} A_\alpha(\mathbf{x})\psi_\alpha$$
$$+\epsilon^2 L^{-1}[\boldsymbol{\xi}\cdot\partial_{\mathbf{x}} A_0 + \boldsymbol{\xi}\cdot\partial_{\mathbf{x}} L^{-1}\cdot\boldsymbol{\xi}\partial_{\mathbf{x}} h^A]. \tag{5.2.18}$$

We must now impose the condition that $\boldsymbol{\xi}\cdot\partial_{\mathbf{x}} A_0 + \boldsymbol{\xi}\cdot\partial_{\mathbf{x}} L^{-1}\sum_{\alpha=1}^{4}\boldsymbol{\xi}\cdot\partial_{\mathbf{x}} A_\alpha(\mathbf{x})\psi_\alpha$ is orthogonal to the collision invariants (otherwise we cannot continue the iteration).

If we repeat the procedure we get more and more terms. We are not interested in obtaining all the terms of the expansion (which presumably is not convergent, in general, but only asymptotic). The important fact is that, if we compute moments of fixed order (such as the stress tensor and the heat flow), a finite number of terms will be needed. To see this we remark that h will depend linearly on $\mathbf{x}$ only through the five quantities $A_\alpha(\mathbf{x})$, or p_1^A, $\mathbf{v}_1^A$, T_1^A (the perturbations of pressure, bulk velocity, and temperature). Then a tensor of a given order can depend on just certain combinations of the derivatives of these quantities (because the rotational invariance of the collision term implies that the Boltzmann gas is an isotropic medium). The heat flow $\mathbf{q}$ is a linear combination with constant coefficients of

$$\partial_{\mathbf{x}} p_1^A,\ \partial_{\mathbf{x}} T_1^A,\ \partial_{\mathbf{x}}\Delta^n p_1^A,\ \partial_{\mathbf{x}}\Delta^n T_1^A,\ \Delta^n \mathbf{v}_1^A \qquad (\mathrm{q}),$$

where Δ is the Laplace operator and $n = 1, 2, 3, \ldots$ (see Problem 5.2.4). The stress deviator $q_{ij} = p_{ij} - p\delta_{ij}$ is a linear combination with constant coefficients (Problem 5.2.5) of

$$\partial_{x_j} u_{1i}^A,\ \partial_{x_j}\Delta^n u_{1i}^A,\ \partial_{x_i x_j} p_1^A,\ \partial_{x_i x_j} T_1^A,\ \partial_{x_i x_j}\Delta^n p_1^A,\ \partial_{x_i x_j}\Delta^n T_1^A \qquad (\mathrm{p}).$$

The coefficients of a term containing derivatives of order m are of order ϵ^m at least. Now we recall the balance equations that must be satisfied as a consequence of the existence of the collision invariants:

$$\mathrm{div}\mathbf{v}_1^A = 0,$$
$$\partial_{x_i} p^A = \sum_{i=1}^{3}\partial_{x_j} q_{ij} \quad (i = 1, 2, 3), \tag{5.2.19}$$
$$\mathrm{div}\mathbf{q} = 0.$$

If we take into account the last relation we see that the left-hand side will contain a linear combination of terms of the following kind:

$$\Delta p_{\text{I}}^{A},\ \Delta T_{\text{I}}^{A},\ \Delta^{n+1} p_{\text{I}}^{A},\ \Delta^{n+1} T_{\text{I}}^{A},$$

because the terms containing $\mathbf{v}_{\text{I}}^{A}$ will disappear since their divergence is zero.

The momentum balance equation will contain in the right-hand side terms of the following kind:

$$\Delta u_{\text{I}i}^{A},\ \Delta^{n+1} u_{\text{I}i}^{A},\ \partial_{x_i} \Delta p_{\text{I}}^{A},\ \partial_{x_i} \Delta T_{\text{I}}^{A},\ \partial_{x_i} \Delta^{n+1} p_{\text{I}}^{A},\ \partial_{x_i} \Delta^{n+1} T_{\text{I}}^{A}.$$

The momentum equation shows that the pressure gradient can be omitted from the variables upon which $\mathbf{q}$ depends, because it can be expressed in terms of other variables appearing in the list (q). Thus ΔT_{I}^{A} is a linear combination of $\Delta^{n+1} p_{\text{I}}^{A},\ \Delta^{n+1} T_{\text{I}}^{A}$. Taking the divergence of the momentum equation, we see that also Δp_{I}^{A} is a linear combinations of $\Delta^{n+1} p_{\text{I}}^{A},\ \Delta^{n+1} T_{\text{I}}^{A}$. Then ΔT_{I}^{A} and Δp_{I}^{A} must vanish; otherwise they would have an exponential change on the scale of a mean free path, which we have excluded. Thus

$$\Delta p_{\text{I}}^{A} = 0, \quad \Delta T_{\text{I}}^{A} = 0, \tag{5.2.20}$$

and the right-hand side of the momentum equation will just contain

$$\Delta u_{\text{I}i}^{A},\ \Delta^{n+1} u_{\text{I}i}^{A}.$$

Taking the Laplacian of the same equation we obtain, since $\Delta p_{\text{I}}^{A} = 0$, that a linear combination of terms of the form $\Delta^{n+1} u_{\text{I}i}^{A}$ $(n = 1, 2, 3, \ldots)$ vanishes. Again, $\Delta^2 u_{\text{I}i}^{A} = 0$, because $\Delta^2 u_{\text{I}j}^{A}$ cannot have an exponential change on the scale of the mean free path. Then the momentum equation reduces to

$$\partial_{x_i} p = \mu \Delta u_{\text{I}i}^{A} \quad (i = 1, 2, 3). \tag{5.2.21}$$

We conclude that the stress deviator and the heat flow have the following form:

$$q_{ij} = -\mu \left(\partial_{x_j} u_{\text{I}i}^{A} + \partial_{x_i} u_{\text{I}j}^{A} \right) + \sigma \partial_{x_i x_j} T_{\text{I}}^{A}, \tag{5.2.22}$$

$$\mathbf{q} = -\kappa \partial_{\mathbf{x}} T_{\text{I}}^{A} + \tau \Delta \mathbf{v}_{\text{I}}^{A}, \tag{5.2.23}$$

where μ, σ, κ, and τ are four coefficients of the order of the mean free path. In particular, the momentum balance equations take on the form given above, Eq. (5.2.21). The most important change with respect to traditional continuum mechanics is the presence of the term with the second derivatives of temperature in the expression of the stress deviator and of the term with the second derivatives of bulk velocity in the expression of the heat flow. These terms were already known to Maxwell.[1] In recent times, their importance has been stressed by Kogan et al.[2] and by Sone et al.[3,4]; approximate values of σ and τ are given in Ref. 4.

In Summary, the solution of the linearized Boltzmann equation reduces to the sum of two terms, one of which, h^B, is important just in the Knudsen layers and the other, h^A, is important far from the boundaries. The latter has a stress deviator and a heat flow given by Eqs. (5.2.22–5.2.23). We remark that, although these constitutive equations are different from those of Navier–Stokes and Fourier, the bulk velocity, pressure, and temperature satisfy the Navier–Stokes equations. In fact, when we take the divergence of the heat flow vector the term proportional to the Laplacian of $\mathbf{v}$ vanishes, thanks to the continuity equation, and thus just the term proportional to the temperature gradient survives; then, taking the divergence of the stress, the term grad (ΔT_1^A) vanishes, because of the energy equation. Yet, the new terms in the constitutive relations may produce physical effects in the presence of boundary conditions different from those of no-slip and no-temperature jump. In fact, as shown by the heuristic derivation of the boundary conditions for the thirteen moment equations (Chapter 2, Section 2.6), we must expect the velocity slip to be proportional to the shear stress and the temperature jump to the heat flow (see Eqs. (2.6.8) and (2.6.9)).

There are however flows where the agreement between the Boltzmann equation and the Navier–Stokes equations does not occur even at small Knudsen numbers. These flows will be discussed in Section 5.6.

Problems

5.2.1 Show that Eq. (5.2.4) holds.

5.2.2 Show that Eq. (5.2.8) holds.

5.2.3 Show that Eq. (5.2.9) holds.

5.2.4 Prove that the heat flow $\mathbf{q}$ for h^A is a linear combination with constant coefficients of $\partial_\mathbf{x} p_1^A$, $\partial_\mathbf{x} T_1^A$, $\partial_\mathbf{x} \Delta^n p_1^A$, $\partial_\mathbf{x} \Delta^n T_1^A$, $\Delta^n \mathbf{v}_1^A$, where Δ is the Laplace operator and $n = 1, 2, 3, \ldots$.

5.2.5 Prove that the stress deviator $q_{ij} = p_{ij} - p\delta_{ij}$ for h^A is a linear combination with constant coefficients of $\partial_{x_j} u_{1i}^A$, $\partial_{x_j} \Delta^n u_{1i}^A$, $\partial_{x_i x_j} p_1^A$, $\partial_{x_i x_j} T_1^A$, $\partial_{x_i x_j} \Delta^n p_1^A$, $\partial_{x_i x_j} \Delta^n T_1^A$.

5.3. Linearized Solutions of Internal Problems

The typical internal problems in more than one dimension are flow in pipes (possibly with annular cross section) or flow between two coaxial rotating cylinders or spheres.

The simplest case is offered by Poiseuille flow in a pipe of arbitrary cross section. The starting point is Eq. (3.4.4). If we adopt the BGK model, we can obtain an integral equation for the bulk velocity $v_2(\mathbf{x})$, where $\mathbf{x}$ is the two-dimensional vector describing the cross section of the pipe divided by the mean free path λ. If we let $v_2(\mathbf{x}) = (RT_0/2)^{1/2}\ell p^{-1}(dp/dy)(1-\phi(\mathbf{x}))$, the integral equation to be solved can be written as follows:

$$\phi(\mathbf{x}) = 1 + \pi^{-1}\int_{\Sigma(\mathbf{x})} T_0(|\mathbf{x}-\mathbf{y}|)|\mathbf{x}-\mathbf{y}|^{-1}\phi(\mathbf{y})d\mathbf{y}. \tag{5.3.1}$$

This equation can be solved numerically for any given cross section; two cases that have been actually considered are the case of a circular cross section[5] and the case of an annulus.[6] Another method is the variational one. The functional to be considered is

$$J(\tilde{\phi}) = \int_{\Sigma}[\tilde{\phi}(\mathbf{x})]^2 d\mathbf{y} - \pi^{-1}\int_{\Sigma}\int_{\Sigma(\mathbf{x})} T_0(|\mathbf{x}-\mathbf{y}|) \\ \times |\mathbf{x}-\mathbf{y}|^{-1}\tilde{\phi}(\mathbf{y})\tilde{\phi}(\mathbf{x})d\mathbf{y}d\mathbf{x} - 2\int_{\Sigma}\tilde{\phi}(\mathbf{x})d\mathbf{x}. \tag{5.3.2}$$

This functional attains its minimum value when $\tilde{\phi} = \phi$ (the solution of Eq. (5.3.1)). This value is

$$J(\phi) = -\int_{\Sigma}\phi(\mathbf{x})\,d\mathbf{x} \tag{5.3.3}$$

a quantity obviously related to the flow rate (Problem 5.3.1).

As in the case of a slab, one can obtain very accurate results by inserting the parabolic profile (which becomes exact in the continuum limit). At variance with the plane case, we have to evaluate double integrals; only one of the quadratures can be performed in closed form, whereas one has to resort to numerical methods to complete the calculations.[7] Table 5.1 gives a comparison for the results for the flow rate as supplied by the two methods (numerical solution[5] and variational method[7]). The agreement is very good. The theoretical results also compare very well with the experimental data (Fig. 5.1). In particular, the flow rate exhibits a minimum for a value of the ratio δ between the radius of the cylinder and the mean free path ℓ close to 0.3, in excellent agreement with experimental data. The values of the flow rate show a maximum deviation of the order of 3% from experimental data.

We remark that the problem of cylindrical Poiseuille flow can be treated by using the ES model[8] (see Section 1.9) to show that the corresponding solution can be obtained from the solution for the BGK model by a simple transformation

Table 5.1. *Comparison between the variational[7] and numerical[5] results for the nondimensional flow rate in cylindrical Poiseuille flow.*

δ	$Q(\delta)^5$	$Q(\delta)^7$
10	3.5821	3.5573
7	2.8440	2.8245
5	2.3578	2.3438
4	2.1188	2.1079
3	1.8850	1.8772
2	1.6608	1.6559
1.6	1.5753	1.5722
1.4	1.5348	1.5321
1.2	1.4959	1.4937
1.0	1.4594	1.4576
0.8	1.4261	1.4247
0.6	1.3982	1.3971
0.4	1.3796	1.3788
0.2	1.3820	1.3815
0.1	1.4043	1.4039
0.01	1.4768	1.4801

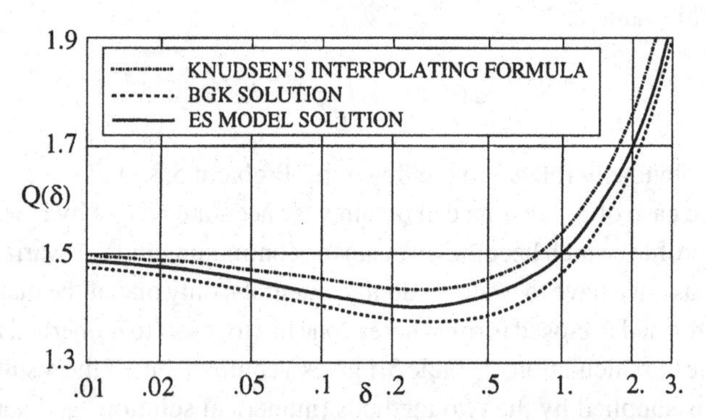

Figure 5.1. Comparison between theory and experiment for cylindrical Poiseuille flow. Here δ is the ratio of the radius a of the cylinder to the mean free path λ and Q is the ratio between the mass flow rate and $-\pi a^2(2RT)^{-1/2}(dp/dz)$.

(Problem 5.3.2). For the flow rate the formula takes the form[9]

$$Q(\delta, \text{Pr}) = Q(\delta\,\text{Pr}, 1) + (1 - \text{Pr})\frac{\delta}{4}, \tag{5.3.4}$$

where $Q(\delta\,\text{Pr}, 1)$ is, of course, given by the solution of the BGK model. If the value Pr = 2/3 is chosen, as is appropriate for a monatomic gas, the disagreement between the experimental data and the theoretical curve reduces to 1 or 2%. This is shown in Fig. 5.1, taken from Ref. 8.

Other important problems that can be solved with little effort by using the linearized BGK model equation include the Couette flow between two concentric cylinders[10] and the heat transfer between two concentric cylinders.[11]

In all the solutions of the problems discussed above, the boundary conditions are those of diffuse reflection according to a Maxwellian distribution.

Problems

5.3.1 Prove that $J(\phi)$ as defined by Eq. (5.3.3) is related to the flow rate.

5.3.2 Using the ES model show that (5.3.4) holds. (Hint: Use the fact that the stress can be calculated explicitly and with a transformation of variables express the ES solution in terms of the BGK solution; see Ref. 7).

5.4. Linearized Solutions of External Problems

If we use the linearized Boltzmann equation, the problem of finding the flow past a body does not have a solution in two dimensions, because it is impossible to match a nonzero flow at infinity since the solutions diverge logarithmically.[12] We shall examine this matter in the next section. However, the solution of three-dimensional problems is amenable to a numerical or variational treatment. A rigorous proof that the linearized solutions supply a good approximation for the nonlinear problem of the flow past a body at low speeds is supplied by the proofs of convergence of the perturbation series by Asano and Ukai.[13]

The most interesting problem is the flow past a sphere. This is a classical problem since Millikan derived experimental data for the drag on a sphere at low speeds, with the purpose of using them in his celebrated oil drop experiments to measure the electron charge.[14] In this experiment a spherical drop of very small size (aerosol) is under the action of an electric field, of gravity, and of the drag exerted by air.

A variational calculation of the drag was performed by Bassanini, Cercignani, and Pagani[15] with the BGK model. The agreement with experimental data is excellent. This is shown in Fig. 5.2, where the results obtained with the Stokes equations (this is the name usually given to the linearized Navier–Stokes equations) and the interpolating formula by Sherman[16] are also shown. The latter is a universal formula that relates a generic quantity $F(\mathrm{Kn})$, a function of the Knudsen number Kn, to its free molecular limit F_{FM} and its value according

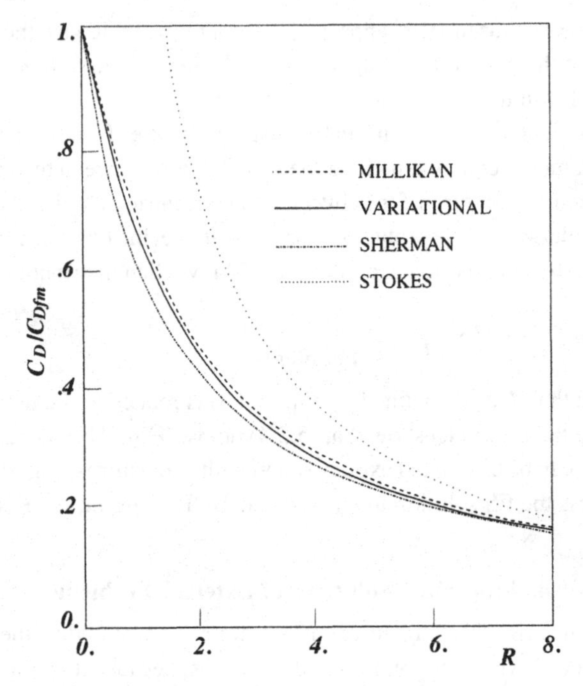

Figure 5.2. Low speed drag coefficient of a sphere versus the nondimensinal radius R. Millikan's curve interpolate his experimental data.

to continuum mechanics $F_C(\text{Kn})$, provided $F_C(\text{Kn})/F_{FM} \to 0$ as $\text{Kn} \to 0$. Sherman's general formula reads as follows:[16]

$$\frac{F(\text{Kn})}{F_{FM}} = \left(1 + \frac{F_{FM}}{F_C(\text{Kn})}\right)^{-1}, \tag{5.4.1}$$

and in the case of the drag on a sphere becomes (Problem 5.4.1)

$$C_D(R) = [1 + (0.685)R]^{-1}, \tag{5.4.2}$$

where R is a suitable inverse Knudsen number, the ratio between the radius of the sphere and the mean free path ℓ.

As Fig. 5.2 shows, the Navier–Stokes equations are completely wrong for values of R of about 10 or less. The variational BGK results[15] are in excellent agreement with the formula provided by Millikan to interpolate his experimental data.[14] Sherman's formula is in reasonably good agreement with the data. The same agreement of this simple interpolation occurs for other typical quantities, with the remarkable exception of the flow rate for Poiseuille flow (no minimum is exhibited by Sherman's formula[16]).

Another problem that can be treated with the variational method is the heat transfer from a sphere.[7] Here, following Sherman, the ratio of the heat flow q to its free molecular limit q_0 is plotted against the ratio of the value of the same heat flow according to continuum mechanics q_c to q_0. The agreement with the data by Takao[17] at $\mathrm{M} = 0$ and by Kavanau[18] at various values of the Mach number ranging from $\mathrm{M} = 0.1$ to $\mathrm{M} = 0.69$ is remarkable.

An exception to the aforementioned difficulty with two-dimensional problems is offered by the flow produced by a rotating cylinder in an infinite expanse of gas, which is at rest at space infinity. In this case the BGK solution can be easily solved numerically;[19] an approximate analytical solution has also been produced.[19] This solution is in a better agreement with the numerical solution than a four-moment solution by Willis.[20]

In all the problems discussed above, the boundary conditions are those of diffuse reflection according to a Maxwellian distribution.

Problem

5.4.1 Prove that (5.4.1) leads to (5.4.2) in the case of the low speed drag coefficient of a sphere.

5.5. The Stokes Paradox in Kinetic Theory

Here we shall discuss the difficulty mentioned at the beginning of the previous section for external flows in two dimensions. The argument begins with the use of the linearized version of an argument we used in Chapter 2 to show that certain problems cannot have a steady solution for certain kinds of boundary conditions.

Let us consider a steady problem past a solid body. The body is finite, at rest, and kept at uniform temperature. The boundary conditions are of the general kind, linear and homogeneous, discussed in Chapter 1, Section 1.11. Thus, unless the boundary conditions are deterministic, there is a Maxwellian distribution M_0, uniquely determined up to a factor, that satisfies the boundary conditions. At space infinity the distribution function is assumed to tend to another Maxwellian distribution M_∞. We linearize the Boltzmann equation with respect to M_0, made unique by choosing its density to be the same as that of M_∞. In the case of specular and reverse reflection we need to fix more things to make the Maxwellian distribution unique; we choose the temperature to be the same as that in M_∞ and choose a vanishing drift velocity. Then we try to find a solution linearized with respect to M_0.

The linearized Boltzmann equation may be written as follows:

$$\boldsymbol{\xi} \cdot \partial_{\mathbf{x}} h = Lh(\mathbf{x}, \boldsymbol{\xi}), \tag{5.5.1}$$

where $\mathbf{x}$ varies in some region Ω outside a finite, smooth region of ordinary space and $\boldsymbol{\xi}$ varies in the entire velocity space.

Because of the assumptions, h will satisfy the same boundary conditions as f; at infinity, h will tend toward a collision invariant (Problem 5.5.1). We shall assume that actually h tends to a collision invariant *uniformly*, as it seems reasonable (far from the boundary the solution of the linearized Boltzmann equation is very smooth, except, possibly, in a zero measure set where the solution is discontinuous, the discontinuity going to zero when $R \to \infty$).

Following Ref. 11, we want to show that if

$$\int_{\partial\Omega_2} \boldsymbol{\xi} \cdot \mathbf{n} h^2 d\boldsymbol{\xi} d\sigma \to 0 \tag{5.5.2}$$

when the closed surface $\partial\Omega_2$ tends to infinity, then h must be a constant.

The linearized version of the Boltzmann inequality reads as follows (Chapter 1, Section 1.10):

$$\int_{\Re^3} hLhd\boldsymbol{\xi} \leq 0, \tag{5.5.3}$$

where the equality sign holds if, and only if, h is a collision invariant.

Let us consider now a bounded region Ω_{12} between two finite surfaces, one inside the other, and both outside Ω. The external surface will be called $\partial\Omega_2$ and the internal one $\partial\Omega_1$.

Let us multiply Eq. (5.5.1) and integrate with respect to both $\boldsymbol{\xi}$ and $\mathbf{x}$, the latter integration being restricted to Ω_{12}. Then, because of Eq. (5.5.3), we have

$$\int_{\partial\Omega_{12}} \boldsymbol{\xi} \cdot \mathbf{n} h^2 d\boldsymbol{\xi} d\sigma \leq 0, \tag{5.5.4}$$

where $\mathbf{n}$ is the outward normal. Here equality holds if, and only if, h is a collision invariant.

We now decompose $\partial\Omega_{12}$ into the two disconnected parts $\partial\Omega_1$ and $\partial\Omega_2$ and orient both normal unit vectors toward space infinity. Then Eq. (5.5.4) may be rewritten as

$$\int_{\partial\Omega_2} \boldsymbol{\xi} \cdot \mathbf{n} h^2 d\boldsymbol{\xi} d\sigma \leq \int_{\partial\Omega_1} \boldsymbol{\xi} \cdot \mathbf{n} h^2 d\boldsymbol{\xi} d\sigma. \tag{5.5.5}$$

If we let the outer surface go to infinity, the corresponding integral will tend to zero, because of our assumption; then we can say, since Eq. (5.5.5) holds whenever $\partial\Omega_2$ encloses $\partial\Omega_1$, that both integrals are nonnegative. If we let $\partial\Omega_1$ tend to $\partial\Omega$, Eq. (5.5.5) then becomes

$$0 \leq \int_{\partial\Omega_2} \boldsymbol{\xi} \cdot \mathbf{n} h^2 d\boldsymbol{\xi} d\sigma \leq \int_{\partial\Omega} \boldsymbol{\xi} \cdot \mathbf{n} h^2 d\boldsymbol{\xi} d\sigma, \tag{5.5.6}$$

where now $\partial\Omega_2$ is any closed surface enclosing $\partial\Omega$. Because of the properties of the kernel appearing in the boundary conditions (see Chapter 1, Section 1.11), the last integral is nonpositive (Problem 5.5.2) and we find that both integrals in Eq. (5.5.6) must vanish. Because of the arbitrariness of $\partial\Omega_2$, this implies (Problem 5.5.3) that the vector $\int_{\mathfrak{R}^3} \boldsymbol{\xi} h^2 d\boldsymbol{\xi}$ is divergence free outside Ω. Then, because of Eqs. (5.5.1) and (5.5.3), we obtain (Problem 5.5.4)

$$\int_{\mathfrak{R}^3} hLh d\boldsymbol{\xi} = 0, \tag{5.5.7}$$

which implies that h is a collision invariant.

We must now impose the condition that this collision invariant is a steady solution of the linearized Boltzmann equation and satisfies the boundary conditions. It is easy to show (Problem 5.5.5) that this is possible if and only if h is a constant.

This result has been obtained under certain assumptions. One of them, expressed by Eq. (5.5.2) is not obvious. The physical meaning of this assumption is that there is no entropy source (or sink) at space infinity (Problem 5.5.6), but the result would hold even if we allow entropy production at space infinity (Problem 5.5.7). Thus there must be an entropy sink there, which swallows the entropy produced near the body.

To investigate whether this is compatible with the behavior of the solutions of the linearized Boltzmann equation, we remark that far from the body the Stokes equations must hold, because of the discussion in Section 5.2. Then the situation in three dimensions is satisfactory, since the Stokes solution is known to be well behaved at infinity,[21] in the sense that it approaches a uniform flow.

The situation is not equally satisfactory in the plane case; in fact, to exemplify, in the case of an axisymmetric body, assuming the temperature to be the same at space infinity and on the surface of the body, the expressions for the pressure and the bulk velocity contain terms of degrees 0 and -1 and terms of lower degrees in the distance r from a point inside the body.[21] The last ones cannot change the limiting behavior of the entropy flow. The constant in the pressure, under our assumptions, is zero (Problem 5.5.8). The viscous stresses, evaluated with the Navier–Stokes formula, obviously turn out to be of the same order as the pressure. The temperature is completely uncoupled from the other quantities. In any case, if the temperature is bounded, the heat flow must decrease more rapidly than r^{-1}, according to Fourier's law (Problem 5.5.9). Thus the leading contribution to the integral in the left-hand side of Eq. (5.5.2) is proportional to a linear combination of the surface integrals of $\sum_{i,j=1}^{3} p_{jk}u_j n_k$ and q_n/T_0 (Problem 5.5.10), which are both of order smaller than r^{-1} and hence produce a vanishing flow when $\partial\Omega_2$ goes to infinity. Then the previous result can be applied to prove that in the two-dimensional case there are no well-behaved solutions for the flow past a body according to the linearized Boltzmann equation. We remark that Ukai and Asano[13] prove their existence theorem in a space of any dimension d, provided $d \geq 3$.

The result just discussed is well known in the theory of the Stokes equations,[21] where it is known under the name of *Stokes paradox*. Here, following Ref. 11, we have extended it to the linearized Boltzmann equation.

In order to avoid a vanishing entropy flow on a large surface, we must allow for a logarithmically diverging bulk velocity. This would have no meaning, if we were solving exact equations; because of the linearization, however, we must consider whether the latter is permitted at large distances. The answer to this question is obviously negative, because at distances much larger than ℓ/M the linearization breaks down (the perturbation becomes singular).[12] Thus there is an outer layer, which must be studied by using a stretched variable $\bar{\mathbf{x}} = \mathbf{x}(\mathrm{M}/\ell)$ and looking for an expansion in powers of the Mach number $\epsilon = \mathrm{M}$. This is equivalent to replacing $\partial_{\mathbf{x}}$ by $\epsilon\partial_{\mathbf{x}}$.[12] Formally, this looks like the Hilbert expansion, which was examined in Chapter 2, with an important difference.

As in the case of the Hilbert expansion, we look for a solution of the Boltzmann equation in the form

$$f = \sum_{n=0}^{\infty} \epsilon^n f_n. \tag{5.5.8}$$

By inserting this formal series into the scaled Boltzmann equation and matching the various orders in ϵ, we obtain equations that one can hope to solve recursively (see Problem 2.3.1):

$$Q(f_0, f_0) = 0, \qquad (5.5.9)_0$$

$$2Q(f_1, f_0) = \boldsymbol{\xi} \cdot \partial_{\mathbf{x}} f_0 \equiv S_0, \qquad (5.5.9)_1$$

$$\ldots,$$

$$2(Q(f_j, f_0) = \boldsymbol{\xi} \cdot \partial_{\mathbf{x}} f_{j-1} - \sum_{i=1}^{j-1} Q(f_i, f_{j-i}) \equiv S_{j-1}, \qquad (5.5.9)_j$$

$$\ldots,$$

where the notation is the same as in Section 2.3. Equation $(5.5.9)_0$ is identical to Eq. (2.3.3) of Chapter 2. The important difference is that we assume f_0 to be the Maxwellian distribution M_0, which is constant. This implies that S_0 is zero. Then Eq. $(5.5.9)_1$ can be written as

$$Lh_1 = 0, \qquad (5.5.10)$$

where L is the linearized Boltzmann operator about M_0, and we let, as usual, $h_j = f_j/M\ s_j = S_j/M$. Then h_1 is a collision invariant:

$$h_1 = \sum_{\alpha=0}^{4} A_\alpha^{(1)}(\mathbf{x})\psi_\alpha. \qquad (5.5.11)$$

The next order equation, Eq. $(5.5.9)_2$ is

$$Lh_2 = \sum_{\alpha=0}^{4} \boldsymbol{\xi} \cdot \partial_{\mathbf{x}} A_\alpha^{(1)} \psi_\alpha - M_0^{-1} Q(M_0 h_1, M_0 h_1). \qquad (5.5.12)$$

But if ψ is a collision invariant (Problem 5.5.11), we have

$$2\psi'\psi'_* - 2\psi\psi_* = -[(\psi')^2 + (\psi'_*)^2 - (\psi)^2 - (\psi_*)^2, \qquad (5.5.13)$$

and hence (Problem 5.5.12)

$$M_0^{-1} Q(M_0 h_1, M_0 h_1) = -\frac{1}{2} L\left(h_1{}^2\right). \qquad (5.5.14)$$

Several terms in the right-hand side vanish because they turn out to be collision invariants; three terms (those not involving the products with ψ_0) survive. Equation (5.5.12) then becomes

$$L\left(h_2 - \frac{1}{2} h_1{}^2\right) = \sum_{\alpha=0}^{4} \boldsymbol{\xi} \cdot \partial_{\mathbf{x}} A_\alpha^{(1)} \psi_\alpha. \qquad (5.5.15)$$

As usual, by the Fredholm alternative for integral equations, this equation has a solution if and only if the integrals of the right-hand side multiplied by the collision invariants and M_0 vanish. Therefore, the solvability condition tells us something about the first five moments of the first-order perturbation of the distribution function, $M_0 h_1$, a little more than in the case of the Hilbert expansion considered in Section 2.3 of Chapter 2. In fact here the bulk velocity has more than one component and cannot be zero. A scalar and a vector equation follow from (5.5.15) (Problem 5.5.13). In terms of $\mathbf{v}_1$ and p_1, the bulk velocity and the pressure in the first approximation, these equations can be written as follows:

$$\operatorname{div} \mathbf{v}_1 = 0, \tag{5.5.16}$$

$$\operatorname{grad} p_1 = 0; \tag{5.5.17}$$

that is, the pressure p_1 is constant and the bulk velocity is divergence free.

We can then construct the solution of Eq. (5.5.15) (Problem 5.5.14):

$$\begin{aligned} h_2 = {} & \frac{1}{2}\sum_{i,j=1}^{3} A_i^{(1)} A_j^{(1)} \xi_i \xi_j + \frac{1}{2} A_4^{(1)\,2}[|\xi|^4 - 35(RT_0)^2] \\ & + \sum_{i=1}^{3} A_i^{(1)} A_4^{(1)} \xi_i (|\boldsymbol{\xi}|^2 - 5RT_0) - \frac{\mu_1}{\rho_0}(\partial_{x_j} A_i^{(1)} + \partial_{x_i} A_j^{(1)})\xi_i \xi_j \\ & - 2\frac{\kappa_1}{\rho_0}\sum_{i=1}^{3} \xi_i (|\boldsymbol{\xi}|^2 - 5RT_0)\partial_{x_i} A_4^{(1)} + \sum_{\alpha=0}^{4} A_\alpha^{(2)}(\mathbf{x})\psi_\alpha, \end{aligned} \tag{5.5.18}$$

where μ_1 and κ_1 are constants, depending on the temperature T_0 appearing in the Maxwellian distribution, given by Eqs. (2.3.13) and ρ_0 is the density in the same Maxwellian.

We can now construct the source $s_2 = \boldsymbol{\xi}\partial_{\mathbf{x}} h_2$, which must also satisfy the solvability conditions. This leads to a system of three equations, two of them being a scalar and the third a vector equation. The first scalar and the vector equation can be written as follows:

$$\operatorname{div} \mathbf{v}_2 = 0\,, \tag{5.5.19}$$

$$\rho_0 \sum_{i=1}^{3} \partial_{x_i}(v_{1i}\mathbf{v}_1) + \operatorname{grad} p_2 = \mu_1 \Delta \mathbf{v}_1, \tag{5.5.20}$$

where $\mathbf{v}_2$ and p_2 are the bulk velocity and the pressure in the second approximation. The third equation governs the temperature in the first approximation and is omitted here. Note that Eqs. (5.5.16) and Eq. (5.5.20) form a system decoupled from the other equations, identical to the Navier–Stokes equations for an incompressible fluid.

It may seem surprising that we obtained nonlinear equations by using a perturbation method. This is due to the fact that we expanded about a nondrifting Maxwellian distribution. If we had expanded about a Maxwellian with a bulk velocity equal to that of the flow at infinity, $\mathbf{v}_\infty$, the result would have been different (and dependent on the order assigned to the various terms). If we are just interested in the deviation of the bulk velocity from its asymptotic value, we can linearize the system formed by (5.5.16) and (5.5.20) about $\mathbf{v}_\infty$, to obtain the Oseen equations (Problem 5.5.15).

How should we solve, then, the problem of the flow past a body in two dimensions? We first should allow the solution of the linearized Boltzmann equation to have a logarithmic term in the bulk velocity, but we should restrict the solution to small values of $r\mathrm{M}/\ell$ (where r is the distance from a suitable point inside the body); then use the Hilbert-like expansion that we used above for large values of $r\mathrm{M}/\ell$, accompanied by a linearization about $\mathbf{v}_\infty$; finally match the two solutions for intermediate values of $r\mathrm{M}/\ell$, using the method of matched asymptotic expansions.[21]

Problems

5.5.1 Prove that, in the linearized problem past a body considered in the text, the linearized conditions at infinity imply that the perturbation h tends to a collision invariant.

5.5.2 Using the properties of the kernel appearing in the boundary conditions show that the second integral in (5.5.6) is nonpositive.

5.5.3 Show that if the first integral in Eq. (5.5.6) vanishes for an arbitrary closed surface $\partial\Omega_2$ outside Ω, the vector $\int_{\mathfrak{R}^3} \boldsymbol{\xi} h^2 d\boldsymbol{\xi}$ is divergence free outside Ω.

5.5.4 Using Eqs. (5.5.1) and (5.5.3), show that if $\int_{\mathfrak{R}^3} \boldsymbol{\xi} h^2 d\boldsymbol{\xi}$ is divergence free outside Ω, then Eq. (5.5.7) holds.

5.5.5 Show that if h is a collision invariant satisfying the steady linearized Boltzmann equation and the boundary conditions then h is a constant.

5.5.6 Show that if Eq. (5.5.2) holds, then there is no entropy source (or sink) at space infinity.

5.5.7 Show that the result that h must be a constant holds even if we allow an entropy production at space infinity.

5.5.8 Prove that the constant in the pressure as computed from the two-dimensional Stokes equations, under the assumptions in the text, is zero.

5.5.9 Show that, if the temperature is bounded, the heat flow must decrease more rapidly than r^{-1}, according to Fourier's law, in a two-dimensional problem.

5.5.10 Show that the leading contribution to the integral in the left-hand side of Eq. (5.5.2) is proportional to a linear combination of the surface integrals of $\sum_{i,j=1}^{3} p_{jk} u_j n_k$ and q_n/T_0.

5.5.11 Show that, if ψ is a collision invariant, then (5.5.13) hods.
5.5.12 Show that if h_1 is a collision invariant, then Eq. (5.5.14) holds. (Hint: Use the result of the previous problem.)
5.5.13 Prove that Eqs. (5.5.16) and (5.5.17) follow from the condition that the right-hand side of Eq. (5.5.15) is orthogonal to the collision invariants.
5.5.14 Show that the solution of Eq. (5.5.15) is given by Eq. (5.5.18).
5.5.15 Show that if one linearizes the system formed by (5.5.16) and (5.5.20) about $\mathbf{v}_\infty$, one obtains the Oseen equations.

5.6. The Continuum Limit

The connection between the Navier–Stokes equations and the Boltzmann equation has been met several times in this and the previous chapters. It seems time to give a general presentation of this aspect of the theory by discussing in more detail a method that originated with Hilbert[22] and Enskog.[23]

The discussion is made complicated by the various possible scalings. For example, if we denote by $(\bar{\mathbf{x}}, \bar{t})$ the microscopic space and time variables (those entering in the Boltzmann equation) and by $(\mathbf{x}, t)$ the macroscopic variables (those entering in the fluid dynamical description), we can study scalings of the following kind:

$$\bar{\mathbf{x}} = \epsilon^{-1}\mathbf{x}, \tag{5.6.1}$$

$$\bar{t} = \epsilon^{-\alpha} t, \tag{5.6.2}$$

where α is an exponent between 1 and 2. For $\alpha = 1$, this reduces to the compressible scaling that was considered in Section 2.3 of Chapter 2 and will be also discussed later in this section. If $\alpha > 1$, we are looking at larger "microscopic" times. We now investigate the limiting behavior of solutions of the Boltzmann equation.

Notice first that the compressible Euler equations,

$$\begin{aligned} \partial_t \rho + \operatorname{div}(\rho \mathbf{v}) &= 0, \\ \partial_t(\rho v_i) + \operatorname{div}\left(\rho \mathbf{v} v_i + \frac{2}{3}\rho e \mathbf{e}_i\right) &= 0, \\ \partial_t\left[\rho\left(e + \frac{1}{2}|\mathbf{v}|^2\right)\right] + \operatorname{div}\left[\rho \mathbf{v}\left(\frac{1}{2}|\mathbf{v}|^2 + \frac{5}{3}e\right)\right] &= 0, \end{aligned} \tag{5.6.3}$$

are invariant with respect to the scaling $t \to \epsilon^{-1}t$, $\mathbf{x} \to \epsilon^{-1}\mathbf{x}$. Here, $\mathbf{e}_i$ denotes the unit vector in the ith direction and p is related to ρ and $e = 3RT/2$ by the state equation for a perfect gas.

To investigate how these equations change under the scalings (5.6.1–5.6.2), let

$$\mathbf{v}^\epsilon(\mathbf{x},t) = \epsilon^\gamma \mathbf{v}(\epsilon^{-1}\mathbf{x}, \epsilon^{-\alpha}t), \ \gamma = \alpha - 1,$$

$$\rho^\epsilon(\mathbf{x},t) = \rho(\epsilon^{-1}\mathbf{x}, \epsilon^{-\alpha}t),$$

$$T^\epsilon(\mathbf{x},t) = T(\epsilon^{-1}\mathbf{x}, \epsilon^{-\alpha}t), \tag{5.6.4}$$

where $(\rho, \mathbf{v}, T)$ solve the compressible Euler equations (5.6.3). We easily obtain

$$\partial_t \rho^\epsilon + \operatorname{div}(\rho^\epsilon \mathbf{v}^\epsilon) = 0, \tag{5.6.5}$$

$$\partial_t \mathbf{v}^\epsilon + (\mathbf{v}^\epsilon \cdot \partial_\mathbf{x})\mathbf{v}^\epsilon = -\frac{1}{\rho^\epsilon}\partial_\mathbf{x} p^\epsilon \cdot \epsilon^{2(1-\alpha)}, \tag{5.6.6}$$

$$\partial_t T^\epsilon + (\mathbf{v}^\epsilon \cdot \partial_\mathbf{x})T^\epsilon + \frac{2}{3}T^\epsilon(\partial_\mathbf{x} \cdot \mathbf{v}^\epsilon) = 0. \tag{5.6.7}$$

The scaling of the bulk velocity field $\mathbf{v}^\epsilon$ in (5.6.4) is done in a dimensionally consistent way.

We expect that the continuum limit of the Boltzmann equation under the scaling (5.6.1–5.6.2) will be given by the asymptotic behavior of $(\rho^\epsilon, \mathbf{v}^\epsilon, T^\epsilon)$, satisfying (5.6.3), in the limit $\epsilon \to 0$. We will now investigate this limit.

To this end, let $\eta = \epsilon^{2(\alpha-1)}$ and expand

$$\mathbf{v}^\eta \equiv \mathbf{v}^\epsilon = \mathbf{v}_0 + \eta \mathbf{v}_1 + \eta^2 \mathbf{v}_2 + \dots,$$

$$\rho^\eta \equiv \rho^\epsilon = \rho_0 + \eta\rho_1 + \eta^2\rho_2 + \dots,$$

$$T^\eta \equiv T^\epsilon = T_0 + \eta T_1 + \eta^2 T_1 + \dots.$$

If we collect the terms of order η^{-1} in (5.6.4), we have

$$\partial_\mathbf{x} p_0 = 0, \tag{5.6.8}$$

and the terms of order η^0 give

$$\partial_t \rho_0 + \operatorname{div}(\rho_0 \mathbf{v}_0) = 0,$$

$$\partial_t \mathbf{v}_0 + (\mathbf{v}_0 \cdot \partial_\mathbf{x})\,\mathbf{v}_0 = -\frac{\partial_\mathbf{x} p_1}{\rho_0},$$

$$\partial_t T_0 + (\mathbf{v}_0 \cdot \partial_\mathbf{x})T_0 + \frac{2}{3}T_0 \operatorname{div} \mathbf{v}_0 = 0. \tag{5.6.9}$$

From (5.6.7) and the perfect gas law $p_0 = \rho_0 R T_0$, which we assume to hold at zeroth order, it follows that $\rho_0 T_0$ is constant as a function of the space variables.

The first and third equations in (5.6.9) now imply

$$\partial_t(\rho_0 T_0) + \operatorname{div}(\rho_0 T_0 \mathbf{v}_0) = -\frac{2}{3}\rho_0 T_0 \operatorname{div} \mathbf{v}_0. \tag{5.6.10}$$

As $\rho_0 T_0$ is only a function of t, say $A(t)$, Eq. (5.6.10) implies that

$$\operatorname{div} \mathbf{v}_0 = -\frac{3}{5}A'(t)/A(t). \tag{5.6.11}$$

Under suitable assumptions, Eq. (5.6.11) implies that $A'(t)=0$ and then $\operatorname{div} \mathbf{v}_0 = 0$. This is the case, for example, if we are in a box with nonporous walls, because then the normal component of $\mathbf{v}_0$ is 0 and we can use the divergence theorem (Problem 5.6.1). A similar argument applies to the case of a box with periodic boundary conditions or if we are in the entire space $\Re^3$ and the difference between $\mathbf{v}_0$ and a constant vector decays fast enough at infinity.

Assuming that we have conditions that imply that $\operatorname{div} \mathbf{v}_0 = 0$, we easily get from the continuity equation that ρ_0 will be independent of $\mathbf{x}$ if the initial value is, and the same for T_0. But with this knowledge Eqs. (5.6.9) then actually entail that ρ_0 and T_0 are constant.

Therefore, if the initial conditions are "well prepared" in the sense that $\mathbf{v}^\epsilon(\mathbf{x}, 0)$ is a divergence-free vector field and $\rho^\epsilon(\mathbf{x}, 0)$, $T^\epsilon(\mathbf{x}, 0)$ are constant, we expect as a first-order approximation for $\rho^\epsilon, \mathbf{v}^\epsilon, T^\epsilon$ the solution of the equations

$$\begin{aligned} \operatorname{div} \mathbf{v}_0 &= 0, \\ \partial_t \mathbf{v}_0 + (\mathbf{v}_0 \cdot \partial_\mathbf{x})\, \mathbf{v}_0 &= -\frac{\partial_\mathbf{x} p_1}{\rho_0}, \end{aligned} \tag{5.6.12}$$

which are the incompressible Euler equations. This limit, known as the "low velocity limit," is well known at the macroscopic level. We refer to Majda's book[24] for references and a detailed discussion. The variable η^{-1} enters into the theory as the square of the speed of sound. If this parameter is large compared to typical speeds of the fluid, then the incompressible model is well suited to describe the time evolution, provided that the initial velocity field is divergence-free and the initial density and temperature are constant.

The incompressible fluid limit was met in the last section in connection with the Stokes paradox. In fact, it seems that this limit and the derivation of the steady incompressible Navier–Stokes equations from the Boltzmann equation were first considered in connection with the flow past a body at small values of the Mach number.[12] We remark that in the steady case $\operatorname{div} \mathbf{v}_0 = 0$ and ρ_0 and T_0 turn out to be constant without using the boundary conditions.

Let us examine the kinetic picture as described by the Boltzmann equation. The above discussion suggests that if $\alpha \in (1, 2)$, then in the scaling (5.6.1–5.6.2), the solutions of the Boltzmann equation will converge to a local Maxwellian distribution whose parameters satisfy the incompressible Euler equations. This assertion can actually be proved rigorously.[25,26]

For $\alpha = 2$ something special happens. Of course, the incompressible Euler equations are invariant under the scaling (5.6.1–5.6.2); however, for $\alpha = 2$ the incompressible Navier–Stokes equations,

$$\partial_t \mathbf{v} + (\mathbf{v} \cdot \partial_{\mathbf{x}})\mathbf{v} = -\partial_{\mathbf{x}} r + \nu \Delta v,$$
$$\partial_{\mathbf{x}} \cdot \mathbf{v} = 0, \tag{5.6.13}$$

(where $r = p/\rho$ and $\nu = \mu/\rho$ is the kinematic viscosity) are also invariant under the same scaling. It is therefore of great interest to understand whether the Boltzmann dynamics "chooses" in this limit the Euler or the Navier–Stokes evolution.

From what we saw in the previous section we can expect that the answer is Navier–Stokes. In other words, considering times larger than those typical for Euler dynamics ($\epsilon^{-2}t$ instead of $\epsilon^{-\alpha}t$, $\alpha < 2$), dissipation becomes nonnegligible. We have seen an illustration of this behavior in the previous section. As another example, consider two parallel layers of fluid moving with velocities $\mathbf{v}$ and $\mathbf{v} + \delta\mathbf{v}$. Suppose that we want to decide whether there is any momentum transfer between these layers (which is expected for the Navier–Stokes equations, but not for the Euler equations). The momentum transfer can in principle be affected by the trend to thermalization typical of the Boltzmann collision term, but a scaling argument shows that it is proportional to $\epsilon^{\alpha-2}\delta\mathbf{v}$, with the consequence that it remains only relevant for $\alpha = 2$ (Problem 5.6.2).

These considerations can be put on a rigorous basis, and, of course, the viscosity coefficient can be computed in terms of kinetic expressions, with the result that we found in Chapter 2.

The incompressible Navier–Stokes equations can be derived from the Boltzmann equation if the time interval is such that smooth solutions of the continuum equations exist. The tool that yields this result is a truncated Hilbert expansion similar to that discussed in Chapter 2 (Section 2.3), and one gets local convergence for the general situation[25] or global and uniform convergence if the data are small in a suitable sense.[26]

Let us consider now the cases in which compressibility is not negligible. If we scale space and time in the same way, the scaled Boltzmann equation becomes

$$\partial_t f^\epsilon + \boldsymbol{\xi} \cdot \partial_{\mathbf{x}} f^\epsilon = \frac{1}{\epsilon} Q(f^\epsilon, f^\epsilon). \tag{5.6.14}$$

We will use the abbreviation $D_t f := \partial_t f + \boldsymbol{\xi} \cdot \partial_{\mathbf{x}} f$. Of course, we expect that

$$\epsilon D_t f^\epsilon \to 0 \text{ as } \epsilon \to 0, \tag{5.6.15}$$

and if

$$f^\epsilon \to f^0, \tag{5.6.16}$$

the limit f^0 must satisfy

$$Q(f^0, f^0) = 0. \tag{5.6.17}$$

This implies, as we know, that f^0 is a local Maxwellian distribution:

$$f^0(\mathbf{x}, \boldsymbol{\xi}, t) \equiv M(\mathbf{x}, \boldsymbol{\xi}, t) = \frac{\rho(\mathbf{x}, t)}{(2\pi RT(\mathbf{x}, t))^{3/2}} \exp\left(-\frac{|\boldsymbol{\xi} - \mathbf{v}(\mathbf{x}, t)|^2}{2RT(\mathbf{x}, t)}\right). \tag{5.6.18}$$

The fields $(\rho, \mathbf{x}, T)$, which characterize the behavior of the local Maxwellian distribution M in space and time, are expected to evolve according to continuum equations that we are going to derive. First, let us emphasize again that these fields are varying *slowly* on the space–time scales that are typical for the gas described in terms of the Boltzmann equation.

From the conservation laws (1.6.36)

$$\int \psi_\alpha Q(f, f) d\boldsymbol{\xi} = 0 \quad (\alpha = 0, \ldots, 4), \tag{5.6.19}$$

we readily obtain, as we know,

$$\int \psi_\alpha (\partial_t f + \boldsymbol{\xi} \cdot \partial_{\mathbf{x}} f) d\boldsymbol{\xi} = 0. \tag{5.6.20}$$

This is a system of equations for the moments of f that is in general not closed. However, if we assume $f = M$ and use the identities (for M they *are* identities; for a general f they are definitions given in (1.6.3–1.6.7), where $e = \frac{3}{2}RT$)

$$\rho = \int M d\boldsymbol{\xi}, \tag{5.6.21$_1$}$$

$$\rho\mathbf{v} = \int M \boldsymbol{\xi} d\boldsymbol{\xi}, \tag{5.6.21$_2$}$$

$$w \equiv \frac{3}{2}\rho RT + \frac{1}{2}\rho|\mathbf{v}|^2 = \frac{1}{2}\int \xi^2 M d\boldsymbol{\xi}, \tag{5.6.21$_3$}$$

we readily obtain from (5.6.20) that

$$\partial_t \rho + \operatorname{div}(\rho \mathbf{v}) = 0, \tag{5.6.22}$$

$$\partial_t (\rho v_i) + \operatorname{div}\left(\int M \boldsymbol{\xi} \xi_i \right) = 0 \quad (i = 1, 2, 3), \tag{5.6.23}$$

$$\partial_t \left(\frac{3}{2} \rho R T + \frac{1}{2} \rho |\mathbf{v}|^2 \right) + \frac{1}{2} \operatorname{div}\left(\int M \boldsymbol{\xi} |\boldsymbol{\xi}|^2 \right) = 0. \tag{5.6.24}$$

These equations are nothing but Eqs. (1.6.18–1.6.20) of Chapter 1, specialized to the case of a Maxwellian distribution. This is of crucial importance if we want to write Eqs. (5.6.23) and (5.6.24) in closed form. To do so we have to express $\int M \boldsymbol{\xi} \xi_i$ and $\int M \boldsymbol{\xi} |\boldsymbol{\xi}|^2$ in terms of the fields $(\rho, \mathbf{v}, T)$. To this end, we use the elementary identities (Problem 5.6.3)

$$\int M(\xi_j - v_j)(\xi_i - v_i) d\boldsymbol{\xi} = \delta_{ij} \rho R T \quad (i, j = 1, 2, 3),$$

$$\int M(\boldsymbol{\xi} - \mathbf{v}) |\boldsymbol{\xi} - \mathbf{v}|^2 d\boldsymbol{\xi} = 0, \tag{5.6.25}$$

which transform Eq. (5.6.23) into

$$\partial_t (\rho v_i) + \operatorname{div}(\rho \mathbf{v} v_i) = -\partial_{x_i} p, \tag{5.6.26}$$

with

$$p = \rho R T. \tag{5.6.27}$$

Equation (5.6.27) is the perfect gas law. Obviously, the quantity p defined here has the meaning of a pressure.

Recalling that the internal energy e is related to temperature T by

$$e = \frac{3}{2} R T \tag{5.6.28}$$

and using this and Eq. (5.6.25) we transform Eq. (5.6.24) into

$$\partial_t \left(\rho \left(e + \frac{1}{2} |\mathbf{v}|^2 \right) \right) + \operatorname{div}\left(\rho \mathbf{v} \left(e + \frac{1}{2} |\mathbf{v}|^2 \right) \right) = -\operatorname{div}(p \mathbf{v}). \tag{5.6.29}$$

Equations (5.6.22), (5.6.26), and (5.6.29) express conservation equations for mass, momentum, and energy respectively and can be rewritten as the Euler equations (5.6.3).

For smooth functions, an equivalent way of writing the Euler equations in terms of the field $(\rho, \mathbf{v}, T)$ is

$$\begin{aligned}
&\partial_t \rho + \operatorname{div}(\rho \mathbf{v}) = 0, \\
&\partial_t \mathbf{v} + (\mathbf{v} \cdot \partial_{\mathbf{x}})\mathbf{v} + \frac{1}{\rho}\partial_{\mathbf{x}} p = 0, \\
&\partial_t T + (\mathbf{v} \cdot \partial_{\mathbf{x}})T + \frac{2}{3} T \partial_{\mathbf{x}} \cdot \mathbf{v} = 0.
\end{aligned} \tag{5.6.30}$$

However, in this form we lose the conservation form as given in (5.6.3), in which the time derivative of a field equals the negative divergence of a current that is a nonlinear function of this field.

Before going on, some comments on our limits are in order, because one might suspect an inconsistency in the passage from a rarefied to a dense gas. Recall that we derived the Boltzmann equation in the Boltzmann–Grad limit ($N\sigma^2 = O(1)$). In the continuum limit, we have to take $N\sigma^2 = 1/\epsilon \to \infty$. This, at first glance, seems contradictory, but there is really no problem. The Boltzmann equation holds for a perfect gas, that is, for a gas such that the density parameter $\delta = N\sigma^3/V$, where V is the volume containing N molecules, tends to zero. The parameter

$$\frac{1}{\mathrm{Kn}} = \frac{N\sigma^2}{V^{\frac{2}{3}}} = N^{\frac{1}{3}}\delta^{\frac{2}{3}} \tag{5.6.31}$$

may tend to zero, tend to ∞, or remain finite in this limit. These are the three cases that occur if we scale N as δ^{-m} ($m \geq 0$), for $m < 2$, $m > 2$, and $m = 2$ respectively. In the first case the gas is in free molecular flow and we can simply neglect the collision term (Knudsen gas). In the second we are in the continuum regime we are treating here, and we cannot simply "omit" the "small" term (i.e., the left-hand side of the Boltzmann equation) because the limit is singular. In the third case the two sides of the Boltzmann equation are equally important (Boltzmann gas) and this is the case dealt with before for solutions close to an absolute Maxwellian distribution.

We already saw, in the case of steady problems in a slab (Chapter 2), that, in spite of the fact that we face a singular perturbation problem, Hilbert[22] proposed an expansion in powers of ϵ. In this way, however, we obtain a Maxwellian distribution at the lowest order, with parameters satisfying the Euler equations and corrections to this solution, which are obtained by solving inhomogeneous linearized Euler equations.[22,27,28] In order to avoid this and to investigate the relationship between the Boltzmann equation and the Navier–Stokes equations, Enskog introduced an expansion, usually called the Chapman–Enskog expansion.[23,27,28,29] The idea behind this expansion is that the functional dependence of f upon the local density, bulk velocity, and internal energy can be expanded into a power series. Although there are many formal similarities with the Hilbert expansion, the procedure is rather different.

As remarked by the author,[27,28] the Chapman–Enskog expansion seems to introduce spurious solutions, especially if one looks for steady states. This is essentially due to the fact that one really considers infinitely many time scales (of orders $\epsilon, \epsilon^2, \ldots, \epsilon^n, \ldots$). The author[27,28] introduced only two time scales (of orders ϵ and ϵ^2) and was able to recover the compressible Navier–Stokes equations. To explain the idea, we remark that the Navier–Stokes equations describe two kinds of processes, convection and diffusion, which act on two different time scales. If we consider only the first scale we obtain the compressible Euler equations; if we insist on the second one we can obtain the Navier–Stokes equations only at the price of losing compressibility. If we want both compressibility and diffusion, we have to keep both scales at the same time and think of f as

$$f(\mathbf{x}, \boldsymbol{\xi}, t) = f(\epsilon\mathbf{x}, \boldsymbol{\xi}, \epsilon t, \epsilon^2 t). \tag{5.6.32}$$

This enables us to introduce two different time variables $t_1 = \epsilon t$ and $t_2 = \epsilon^2 t$ and a new space variable $\mathbf{x}_1 = \epsilon\mathbf{x}$ such that $f = f(\mathbf{x}_1, \boldsymbol{\xi}, t_1, t_2)$. The fluid dynamical variables are functions of $\mathbf{x}_1$, t_1, and t_2, and for both f and the fluid dynamical variables the time derivative is given by

$$\frac{\partial}{\partial t} = \epsilon\frac{\partial f}{\partial t_1} + \epsilon^2\frac{\partial f}{\partial t_2}. \tag{5.6.33}$$

In particular, the Boltzmann equation can be rewritten as

$$\epsilon\frac{\partial f}{\partial t_1} + \epsilon^2\frac{\partial f}{\partial t_2} + \epsilon\boldsymbol{\xi}\cdot\frac{\partial}{\partial \mathbf{x}_1} f = Q(f, f).$$

If we expand f formally in a power series in ϵ, we find that at the lowest order f is a Maxwellian distribution. The compatibility conditions at the first order give that the time derivatives of the fluid dynamic variables with respect to t_1 are determined by the Euler equations, but the derivatives with respect to t_2 are determined only at the next level and are given by the terms of the compressible Navier–Stokes equations describing the effects of viscosity and heat conductivity. The two contributions are, of course, to be added as specified by (5.6.33) in order to obtain the full time derivative and thus write the compressible Navier–Stokes equations.

It is not among the aims of this book to indicate the techniques applied to and the results obtained from the computations of the transport coefficients, such as the viscosity and heat conduction coefficients, for a given molecular interaction. For this we refer to standard treatises.[30–31]

The results discussed in this section show that there is a qualitative agreement between the Boltzmann equation and the Navier–Stokes equations for sufficiently low values of the Knudsen number.

There are however flows where this agreement does not occur. They have been especially studied by Sone.[32] These effects arise because the no-slip and no temperature jump boundary conditions do not hold. In addition to the thermal creep induced along a boundary with a nonuniform temperature, discovered by Maxwell (Chapter 3, Section 3.5), two new kinds of flow are induced over boundaries kept at uniform temperatures. They are related to the presences of thermal stresses in the gas.

The first effect[32–34] is present even for small Mach numbers and small temperature differences and follows from the fact that there are stresses related to the second derivatives of the temperature (see Eq. (5.2.22)). Although these stresses do not change the Navier–Stokes equations, they change the boundary conditions; the gas slips on the wall, and thus a movement occurs even if the wall is at rest. This effect is particularly important in small systems, such as micromachines, since the temperature differences are small but may have relatively large second derivatives; it is usually called the thermal stress slip flow.[32–34]

The second effect is nonlinear[35,36] and occurs when two isothermal surfaces do not have constant distance (thus in any situation with large temperature gradients, in the absence of particular symmetries). In fact, if we assume that in the Hilbert expansion the velocity vanishes at the lowest order (i.e., the speed is of the order of the Knudsen number), the terms of second order in the temperature show up in the momentum equation. These terms are associated with thermal stresses and are of the same importance as those containing the pressure and the viscous stresses (Problem 5.6.4). A solution in which the gas does not move can be obtained if and only if (Problem 5.6.5)

$$\text{grad}\, T \wedge \text{grad}(|\text{grad}\, T|^2) = 0. \tag{5.6.34}$$

Since $|\text{grad}\, T|$ measures the distance between two nearby isothermal lines, if this quantity has a gradient in the direction orthogonal to grad T (and hence (5.6.16) does not hold), the distance between two neighboring isothermal lines varies and we must expect that the gas moves.

These effects may occur even for sufficiently large values of the Knudsen number; they cannot be described, however, in terms of the local temperature field. Rather, they depend on the configuration of the system. They should not be confused with flows due to the presence of a temperature gradient along the wall, such as the transpiration flow[37] and the thermophoresis of aerosol particles.[38]

Numerical examples of simulations of this kind of flow will be given in Chapter 7.

Problems

5.6.1 Prove that if the gas is in a bounded region with no matter exchange with the exterior, then Eq. (5.6.10) implies that $A'(t) = 0$.

5.6.2 Consider two parallel layers of fluid moving with velocities $\mathbf{v}$ and $\mathbf{v} + \delta\mathbf{v}$. Show, by a scaling argument, that there is a momentum transfer between these layers, proportional to $\epsilon^{\alpha-2}\delta\mathbf{v}$.

5.6.3 Prove that Eqs. (5.6.25) hold for any Maxwellian distribution.

5.6.4 Show that if in the Hilbert expansion the velocity vanishes at the lowest order (i.e., the speed is of the order of the Knudsen number), the terms of second order in the temperature show up in the momentum equation as terms of the same importance as those containing the pressure and the viscous stresses.

5.6.5 Show that, in the situation envisaged in the previous problem, a solution in which the gas does not move can be obtained if and only if (5.6.34) holds.

5.7. Free Molecular Flows

We discussed free molecular flows in a slab geometry in Chapter 2. As we know, this is the case exactly opposite to the continuum limit. The Knudsen number tends to infinity and the collisions are negligible. Thus the Boltzmann equation (in the absence of a body force) reduces to the simple form

$$D_t f = \partial_t f + \boldsymbol{\xi} \cdot \partial_{\mathbf{x}} f = 0. \tag{5.7.1}$$

Since the molecular collisions are negligible, the gas–surface interaction discussed in Section 1.11 plays a major role. This situation is typical for artificial satellites, since the mean free path is 50 meters at 200 kilometers of altitude.

The general solution of Eq. (5.7.1) is in terms of an arbitrary function of two vectors $g(\cdot, \cdot)$ (Problem 5.7.1):

$$f(\mathbf{x}, \boldsymbol{\xi}, t) = g(\mathbf{x} - \boldsymbol{\xi} t, \boldsymbol{\xi}). \tag{5.7.2}$$

In the steady case, Eq. (5.7.1) reduces to

$$\boldsymbol{\xi} \cdot \partial_{\mathbf{x}} f = 0, \tag{5.7.3}$$

and the general solution becomes (Problem 5.7.2)

$$f(\mathbf{x}, \boldsymbol{\xi}, t) = g(\mathbf{x} \wedge \boldsymbol{\xi}, \boldsymbol{\xi}). \tag{5.7.4}$$

Frequently it is easier to work with the property that f is constant along the molecular trajectories than with the explicit solutions given by Eqs. (5.7.3–5.7.4).

The easiest problem to deal with (in a sense, easier even than the plane Couette flow discussed in Chapter 2) is the flow past a convex body. In this case, in fact, the molecules arriving at the surface of the wall have an assigned distribution function f_∞, usually a Maxwellian distribution with the density ρ_∞, bulk velocity $\mathbf{v}_\infty$, and temperature T_∞, prevailing far away from the body, and the distribution function of the molecules leaving the surface is given by the boundary conditions. The distribution function at any other point P, if needed, is simply obtained by the following rule: If the straight line through P having the direction of $\boldsymbol{\xi}$ intersects the body at a point Q and $\boldsymbol{\xi}$ points from Q toward P, then the distribution function at P is the same as that at Q; otherwise it equals f_∞ (Problem 5.7.3).

Interest is usually confined to the total momentum and energy exchanged between the molecules and the body, which, in turn, easily yield the drag and lift exerted by the gas on the body and the heat transfer between the body and the gas.

An interesting result concerns the recovery factor defined as

$$r = \frac{T_{\text{eq}} - T_\infty}{T_0 - T_\infty}, \tag{5.7.5}$$

where T_{eq} is the *uniform* temperature at which a finite body is to be kept in order to have a zero *global* heat transfer and, in terms of the Mach number M_∞ and of the ratio of specific heats γ (5/3 for a monatomic gas), the stagnation temperature T_0 is defined by

$$T_0 = T_\infty \left(1 + \frac{\gamma - 1}{2} \mathrm{M}_\infty^2\right). \tag{5.7.6}$$

According to a known result in boundary layer theory[39] r is always less than unity. In the theory of free molecular flows, r for an element at a given temperature turns out to be larger than 5/4 for a monatomic gas (Problem 5.7.4) in the case of perfect accommodation. As a consequence $r > 1$ is expected for bodies of sufficiently symmetrical shape for cancellations not to occur. This result is confirmed by both calculations and experiments. We remark, however, that this is not a rigorous consequence obtained from the scattering kernel theory, but only from the simplified theory based on the accommodation coefficients. A theoretical counterexample has been given.[40]

In practice, of course, the temperature of a body is determined by a balance of all forms of heat transfer at the body surface. For an artificial satellite, a considerable portion of the heat energy is lost by radiation and this process must be duly taken into account in the balance.

The results take a particularly simple form in the case of a large Mach number since we can let the latter go to infinity in the various formulas (Problems 5.7.5). One must, however, be careful, because the speed ratio is multiplied by $\sin\theta$ in many terms and thus the aforementioned limit is not uniform in θ. Therefore the limiting formulas can be used if and only if the area where $S\sin\theta \le 1$ is small.

The standard treatment is based on the definition of accommodation coefficients, but calculations based on the CL model are available[28,41,42] (Problem 5.7.6).

The case of nonconvex boundaries is, of course, more complicated and one must solve an integral equation to obtain the distribution function at the boundary (Problem 5.7.7). If one assumes diffuse reflection according to a Maxwellian, the integral equation simplifies considerably, because just the mass flow at the boundary must be computed[28] (Problem 5.7.8).

In particular the latter equation can be used to study free molecular flows in pipes of arbitrary cross section with a typical diameter much smaller than the mean free path (capillaries). If the cross section is circular the equation becomes particularly simple (Problem 5.7.9) and is known as Clausing's equation.[28]

Problems

5.7.1 Show that the general solution of Eq. (5.7.1) is given by (5.7.2), where $g(\cdot,\cdot)$ is an arbitrary function of two vectors.

5.7.2 Show that the general solution of Eq. (5.7.3) is given by (5.7.4), where $g(\cdot,\cdot)$ is an arbitrary function of two vectors orthogonal to each other.

5.7.3 Show that the distribution function at any point P outside a convex body in free molecular flow is obtained by the following rule: If the straight line through P having the direction of $\boldsymbol{\xi}$ intersects the body at a point Q and $\boldsymbol{\xi}$ points from Q toward P, then the distribution function at P is the same as that at Q; otherwise it equals f_∞.

5.7.4 Find the total momentum and energy exchanged between the molecules of a monatomic gas and a body in free molecular flow and, as a consequence, the drag and lift exerted by the gas on the body and the heat transfer between the body and the gas. Assume perfect accommodation, for simplicity. As a consequence show that the recovery factor, defined in (5.7.5) is larger than 5/4. (See Ref. 28.)

5.7.5 Show that the results of Problem 5.7.4 take a particularly simple form in the case of a large Mach number since we can let the latter go to infinity in the various formulas. Show in particular that the drag coefficient turns out to be 2. (See Ref. 28.)

5.7.6 Repeat the calculations of Problem 5.7.4 for the CL model (see Ref. 27.)

5.7.7 Discuss the case of free molecular flow in the presence of nonconvex boundaries and find the integral equation giving the distribution function at the boundary (see Ref. 28.)

5.7.8 Assume diffuse reflection according to a Maxwellian and show that the integral equation of Problem 5.7.6 simplifies.

5.7.9 Show that the integral equation of Problem 5.7.8 can be used to study free molecular flows in pipes of arbitrary cross section. Show that if the cross section is circular the equation becomes particularly simple (Clausing's equation). (See Ref. 28.)

5.8. Nearly Free Molecular Flows

We have already examined the perturbation of free molecular flows in the case of a gas between two plates (Chapter 2, Section 2.4). The discussion in more than one dimension is similar. The only remark is that if one tries the naïve Knudsen iteration, the singularity in the first iterate cancels when integrating to obtain moments (a first-order pole is milder, if the dimension is higher). The singularity is still present and, although it is mild, it can build up a worse singularity when computing subsequent steps. The difficulties are enhanced in unbounded domains where the subsequent terms diverge at space infinity. The reason for the latter fact is that the ratio between the mean free path ℓ and the distance d of any given point from the body is a local Knudsen number that tends to zero when d tends to infinity; hence collisions certainly arise in an unbounded domain and tend to dominate at large distances. On this basis we are led to expect that a continuum behavior takes place at infinity, even when the typical lengths characterizing the size of the body are much smaller than the mean free path; this is confirmed by the discussion of the Stokes paradox for the steady linearized Boltzmann equation, which was considered in Section 5.5. There we found that far from the body, continuum equations always apply, at least for small Mach numbers.

Both difficulties are removed, as we saw in Chapter 2, by the collision iteration. The main difference is that logarithmic terms in the Knudsen number may occur but multiplied by a power of Kn typically equal to the number of space dimensions relevant for the problem under consideration in a bounded domain (Problem 5.8.1). In particular the dependence upon coordinates will show the same singularity (we can think of local Knudsen numbers based on the distance from the nearest wall); as a consequence first derivatives will diverge at the boundary in one dimension and the same will occur for second, or third derivatives, in two, or three, space dimensions, respectively. In an external domain we have, in addition to the low speed effects, the effect of particles coming from infinity, which actually dominates. In particular in one dimension (half-space problems) the terms coming from iterations are of the same order as the the lowest order terms; actually for a half-space problem we can hardly define a Knudsen number (the local one is an exception). In two dimensions the corrections in the moments are of order $\mathrm{Kn}^{-1} \log \mathrm{Kn}$. In three dimensions a correction of order $\mathrm{Kn}^{-2} \log \mathrm{Kn}$ is preceded by a correction of order Kn^{-1} (Problem 5.8.2).

Care must be exercised when applying the aforementioned results to a concrete numerical evaluation, as already mentioned in Chapter 2. In fact, for large but not extremely large Knudsen numbers (say $10 \leq \mathrm{Kn} \leq 100$) $\log \mathrm{Kn}$ is a relatively small number, although $\log \mathrm{Kn} \to \infty$ for $\mathrm{Kn} \to \infty$. Hence terms of order $\log \mathrm{Kn}/\mathrm{Kn}$, though mathematically dominating over terms of order $1/\mathrm{Kn}$ are of the same order as the latter for practical purposes. As consequence, the two kinds of terms must be computed together if numerical accuracy is desired for the aforementioned range of Knudsen numbers.

Related to this remark is the fact that any factor appearing in front of Kn in the argument of the logarithm is meaningless unless the term of order Kn^{-1} is also computed. This is particularly important when the factor under consideration depends upon a parameter that can take very large (or very small) values (typically a speed ratio). Thus Hamel and Cooper[43,44] have shown that the first iterate of the integral iteration is incapable of describing the correct dependence upon the speed ratio and have applied the method of matched asymptotic expansions[21] to regions near a body and far from a body. Specifically, for the hypersonic flow of a gas of hard spheres past a two-dimensional strip, they find for the drag coefficient

$$C_D = C_{D\mathrm{f.m.}} \left[1 + \frac{\epsilon \log \epsilon}{2\pi} \right], \tag{5.8.1}$$

where the inverse Knudsen number ϵ is based on the mean free path $\lambda = \pi^{3/2}\sigma^2 n_\infty S_w$ (σ is the molecular diameter and $S_w = S_\infty(T_w/T_\infty)$, whereas n_∞ and S_∞ are the number density and the speed ratio at infinity).

If we consider infinite-range intermolecular potentials, then the considerations of Chapter 2 apply and we have fractional powers rather than logarithms.

All the considerations of this section have the important consequence that approximate methods of solution that are not able to allow for a nonanalytic behavior for $\mathrm{Kn} \to \infty$ produce poor results for large Knudsen numbers. A typical example is the moment method of Liu ad Lees when applied to the Poiseuille flow between parallel plates;[45] the method, which looks a priori so promising because it embodies the main features of the behavior at both small and large Knudsen numbers, turns out to produce the Knudsen minimum in the flow rate but gives a wrong asymptotic behavior in the region of small Knudsen numbers when logarithmic terms become important. The same difficulty can show up in the variational method when applied to the integro-differential form of the linearized Boltzmann equation, unless the trial function is chosen in such a way as to be able to produce the logarithmic term.

Problems

5.8.1 Use the method of collision iteration to compute the leading singularity in a bounded domain in one, two, and three dimensions. (See Ref. 28.)

5.8.2 Perform the same calculation as in Problem 1 for an external domain. (See Ref. 28.)

5.9. Expansion of a Gas into a Vacuum

A fundamental theoretical and experimental problem of rarefied gas dynamics is the free expansion of a gas into a vacuum. This expansion occurs, for example, in the discharge from an orifice into a low pressure chamber. This problem embodies, within a simple framework, a transition from continuum almost to free molecular flow without the usual complicating effects of gas–surface interaction.

As remarked by Ashkenas and Sherman,[46] the flow along the axis of the expanding jet can be to some extent simulated by the spherically symmetric expansion of a monatomic gas from a near continuum source. We shall say more on expanding jets in Section 8.7 of the last chapter, devoted to evaporation and condensation phenomena. In fact a jet can be produced not only by discharging a gas from an orifice, but also by producing an intense evaporation from a disk-shaped region in a plane. Here we shall restrict ourselves to the simpler problem of a spherical source flow into a vacuum, which is theoretically simpler and has been studied by many authors.

According to inviscid gas theory, the expansion from a spherical source into a vacuum should accelerate the gas to any Mach number and to an arbitrarily low density. However, at large distances from the source, the density will be so low that collisions will be unable to support the continuum expansion. In fact the average energy due to random motion perpendicular to the streamlines is continually decreasing and is being fed into the mean motion of the gas as well as into the random motion parallel to the streamlines. Since random motion is connected with temperature, this circumstance is often referred to as the "freezing" of the parallel temperature. A better name than parallel temperature would be parallel thermal energy. Since, however, there is a linear relation between internal energy and temperature for a perfect gas, this abuse of terminology is acceptable and will be used in the following.

The first careful attempts at quantifying the picture given above for a spherical expansion are due to Hamel and Willis[47] and Edwards and Cheng,[48] who both used moment equations and the so-called hypersonic approximation. The latter is based on an expansion in negative powers of the speed ratio of all moments of the distribution function. Thus a rational truncation of the infinite set of moment equations is achieved.

The mathematical theory was formalized by Freeman[49] and applied to unsteady problems by Freeman and Grundy.[50] The technique can be also extended to the case of an expanding jet, as shown by Grundy,[51] who supported the assumption that the free jet can be approximated by a spherical source. According to these authors, solutions of the Boltzmann equation for these problems can be found that asymptotically approach the isentropic flow solution for small values of a suitably scaled radial distance. This is essentially the Hilbert expansion, which breaks down at large distances from the source. However, it is possible to re-scale the Boltzmann equation in this outer region, and the resulting moment equations form a closed set.

To see this we remark that the conservation of mass implies that the radial velocity v_r is constant to order $O(r^{-2})$ and hence $\rho = \rho_1(r_1/r)^2$, where the subscript 1 designates that a quantity is evaluated at a reference point upstream where the flow is supersonic but still isentropic. Bernoulli's theorem immediately gives that v_r tends to a constant V_∞ (the limiting speed) when the distance from the source tends to infinity.

The temperature is proportional to $\rho^{2/3}$ and hence to $r^{-4/3}$. It is thus natural to re-scale the Boltzmann equation by letting $\bar{\mathbf{x}} = \epsilon\mathbf{x}$, where ϵ is the mean free path (as in the Hilbert expansion), and, at the same time, letting the peculiar molecular velocity $\mathbf{c} = \mathbf{C}\epsilon^{2/3}$. Then, if we expand $F = \epsilon^{-2} f$ in powers of $\epsilon^{2/3}$, we find the behavior at large distances. To the lowest order the radial peculiar velocity does not appear explicitly in the equation for F_0. If we use as variables $\overline{r} = |\overline{\mathbf{x}}|$, $\overline{C}_\theta = \overline{r}C_\theta$, and $\overline{C}_\varphi = \overline{r}C_\varphi$ for the radial coordinate and the transverse components of the peculiar velocity, it turns out that F_0 formally satisfies the *space-homogeneous* Boltzmann equation with the time variable t replaced by $\overline{r}/V_\infty$. The variables in the collision term must, however, be modified accordingly (Problem 5.9.1).

The set of moment equations to be solved is obtained by multiplying by C_r^2 and $\overline{C}_\theta^2 + \overline{C}_\varphi^2$ and integrating. An equivalent set is the following (where we have omitted the bar over r) (Problem 5.9.2):

$$\frac{d}{dr}(r^2 p_\parallel) = \frac{\tilde{\nu}}{V_\infty} r^2(\rho RT - p_\parallel),$$

$$r\frac{d}{dr}(RT) = -\frac{2}{3}\left(3RT - \frac{p_\parallel}{\rho}\right). \tag{5.9.1}$$

Here $p_\parallel$ is the normal stress in the radial direction and $\tilde{\nu}$ is an average collision frequency.

Since $\tilde{\nu}$ is proportional to $\rho = \rho_1(r_1/r)^2$, Eqs. (5.9.1) can be rewritten as follows:

$$\frac{dT_\parallel}{dr} = \frac{\beta}{r^2}(T - T_\parallel),$$

$$r\frac{dT}{dr} = -\frac{2}{3}(3T - T_\parallel). \tag{5.9.2}$$

Here

$$T_\parallel = \frac{p_\parallel}{\rho R} \tag{5.9.3}$$

is the parallel temperature and

$$\beta = \frac{\rho_1 {r_1}^2}{V_\infty}\frac{\tilde{\nu}}{\rho} \tag{5.9.4}$$

is a constant for Maxwell molecules and can be assumed to be a power of T for inverse power potentials:

$$\beta = \beta_\infty\left(\frac{T}{T_\infty}\right)^\alpha, \tag{5.9.5}$$

where α is an exponent obviously related to the force exponent ($\alpha = 0$ for Maxwell's molecules).

The parallel temperature $T_\parallel$ can be eliminated between Eqs. (5.9.2) to yield

$$r^2\frac{d^2T}{dr^2} + (3r + \beta)\frac{dT}{dr} + \frac{4}{3r}\beta T = 0. \tag{5.9.6}$$

We assume an inverse power potential. Then (5.9.5) applies and we let

$$\theta = \frac{T}{T_\infty}, \qquad s = \frac{r}{\beta_\infty}. \tag{5.9.7}$$

Equation (5.9.6) can be rewritten as

$$s^2 \frac{d^2\theta}{ds^2} + (3s + \theta^\alpha)\frac{d\theta}{ds} + \frac{4}{3s}\theta^{1+\alpha} = 0. \tag{5.9.8}$$

If $\alpha = 0$ (Maxwell's molecules), Eq. (5.9.8) can be solved in terms of confluent hypergeometric functions and a unique solution can be obtained that satisfies the conditions $\theta(s) \to 1$ for $s \to \infty$ and $\theta = O(s^{-4/3})$ when $s \to 0$ (approach to the isentropic solution). Otherwise, Eq. (5.9.8) can be integrated numerically.[47,48] Some results for the case of Maxwell's molecules are given in Figs. 5.3 and 5.4, where plots of $T_\parallel$ and

$$T_\perp = \frac{3T - T_\parallel}{2} \tag{5.9.9}$$

are given.

The asymptotic behavior for θ can be obtained by letting

$$\theta = 1 + \frac{c}{s} + \cdots. \tag{5.9.10}$$

Inserting Eq. (5.9.10) into (5.9.8) and comparing the coefficients of the terms of order s^{-1} yields $c = 4/3$. We may now compute the behavior of the transverse

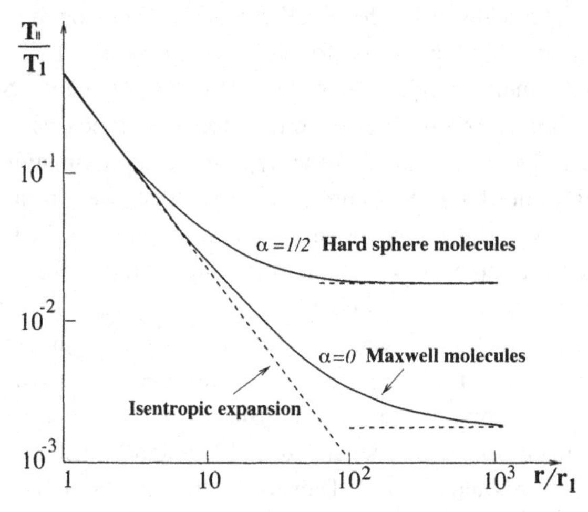

Figure 5.3. The results of Hamel and Willis for the parallel temperature in the expansion from a spherical source.

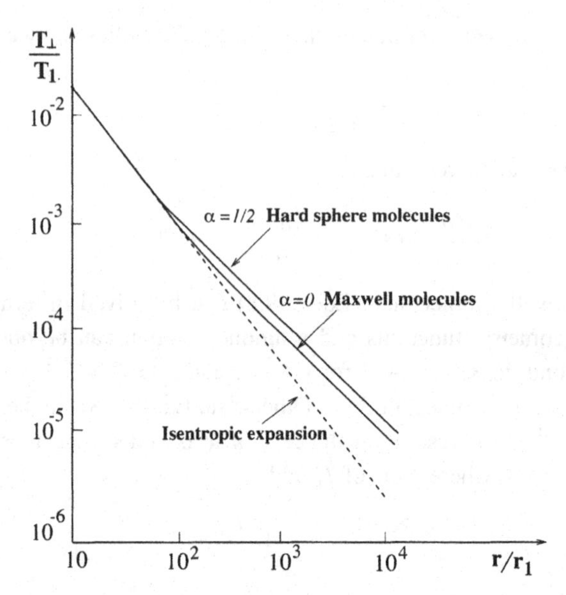

Figure 5.4. The results of Hamel and Willis for the transverse temperature in the expansion from a spherical source.

temperature defined in (5.9.9). Equation (5.9.2) shows that

$$\theta = \frac{T_\perp}{T_\infty} = -\frac{3}{4}s\frac{d\theta}{ds} = \frac{3}{4}\frac{c}{s} + \cdots = \frac{1}{s} + \cdots. \tag{5.9.11}$$

Hence for a spherical source, the parallel temperature is found to freeze; no such freezing is obtained, however, for a cylindrical source, in contrast with the previous, extremely simplified analysis of Brook and Oman.[52] No accurate prediction of the observed freezing phenomena for the spherical case is obtained from the Navier–Stokes equations in the hypersonic approximation, since the solution starting out along the isentrope at small distances from the source eventually violates the hypersonic assumption on which it is based.

When the kinetic theory descriptions of the translational freezing process were first published, they appeared to contradict molecular-beam findings concerning the distribution of random velocities perpendicular to the streamlines. Specifically, according to Eq. (5.9.11), it was predicted that the energy in these degrees of freedom eventually diminished as r^{-1}, whereas experiments seemed clearly to indicate an r^{-2} dependence.[53] This apparent contradiction was elucidated by Edwards and Cheng.[54] They found that, as $r \to \infty$ the distribution of random velocities perpendicular to the streamlines has a fairly narrow central spike with rather thick wings. The spike width decreases as r^{-1} and, since the temperature measurements are based on the spike width, it explained why they indicate a temperature decay proportional r^{-2}. However, experiments measuring the energy content of the wings, if possible, should indicate the r^{-1} behavior. A study of the fourth-order moments of the distribution function for Maxwell's molecules was performed by Freeman and Thomas.[55] Their results were in a qualitative agreement with the previous results of Edwards and Cheng.[54]

The same conclusions are indirectly confirmed by the results of Miller and Andres[56] who, in order to treat the intermolecular potential in a realistic fashion, assumed an ellipsoidal distribution function of the kind used by Holway to study the shock wave structure (see Chapter 4). This assumption, according to Willis, Hamel, and Lin,[57] is valid only for $T_{\|}/T_{\perp} \leq 2$. The resulting behavior of $T_{\perp}$ does not follow the r^{-1} law predicted by Eq. (5.9.11).

Problems

5.9.1 Prove that if we re-scale the Boltzmann equation by letting $\bar{\mathbf{x}} = \epsilon \mathbf{x}$, where ϵ is the mean free path, and $\mathbf{c} = \mathbf{C}\epsilon^{2/3}$ and expand $F = \epsilon^{-2} f$ in powers of $\epsilon^{2/3}$, to the lowest order the radial peculiar velocity does not appear explicitly in the equation for F_0 and, if we use as variables $\bar{r} = |\bar{\mathbf{x}}|$, $\overline{C}_\theta = \bar{r} C_\theta$, and $\overline{C}_\varphi = \bar{r} C_\varphi$ for the radial coordinate and the transverse components of the peculiar velocity, F_0 satisfies an equation similar to the space-homogeneous Boltzmann equation.

5.9.2 Obtain the moment equations (5.9.2).

5.9.3 Obtain Eq. (5.9.6) from the system (5.9.2).

5.9.4 Check that $c = 4/3$ in (5.9.10).

5.10. Concluding Remarks

The techniques discussed in this chapter have a wide applicability and can originate many interesting investigations. In particular, when put together with the study of the kinetic layers discussed in Chapter 3 and the shock waves discussed in Chapter 4, they shed light on many qualitative aspects of rarefied flows.

Other techniques of the same nature might have been discussed. We mention, in particular, the "method of boundary conditions," which seems to have some connection with the variational method but might be used for nonlinear problems as well. It consists of using the Navier–Stokes equations (for any Knudsen number!) with suitable boundary conditions that simulate the effect of Knudsen layers in a rather detailed way. By its nature the method is useful for computing overall quantities. Although some unnoticed numerical errors produced results that induced the originators of the method[58,59] to dismiss it, the subsequent contribution by Loyalka[60] seems to suggest that the method might have deserved further study. It was rediscovered by Kuščer and Miklavčič in 1980.[61]

We should also mention that our treatment is not meant to describe exhaustively the work that has been done on the various problems. Rarefied flows through tubes and nozzles for instance are described in about one hundred or so papers. Only some of the many other interesting problems have been mentioned here. It would be difficult to do justice to the various authors. Thus we prefer to refer to a survey[62] dealing with the research on this topic performed before 1976.

So far, our treatment has been mainly based on perturbation methods and approximate analytical techniques. Even when we used numerical methods, they were meant to illustrate or complete these techniques. It seems clear, however, that numerical techniques are required to obtain accurate results for the velocity, pressure, and temperature fields. Thus, after modifying the basic concepts and equations to allow us to cope with mixtures and polyatomic gases in the next chapter, we shall devote the entire Chapter 7 to a study of the basic ideas and results of numerical simulations.

References

[1] J. C. Maxwell, "On stresses in rarified gases arising from inequalities of temperature," *Phil. Trans. Roy. Soc.* **170**, 231–256 (1879).

[2] M. N. Kogan, V. S. Galkin, and O. G. Fridlender, "Stresses produced in gases by temperature and concentration inhomogeneities. New types of free convection," *Sov. Phys. Usp.* **19**, 420–428 (1976).

[3] Y. Sone, K. Aoki, H. Sugimoto, and A. Bobylev, "Inappropriateness of the heat-conduction equation for description of a temperature field of a stationary gas in the continuum limit: Examination by asymptotic analysis and numerical computation of the Boltzmann equation," *Phys. Fluids* **8**, 628–638 (1996).

[4] Y. Sone, "Asymptotic theory of a steady flow of a rarefied gas past bodies for small Knudsen numbers," in *Advances in Kinetic Theory and Continuum Mechanics*, R. Gatignol and Soubbaramayer, eds., 19–31, Springer-Verlag, Berlin (1991).

[5] C. Cercignani and F. Sernagiotto, "Cylindrical Poiseuille flow of a rarefied gas," *Phys. Fluids* **9**, 40–44 (1966).

[6] P. Bassanini, C. Cercignani, and F. Sernagiotto, "Flow of a rarefied gas in a tube of annular section," *Phys. Fluids* **9**, 1174–1178 (1966).

[7] C. Cercignani and C. D. Pagani, "Variational approach to rarefied flows in cylindrical and spherical geometry," in *Rarefied Gas Dynamics*, C. L. Brundin, ed., Vol. I, 555–573, Academic Press, New York (1967).

[8] C. Cercignani and G. Tironi, "Alcune applicazioni di un nuovo modello linearizzato dell'equazione di Boltzmann," *Atti del Congresso Nazionale AIDA-AIR 1965*, 174–181, AIDA-AIR (1967).

[9] C. Cercignani, "Reply to the comments by A. S. Berman," *Phys. Fluids* **10**, 1859–1860 (1967).

[10] C. Cercignani and F. Sernagiotto, "Cylindrical Couette flow of a rarefied gas," *Phys. Fluids* **10**, 1200–1204 (1967).

[11] P. Bassanini, C. Cercignani, and C. D. Pagani, "Influence of the accommodation coefficient on the heat transfer in a rarefied gas," *Int. J. Heat Mass Transfer* **11**, 1359–1369 (1968).

[12] C. Cercignani, "Stokes paradox in kinetic theory," *Phys. Fluids* **11**, 303–308 (1968).

[13] S. Ukai and K. Asano, "Steady solutions of the Boltzmann equation for a gas flow past an obstacle. I. Existence," *Archve Rational Mech. Anal.* **84**, 249–291 (1983).

[14] R. A. Millikan, "The general law of fall of a small spherical body through a gas, and its bearing upon the nature of molecular reflection from surfaces," *Phys. Rev.* **22**, 1–23 (1923).

[15] C. Cercignani, C. D. Pagani, and P. Bassanini, "Flow of a rarefied gas past an axisymmetric body. II. Case of a sphere," *Phys. Fluids* **11**, 1399–1403 (1968).

[16] F. S. Sherman, "A survey of experimental results and methods for the transition regime of rarefied gas dynamics," in *Rarefied Gas Dynamics*, J. A. Laurman, ed., Vol. II, 228–260, Academic Press, New York (1963).

[17] K. Takao, "Heat transfer from a sphere in a rarefied gas," in *Rarefied Gas Dynamics*, J. A. Laurmann, ed., Vol. II, 102–111, Academic Press, New York (1963).

[18] L. Kavanau, "Heat transfer from spheres to a rarefied gas in subsonic flow," *Trans. ASME* **77**, 617–623 (1955).

[19] P. Bassanini, C. Cercignani, and P. Schwendimann, "The problem of a cylinder rotating in a rarefied gas," in *Rarefied Gas Dynamics*, C. L. Brundin ed., Vol. I, 505–516, Academic Press, New York (1967).

[20] D. R. Willis, "Heat transfer and shear between coaxial cylinders for large Knudsen numbers," *Phys. Fluids* **8**, 1908–1910 (1965).

[21] M. Van Dyke, *Perturbation Methods in Fluid Mechanics*, Academic Press, New York (1964).

[22] D. Hilbert, "Begründung der kinetischen Gastheorie," *Mathematische Annalen* **72**, 562–577 (1912).

[23] D. Enskog, *Kinetische Theorie der Vorgänge in Mässig Verdünnten Gasen. I. Allgemeiner Teil*, Almqvist and Wiksell, Uppsala (1917).

[24] A. Majda, *Compressible Fluid Flow and Systems of Conservation Laws in Several Space Variables,* Springer-Verlag, Appl. Math. Ser. 53 (1984).

[25] A. De Masi, R. Esposito, and J. L. Lebowitz, "Incompressible Navier–Stokes equations and Euler limit of the Boltzmann equation," *Commun. Pure Appl. Math.* **42**, 1189–1214 (1989).

[26] C. Bardos and S. Ukai, "The classical incompressible Navier–Stokes limit of the Boltzmann equation," *Math. Models Methods Appl. Sci.* **1**, 235–257 (1991).

[27] C. Cercignani, *Mathematical Methods in Kinetic Theory*, Plenum Press, New York (1969; 1990, 2nd ed.).

[28] C. Cercignani, *The Boltzmann Equation and Its Applications*, Springer-Verlag, New York (1988).

[29] S. Chapman, and T. G. Cowling: *The Mathematical Theory of Non-Uniform Gases*, Cambridge Univ. Press, London (1940).

[30] J. O. Hirschfelder, C. F. Curtiss, and R. B. Bird, *Molecular Theory of Gases and Liquids*, Wiley, New York (1954).

[31] J. H. Ferziger and H. G. Kaper, *Mathematical Theory of Transport Processes in Gases*, North Holland, Amsterdam (1972).

[32] Y. Sone, "Asymptotic theory of flow of rarefied gas over a smooth boundary I," in *Rarefied Gas Dynamics*, D. Dini et al., ed., Vol. I, 243–253, Editrice Tecnico Scientifica, Pisa (1974).

[33] Y. Sone, "Flow induced by thermal stress in rarefied gas," *Phys. Fluids* **15** 1418–1423 (1972).

[34] T. Ohwada and Y. Sone, "Analysis of thermal stress slip flow and negative thermophoresis using the Boltzmann equation for hard-sphere molecules," *Euro. J. Mech. B* **11** 389–414 (1992).

[35] M. N. Kogan, V. S. Galkin, and O. G. Fridlender, "Stresses produced in gases by temperature and concentration inhomogeneities. New type of free convection," *Sov. Phys. Usp.* **19**, 420–438 (1976).

[36] Y. Sone, K. Aoki, S. Tanaka, H. Sugimoto, and A. V. Bobylev, "Inappropriateness of the heat-conduction equation for the description of a temperature field of a stationary gas in the continuum limit: Examination by asymptotic analysis and numerical computation of the Boltzmann equation," *Phys. Fluids A* **8**, 628–638 (1996).

[37] T. Ohwada, Y. Sone, and K. Aoki, "Numerical analysis of the Poiseuille and thermal transpiration flows between two parallel plates on the basis of the Boltzmann equation for hard-sphere molecules," *Phys. Fluids A*, **1** 2042–2049 (1989); "Erratum," *Phys. Fluids A*, **2** 639 (1990).

[38] S. Takata and Y. Sone, "Flow induced around a sphere with a nonuniform surface temperature in a rarefied gas, with application to the drag and thermal force problems of a spherical particle with an arbitrary thermal conductivity," *Euro. J. Mech. B* **14** 487–518 (1995).

[39] H. Schlichting, *Boundary Layer Theory*, McGraw-Hill, New York (1958).

[40] C. Cercignani and M. Lampis, "On the recovery factor in free molecular flow," *J. Appl. Math. Phys. (ZAMP)* **27**, 733–738 (1976).

[41] C. Cercignani and M. Lampis, "Influence of gas-surface interaction on drag and lift in free molecular flow," *Entropie* **44**, 40–46 (1972).

[42] C. Cercignani and M. Lampis, "Free molecular flow past a flat plate in the presence of a nontrivial gas–surface interaction," *J. Appl. Math. Phys. (ZAMP)* **23**, 713–728 (1972).

[43] A. L. Cooper and B. B. Hamel, "Asymptotic theory of the Boltzmann equation at large Knudsen numbers," *Phys. Fluids* **16**, 35–42 (1973).

[44] B. B. Hamel and A. L. Cooper, "Nature of the first iterate for nearly-free molecular flow," *Phys. Fluids* **16**, 43–44 (1973).

[45] C. Y. Liu, "Plane Poiseuille flow of a rarefied gas," *Phys. Fluids* **11**, 481–485 (1968).

[46] H. Ashkenas and F. S. Sherman, "The structure and utilization of supersonic free jets in low density wind tunnels," in *Rarefied Gas Dynamics*, J. H. de Leeuw, ed., Vol. 2, 84–105, Academic Press, New York (1965).

[47] B. B. Hamel and D. R. Willis, "Kinetic theory of source flow expansion with application to the free jet," *Phys. Fluids* **9**, 829–841 (1966).

[48] R. H. Edwards and H. K. Cheng, "Steady expansion of a gas ito a vacuum," *AIAA J.* **4**, 558–561 (1966).

[49] N. C. Freeman, "Solution of the Boltzmann equation for expanding flows," *AIAA J.* **5**, 1696–1698 (1967).

[50] N. C. Freeman and R. E. Grundy, "On the solution of the Boltzmann equation for an unsteady cylindrically symmetric expansion of a monatomic gas into a vacuum," *J. Fluid Mech.* **31**, 723–736 (1968).

[51] R. E. Grundy, "Axially symmetric expansion of a monatomic gas from an orifice into vacuum," *Phys. Fluids* **12**, 2011–2018 (1969).

[52] J. W. Brook and R. A. Oman, "Steady expansions at high speed ratio using the BGK model," in *Rarefied Gas Dynamics*, J. H. de Leeuw, ed., Vol. 1, 125–139, Academic Press, New York (1965).

[53] N. Abuaf, J. B. Anderson, R. D. Andres, J. B. Fenn, and D. R. Miller, "Studies in low density supersonic jets," in *Rarefied Gas Dynamics*, C. L. Brundin, ed., Vol. 2, 1317–1336, Academic Press, New York (1967).

[54] R. H. Edwards and H. K. Cheng, "Distribution function and temperature in a monatomic gas under steady expansion into a vacuum," in *Rarefied Gas Dynamics*, C. L. Brundin, ed., Vol. 1, 819–835, Academic Press, New York (1967).

[55] N. C. Freeman and D. R. Thomas, "On spherical expansions of monatomic gases into vacuum," in *Rarefied Gas Dynamics*, L. Trilling and H. Y. Wachman, Vol. 1, 163–171, Academic Press, New York (1969).

[56] D. R. Miller and R. P. Andres, "Translational relaxation in low density supersonic jets," in *Rarefied Gas Dynamics*, L. Trilling and H. Y. Wachman, Vol. 2, 1385–1402, Academic Press, New York (1969).

[57] D. R. Willis, B. B. Hamel, and J. T. Lin, University of Calif. Berkeley Report No. AS-70-8 (1970).

[58] C. Cercignani and G. Tironi, "New boundary conditions in the transition regime," *J. Plasma Phys.* **2**, 293–310 (1968).

[59] C. Cercignani and G. Tironi, "Some applications to the transition regime of a new set of boundary conditions for Navier–Stokes equations," in *Rarefied Gas Dynamics*, L. Trilling and H. Y. Wachman, Vol. 1, 281–290, Academic Press, New York (1969).

[60] S. K. Loyalka, "On boundary conditions method in the kinetic theory of gases," *Z. Naturforsch.* **26a**, 1708–1712 (1971).

[61] I. Kuščer and M. Miklavčič, "Generalized Maxwell method for solving kinetic boundary value problems," in *Mathematical Problems in the Kinetic Theory of Gases*, D. C. Pack and H. Neunzert, eds., 113–128, P. Lang, Frankfurt (1980).

[62] R. H. Edwards, "Low-density flows through tubes and nozzles," in *Rarefied Gas Dynamics*, J. L. Potter, ed., Part I, 199–223, AIAA, New York (1977).

6

Polyatomic Gases, Mixtures, Chemistry, and Radiation

6.1. Introduction

As is well known, air at room pressure and temperature is a mixture, its main components being two diatomic gases, nitrogen and oxygen. This immediately calls for a change in our basic equation, Eq. (1.4.1) of Chapter 1, which is only suitable for a single monatomic gas.

A remarkable feature of aerodynamics at the molecular level is that the evolution equations change in a significant way when involving polyatomic rather than monatomic gases. This is not the case when the gas is treated as a continuum, where, at least at room conditions, only a few changes in the equations occur, the most remarkable being the change of the ratio of specific heats γ. Incidentally, the value of γ has caused some problems from the very beginning of kinetic theory. In fact, the value for a monatomic gas is easily found to be $5/3$ (the first calculation goes back to Clausius). Now, at that time, no monatomic gas was known, except mercury vapor. Indeed the data for this gas existed thanks to the experiments of Kundt and Warburg,[1] but these data were old and soon forgotten. Maxwell assumed the molecules to be solid bodies different from perfectly smooth hard spheres and found the value $4/3$. Neither value agreed with the ratio for the most common gases, such as nitrogen, oxygen, hydrogen, etc. This caused a difficulty, until Boltzmann pointed out that a diatomic molecule, modeled as two point masses at a fixed distance or, more generally, as a solid body with rotational symmetry, and hence having all the mass placed symmetrically with respect to an axis, had only five degrees of freedom, which led to the value $\gamma = 7/5$, in agreement with the experimental data. Later Rayleigh and Ramsay discovered the first rare gases and the value $5/3$ was found to apply to the ratio of their specific heats. More problems arise when vibrational degrees of freedom are considered, which certainly must be the case at high temperatures.

We shall first deal with the case of a mixture of monatomic gases, and then with the case of a single monatomic gas; subsequently we shall consider polyatomic gases and other complications, that occur at higher temperatures (chemical reactions, ionization, radiation). The case of a mixture of polyatomic gases easily follows by combining the theory for a polyatomic gas with the discussion of mixtures of monatomic gases.

The subject matter of this chapter is clearly a very important area for Monte Carlo simulation and development of new models.

Problem

6.1.1 Compute the ratio of specific heats $\gamma = c_p/c_v$ for a molecule of a perfect gas with n degrees of freedom ($c_v = e_{\text{thm}}/T$, where e_{thm} is the thermal energy, and $c_p = c_v + R$ by a well-known thermodynamic relation.)

6.2. Mixtures

If we consider a mixture of monatomic gases, the differences between the various species occurs in the values of the masses and in the law of interaction between molecules of different species; in the simplest case, when the molecules are pictured as hard spheres, the second difference is represented by unequal values of the molecular diameters. In the mathematical treatment, a first difference will be in the fact that we shall need n distribution functions f_i $(i = 1, 2, \dots, n)$ if there are n species. The notation becomes complicated, but there is no new idea, except, of course, for the fact that we must derive a system of n coupled Boltzmann equations for the n distribution functions. The arguments are exactly the same, with obvious changes, and the result is

$$\frac{\partial f_i}{\partial t} + \boldsymbol{\xi} \cdot \frac{\partial f_i}{\partial \mathbf{x}} + \mathbf{X}_i \cdot \frac{\partial f_i}{\partial \boldsymbol{\xi}}$$
$$= \sum_{k=1}^{n} \int_{\mathbf{R}^3} \int_{\mathcal{B}_+} (f_i' f_{k*}' - f_i f_{k*}) \mathcal{B}_{ik}(\mathbf{n} \cdot \mathbf{V}, |\mathbf{V}|) d\boldsymbol{\xi}_* d\mathbf{n}, \qquad (6.2.1)$$

where $\mathcal{B}_{ik}$ is computed from the interaction law between the ith and kth species, while in the kth term in the left-hand side, $\mathbf{V} = \boldsymbol{\xi} - \boldsymbol{\xi}_*$ is the relative velocity of the molecule of the ith species (whose evolution we are following) with respect to a molecule of the kth species (against which the former is colliding). The arguments $\boldsymbol{\xi}'$ and $\boldsymbol{\xi}_*'$ are computed, as before, from the laws of conservation of mass and energy in a collision with the following result:

$$\xi' = \xi - \frac{2\mu_{ik}}{m_i}\mathbf{n}[(\xi - \xi_*) \cdot \mathbf{n}],$$
$$\xi'_* = \xi_* + \frac{2\mu_{ik}}{m_k}\mathbf{n}[(\xi - \xi_*) \cdot \mathbf{n}], \quad (6.2.2)$$

where $\mu_{ik} = m_i m_k/(m_i + m_k)$ is the reduced mass.[2]

To prepare some material for the description of polyatomic gases and chemical reactions, we remark that Eq. (6.2.1) can be rewritten as follows:[3]

$$\frac{\partial f_i}{\partial t} + \xi \cdot \frac{\partial f_i}{\partial \mathbf{x}} + \mathbf{X}_i \cdot \frac{\partial f_i}{\partial \xi}$$
$$= \sum_{k=1}^{n} \int_{\mathbf{R}^3 \times \mathbf{R}^3 \times \mathbf{R}^3} (f_i' f_{k*}' - f_i f_{k*}) W_{ik}(\xi, \xi_* | \xi', \xi_*') d\xi_* d\xi' d\xi_*'. \quad (6.2.3)$$

Now ξ', ξ'_*, ξ, and ξ_* are independent variables (i.e., they are not related by the conservation laws) and

$$W_{ik}(\xi, \xi_* \,|\, \xi'\xi'_*) = S_{ik}(\mathbf{n} \cdot \mathbf{V}, |\mathbf{V}|)\delta(m_i\xi_* + m_k\xi_* - m_i\xi' - m_k\xi'_*)$$
$$\times\, \delta(m_i|\xi_*|^2 + m_k|\xi_*|^2 - m_i|\xi'|^2 - m_k|\xi'_*|^2), \quad (6.2.4)$$

where $\mathbf{n} = (\xi - \xi')/|\xi - \xi'|$ and

$$S_{ik}(\mathbf{n} \cdot \mathbf{V}, |\mathbf{V}|) = \frac{\mathcal{B}_{ik}(\mathbf{n} \cdot \mathbf{V}, |\mathbf{V}|)}{2\mathbf{n} \cdot \mathbf{V}}(m_i + m_k)^3 m_i m_k. \quad (6.2.5)$$

Conservation of momentum and energy is now taken care of by the delta functions appearing in Eq. (6.2.4)[3] (Problem 6.2.1).

With a slight modification, Eq. (6.2.3) can be extended to the case of a mixture in which a collision can transform the two colliding molecules of species j, l into two molecules of different species k, i (a very particular kind of chemical reaction). In this case the relations between the velocities before and after the encounter are different from the ones used so far, but we may still write a set of equations for the n species:

$$\frac{\partial f_i}{\partial t} + \xi \cdot \frac{\partial f_i}{\partial \mathbf{x}} + \mathbf{X}_i \cdot \frac{\partial f_i}{\partial \xi}$$
$$= \sum_{k,l,j=1}^{n} \int_{\mathbf{R}^3 \times \mathbf{R}^3 \times \mathbf{R}^3} (f_l' f_{j*}' - f_i f_{k*}) W_{ik}^{lj}(\xi, \xi_* \,|\, \xi', \xi'_*) d\xi_* d\xi' d\xi'_*. \quad (6.2.6)$$

where W_{ik}^{lj} gives the probability density that a transition from velocities $\boldsymbol{\xi}', \boldsymbol{\xi}'_*$ to velocities $\boldsymbol{\xi}, \boldsymbol{\xi}_*$ takes place in a collision that transforms two molecules of species l, j, respectively, into two molecules of species i, k, respectively. It is clear how the previous model is included into the new one when the species change does not occur.

The idea of kinetic models analogous to the BGK model can be naturally extended to mixtures and polyatomic gases.[4,5,6] A typical collision term of the BGK type will read

$$J_i(f_r) = \sum_{j=1}^{n} J_{ij}(f_r) = \sum_{j=1}^{n} \nu_{ij}[\Phi_{ij}(\boldsymbol{\xi}) - f_i(\boldsymbol{\xi})], \tag{6.2.7}$$

where ν_{ij} are the collision frequencies and Φ_{ij} is a Maxwellian distribution to be determined by suitable conditions that generalize Eq. (1.9.4).

There are some important changes concerning the collision invariants and the definition of macroscopic functions in the case of mixtures. First, the collision invariants in the case of n species have n components and are defined as follows: ψ_i $(i = 1, \ldots, n)$ is a collision invariant if and only if

$$\sum_i \int_{\mathbf{R}^3} \psi_i Q_i d\boldsymbol{\xi} = 0, \tag{6.2.8}$$

where Q_i denotes the right-hand side of Eq. (6.2.1).

There are $n + 4$ rather than 5 linearly independent collision invariants. There are three invariants related to momentum conservation, $\psi_{(n+\alpha-1)i} = m_i \xi_\alpha$ $(\alpha = 1, 2, 3)$ and one related to energy conservation, $\psi_{(n+3)i} = m_i|\boldsymbol{\xi}|^2$; the remaining n invariants are related to the conservation of the number of particles of each species $\psi_{ij} = \delta_{ij}$ $(i, j = 1, \ldots, n)$. This, of course, applies when there are no chemical reactions.

In the case of mixtures it is more convenient to normalize the distribution function as a number density (this has been already taken into account when giving the expression of the collision invariants). Then the number densities of the single species are given by

$$n^{(i)} = \int f_i d\boldsymbol{\xi} \quad (i = 1, \ldots, n) \tag{6.2.9}$$

and the mass density $\rho^{(i)}$ is given by $m_i n^{(i)}$. The number and mass densities for the mixture are given by

$$n = \sum_{i=1}^{n} n^{(i)} \tag{6.2.10}$$

and

$$\rho = \sum_{i=1}^{n} \rho^{(i)}. \tag{6.2.11}$$

It is convenient to define the bulk velocities of the single species and the bulk velocity of the mixture as follows:

$$\mathbf{v}^{(i)} = \frac{\int_{R^3} \boldsymbol{\xi} f_i d\boldsymbol{\xi}}{\int_{R^3} f_i d\boldsymbol{\xi}} \quad (i = 1, \ldots, n), \tag{6.2.12}$$

$$\rho\mathbf{v} = \sum_{i=1}^{n} \rho^{(i)}\mathbf{v}^{(i)} \tag{6.2.13}$$

It is usual to define the peculiar velocity

$$\mathbf{c} = \boldsymbol{\xi} - \mathbf{v}. \tag{6.2.14}$$

The stress tensor for the ith species is given by

$$p_{jk}^{(i)} = m_i \int_{R_3} \xi_j \xi_k f_i d\boldsymbol{\xi}; \qquad (i = 1, \ldots, n; \; j, k = 1, 2, 3), \tag{6.2.15}$$

and the stress tensor for the mixture is the sum of the various stresses:

$$p_{jk} = \sum_{i=1}^{n} p_{jk}^{(i)} \quad (j, k = 1, 2, 3). \tag{6.2.16}$$

Although these definitions are the most common and natural, they are not used by all authors. One might, for example, define a peculiar velocity for each species and use it to define the partial stresses. Then it is no longer true that the stress tensor for the mixture is the sum of the partial stresses.

Similarly, the thermal energy per unit mass (associated with random motions) is defined for each species by

$$n^{(i)}e^{(i)} = \frac{1}{2} \int_{R^3} |\mathbf{c}|^2 f_i d\boldsymbol{\xi} \quad (i = 1, \ldots, n) \tag{6.2.17}$$

and for the mixture by

$$\rho e = \sum_{i=1}^{n} \rho^{(i)} e^{(i)}. \tag{6.2.18}$$

A similar procedure can be applied to the heat flow.

The pressure is, as usual, $1/3$ of the trace of the stress matrix and is related to the temperature by $p = nk_BT$. Please note that *there is not* a constant R such that $p = \rho RT$.

Problem

6.2.1 Show that conservation of momentum and energy follow from Eq. (6.2.4) because of the delta functions contained in the kernel. (See Ref. 3.)

6.3. Polyatomic Gases

A possible picture of a molecule of a polyatomic gas, suggested by quantum mechanics, is as follows:[7] The molecule is a mechanical system, which differs from a point mass by having a sequence of internal states, which can be identified by a label, assuming integral values. In the simplest cases these states differ from each other because the molecule has, besides kinetic energy, an internal energy taking different values E_i in each of the different states. A collision between two molecules, besides changing the velocities, can also change the internal states of the molecules and, as a consequence, the internal energy enters in the energy balance. From the viewpoint of writing evolution equations for the statistical behavior of the system, it is convenient to think of a single polyatomic gas as a mixture of different monatomic gases. Each of these gases is formed by the molecules corresponding to a given internal energy, and a collision changing the internal state of at least one molecule is considered as a reactive collision of the kind considered above. $W_{ik}^{lj}(\boldsymbol{\xi}, \boldsymbol{\xi}_* \mid \boldsymbol{\xi}', \boldsymbol{\xi}'_*)$ gives the probability density of a collision transforming two molecules with internal states l, j, respectively, and velocities $\boldsymbol{\xi}'$, $\boldsymbol{\xi}'_*$, respectively, into molecules with internal states i, k, respectively, and velocities $\boldsymbol{\xi}$, $\boldsymbol{\xi}_*$, respectively.

This model is amply sufficient to discuss aerodynamic applications. We want to mention, however, that it requires nondegenerate levels of internal energy, if there are, for example, strong magnetic fields that can act on the internal variables such as (typically) the spin of the molecules. In that case, if the molecule has spin s, the distribution function f becomes a square matrix of order $2s + 1$ and the kinetic equation reflects the fact that matrices in general do not commute and, as remarked by Waldmann[8,9] and Snider,[10] the collision term contains not simply the cross section but the scattering amplitude, which may not commute with f.

It is appropriate now to enquire why we started talking about quantum rather than classical mechanics. The main reason is not related to practice, but rather to history. Classical models of polyatomic molecules have been regarded with suspicion since 1887 when Lorentz found a mistake in the proof of the H-theorem of Boltzmann[11] for general polyatomic molecules. The question arises from the fact that when one proves the H-theorem for a monatomic gas one usually does not explicitly underline (because it is irrelevant in that case) that the velocities $\boldsymbol{\xi}'$ and $\boldsymbol{\xi}'_*$ are *not* the velocities into which a collision transforms the velocities $\boldsymbol{\xi}$ and $\boldsymbol{\xi}_*$ but are the velocities that are transformed by a collision in the latter ones; this is conceptually very important, but the lack of a detailed discussion does not lead to any inconvenience because the expressions for $\boldsymbol{\xi}'_*$ and $\boldsymbol{\xi}'$ are invariant with respect to a change of sign of the unit vector $\mathbf{n}$, which permits an equivalence between velocity pairs that are carried into the pair $\boldsymbol{\xi}_*$, $\boldsymbol{\xi}$ and those which originate from the latter pair, as a consequence of a collision. The remarkable circumstance that we have just recalled is related to the particular symmetry of a collision described by a central force, which allows us to associate to a collision $[\boldsymbol{\xi}_*, \boldsymbol{\xi}] \to [\boldsymbol{\xi}'_*, \boldsymbol{\xi}']$ another collision, the "inverse collision" $[\boldsymbol{\xi}'_*, \boldsymbol{\xi}'] \to [\boldsymbol{\xi}_*, \boldsymbol{\xi}]$, which differs from the former just because of the transformation of the unit vector $\mathbf{n}$ into $-\mathbf{n}$. When polyatomic molecules are considered, the states before and after a collision require more than just the velocities of the mass centers to be described (the angular velocity, for example, if the molecule is pictured as a solid body). Let us symbolically denote by $[A, B]$ the state of the pair of molecules. Then there is no guarantee that one can correlate an "inverse collision" $[A', B'] \to [A, B]$, differing from the previous one just because of the change of $\mathbf{n}$ into $-\mathbf{n}$, with the collision $[A, B] \to [A', B']$. Now in the original proof of the H-theorem for polyatomic molecules proposed by Boltzmann,[11] the assumption was implicitly made that there is always such a collision. Lorentz remarked[12–14] that this is not true in general. Boltzmann recognized his blunder and proposed another proof based on the "closed cycles of collisions";[12–15] the initial state $[A, B]$ is reached not through a single collision but through a sequence of collisions. This proof, although called unobjectionable by Lorentz[13] and Boltzmann,[14] never satisfied anybody.[17,9]

The matter was forgotten until a quantum mechanical proof showed that the required property followed from the unitarity of the S matrix.[9] A satisfactory proof of the inequality required to prove the H-theorem for a purely classical, but completely general, model was given only in 1981[17] and will be discussed in the next section.

For aerodynamic applications none of these aspects are particularly relevant and, in fact, the main problem is to find a sufficiently handy model for practical calculations.

Lordi and Mates[18] studied the "two centers of repulsion" model and found that a rather complicated numerical solution was required for a given set of impact parameters. The lack of closed-form expressions makes the model impractical for applications, where the numerical solution describing the collision should be repeated many millions of times. Curtiss and Muckenfuss[19] developed the collision mechanics of the so-called sphero-cylinder model, consisting of a smooth elastic cylinder with two hemispherical ends. Whether or not two molecules collide depends on more than one parameter; in addition, there are several "chattering" collisions in a single collision event. The loaded-sphere model had been already developed by Jeans[20] in 1904 and was subsequently developed by Dahler and Sather[21] and Sandler and Dahler.[22] Although it is spherical in geometry, the molecules rotate about the center of mass, which does not coincide with the center of the sphere, with the consequence that it has essentially the same disadvantages as the sphero-cylinder model.

The only exact model that is amenable to explicit calculations is the perfectly rough sphere model, first suggested by Bryan[23] in 1894. The name results from the fact that the relative velocity at the point of contact of the two molecules is reversed by the collision. This model has some obvious disadvantages. First, a glancing collision may result in a large deflection; second, all collisions can produce a large interchange of rotational and translational energy, with the consequence that the relaxation time for rotational energy is unrealistically short; third, the number of internal degrees of freedom is three, rather than two, which makes the model inappropriate for a description of the main components of air, which are diatomic gases. One can disregard the first disadvantage and remedy the second by assuming that a fraction of collisions follow the smooth-sphere rather that rough-sphere dynamics; but there is obviously no escape from the third difficulty.

In practical calculations, one has learned, over the years, that one must compromise between the faithful adherence to a microscopic model and the computational time required to solve a concrete problem. This was true in the early days of rarefied gas dynamics (and may still be true nowadays when one tries to find approximate closed-form solutions or spare computer time) even for monatomic gases, as we discussed in Chapter 1 (in connection with the BGK model) and shall discuss in the next chapter (in connection with Direct Simulation Monte Carlo). As a matter of fact, when trying to solve the Boltzmann equation, one of the major shortcomings is the complicated structure of the collision term; even in the simplest case, one adds the complication of the presence of internal degrees of freedom, and so any practical problem becomes intractable, unless one is ready to accept the aforementioned compromise. Fortunately, when one is not interested in fine details, it is possible to obtain reasonable results by replacing the collision integral by a phenomenological collision model, (e.g., a simpler expression that retains only the qualitative and average properties of the true collision term). As computers become more and more powerful, the amount of phenomenological simplification diminishes and the calculations may more closely mimic the microscopic models.

For polyatomic gases, the basic new fact with respect to the monatomic ones is that the total energy is redistributed between translational and internal degrees of freedom at each collision. Those collisions for which this redistribution is negligible are called elastic, while the others are called inelastic. The simplest approach would be to calculate the effect of collisions as a linear combination of totally elastic and completely inelastic collisions, the second contribution being described by a model analogous to the BGK model briefly described in Chapter 1.

There are, of course, problems related to the molecule spins and their alignment; these are particularly important when we put the molecules in a magnetic field and peculiar phenomena, which go under the general name of Senftleben–Beenakker effects, arise. There is an entire book devoted to this topic[24] and we shall not deal with these problems, except for the conceptually important aspect of the H-theorem, to be discussed in the next section, and the curious Scott effect,[25] briefly discussed in Section 6.7.

Let us now consider in more detail the case of a continuous internal-energy variable. In this case, it is convenient to take the unit vector $\mathbf{n}$ in the center-of-mass system and use the internal energy E_i and E_{i*} of the colliding molecules. As usual, the values before a collision will be denoted by a prime. Equation (6.1.2) is replaced by

$$Q(f,f) = \int_{\mathbf{R}^3} d\boldsymbol{\xi}_* \int_0^\infty E_{i*}^{\frac{n-2}{2}}\, dE_{i*} \int_{\mathbf{S}^2} \int_0^E (E_i')^{\frac{n-2}{2}}\, dE_i' \times \int_0^{E-E_i'} (E_{i*}')^{\frac{n-2}{2}}\, dE_{i*}' (f'f_*' - ff_*)\mathcal{B}(E; \mathbf{n}\cdot\mathbf{n}'; E_i', E_{i*}' \to E_i, E_{i*}). \tag{6.3.1}$$

Here $E = m|\boldsymbol{\xi}|^2/4 + E_i + E_{i*}$ is the total energy in the center-of-mass system, which is conserved in a collision. The kernel $\mathcal{B}$ satisfies the reciprocity relation

$$|\mathbf{V}|\mathcal{B}(E; \mathbf{n}\cdot\mathbf{n}'; E_i', E_{i*}' \to E_i, E_{i*}) = |\mathbf{V}'|\mathcal{B}(E; \mathbf{n}\cdot\mathbf{n}'; E_i, E_{i*} \to E_i', E_{i*}'). \tag{6.3.2}$$

Here we follow a paper by Kuščer[25] and look for a one-parameter family of models, assuming that the scattering is isotropic in the center-of-mass system. The second crucial assumption will be that the redistribution of energy among the various degrees of freedom only depends upon the ratios of the various energies to the total energy E, $\epsilon_i = E_i/E$, etc. This assumption is valid for collisions of rigid elastic bodies and can be considered as a good approximation for steep repulsive potentials. It is then possible to write $\mathcal{B}$ in the following form:

$$|\mathbf{V}|\mathcal{B}(E;\mathbf{n}\cdot\mathbf{n}';E_i',E_{i*}'\to E_i,E_{i*}) = \frac{|\mathbf{V}'|^2}{|\mathbf{V}|}\frac{\sigma_{tot}(E)}{4\pi E^n}\theta(\epsilon_i',\epsilon_{i*}'\to\epsilon_i,\epsilon_{i*};\tau). \tag{6.3.3}$$

The denominator on the right takes care of normalization. Then the function θ satisfies the following relations:

$$\int_0^1 \epsilon^{\frac{n-2}{2}}d\epsilon\int_0^{1-\epsilon}\epsilon_*^{\frac{n-2}{2}}d\epsilon_*\theta(\epsilon_i',\epsilon_{i*}'\to\epsilon_i,\epsilon_{i*};\tau)=1, \tag{6.3.4}$$

$$(1-\epsilon_i'-\epsilon_{i*}')\theta(\epsilon_i',\epsilon_{i*}'\to\epsilon_i,\epsilon_{i*};\tau) = (1-\epsilon_i-\epsilon_{i*})\theta(\epsilon_i,\epsilon_{i*}\to\epsilon_i',\epsilon_{i*}';\tau). \tag{6.3.5}$$

The dependence of σ_{tot} on E makes it possible to adjust the model to the correct dependence of the viscosity on temperature. The parameter τ will be chosen in such a way as to represent the degree of inelasticity of the collisions. $\tau = 0$ will correspond to elastic collisions:

$$\theta(\epsilon_i',\epsilon_{i*}'\to\epsilon_i,\epsilon_{i*};0) = \epsilon^{-\frac{n-2}{2}}\epsilon_*^{-\frac{n-2}{2}} = \delta(\epsilon_i-\epsilon_i')\delta(\epsilon_i'-\epsilon_{i*}'). \tag{6.3.6}$$

$\tau = \infty$ will correspond to maximally inelastic collisions:

$$\theta\left(\epsilon_i',\epsilon_{i*}'\to\epsilon_i,\epsilon_{i*};\infty\right) = \epsilon^{-\frac{n-2}{2}}\epsilon_*^{-\frac{n-2}{2}}\frac{\Gamma(n+2)}{(\Gamma(n))^2}(1-\epsilon_i-\epsilon_{*i}), \tag{6.3.7}$$

where $\Gamma(n)$ denotes the Gamma function, which coincides with $(n-1)!$ if n is an integer.

A mixture of the two extreme cases gives the model first proposed by Borgnakke and Larsen[27] in 1975:

$$\begin{aligned}\theta(\epsilon_i',\epsilon_{i*}'\to\epsilon_i,\epsilon_{i*};\tau) &= e^{-\tau}\theta(\epsilon_i',\epsilon_{i*}'\to\epsilon_i,\epsilon_{i*};0)\\ &\quad+(1-e^{-\tau})\theta(\epsilon_i',\epsilon_{i*}'\to\epsilon_i,\epsilon_{i*};\infty).\end{aligned} \tag{6.3.8}$$

Kuščer[25] notices an analogy between this model and Maxwell's model for gas–surface interaction, as discussed in Chapter 1, and introduces another model, called the theta model, which would correspond to the Cercignani–Lampis model in this analogy.

The Larsen–Borgnakke model has become a customary tool in numerical simulations of polyatomic gases. It can be also applied to the vibrational modes either through a classical procedure that assigns a continuously distributed vibrational energy to each molecule or through a quantum approach that assigns discrete vibrational levels to each molecule. It would be out of place here to discuss this point in more detail, for which we refer to the book of Bird.[28] We also refrain from discussing the interesting recent developments[29,30] of an old idea of Boltzmann[31,32] to interpret, in the frame of classical statistical mechanics, the circumstance that at low temperatures the internal degrees of freedom appear to be frozen, as due to the extremely long relaxation times for the energy transfer process.

The Larsen–Borgnakke model suffers from the limitation that it considers all the collisions as a mixture of the elastic or completely inelastic collisions, disregarding the possibility of a partially inelastic one. In order to construct a more general model we use a procedure already indicated in Chapter 1 to deal with models for boundary conditions. We start from a sensible approximate kernel $\theta_0(\epsilon_i', \epsilon_{i1}' \to \epsilon_i, \epsilon_{i1})$, which is chosen on the basis of intuition but does not satisfy the basic properties (6.3.4) and (6.3.5). At this point we add to it some other terms, which ensure that the three fundamental properties are satisfied, as follows:

$$\theta(\epsilon_i', \epsilon_{i1}' \to \epsilon_i, \epsilon_{i1}) = \theta_0(\epsilon_i', \epsilon_{i1}' \to \epsilon_i, \epsilon_{i1}) + \frac{\Gamma(n+2)}{(\Gamma(n))^2}(1 - \epsilon_i - \epsilon_{i1}) \times (1 - H(\epsilon_i, \epsilon_{i1}))\,(1 - H(\epsilon_i', \epsilon_{i1}'))/I, \tag{6.3.9}$$

$$H(\epsilon_i', \epsilon_{i1}') = \int_0^1 \int_0^{1-\epsilon_i} \theta_0(\epsilon_i', \epsilon_{i1}' \to \epsilon_i, \epsilon_{i1}) \epsilon_i^{\frac{n-2}{2}} \epsilon_{i1}^{\frac{n-2}{2}} \, d\epsilon_i d\epsilon_{i1}, \tag{6.3.10$_1$}$$

$$I = 1 - \frac{\Gamma(n+2)}{(\Gamma(n))^2} \int_0^1 \int_0^{1-\epsilon_i} (1 - \epsilon_i - \epsilon_{i1}) H(\epsilon_i, \epsilon_{i1}) \epsilon_i^{\frac{n-2}{2}} \epsilon_{i1}^{\frac{n-2}{2}} \, d\epsilon_i d\epsilon_{i1}. \tag{6.3.10$_2$}$$

As in the aforementioned case, Eq. (6.3.9) may be interpreted as the linear combination of two normalized kernels $\theta_1(\ldots)$ and $\theta_2(\ldots)$, while $H(\epsilon_i', \epsilon_{i1}')$ is a sort of accommodation coefficient depending on the energies of the impinging molecules ($H(\epsilon_i', \epsilon_{i1}')$ must lay in the interval [0,1]); we write

$$\begin{aligned}
\theta(\ldots) &= H(\epsilon_i', \epsilon_{i1}')\theta_1(\ldots) + (1 - H(\epsilon_i', \epsilon_{i1}'))\theta_2(\ldots) \\
\theta_1(\ldots) &= \theta_0(\ldots)/H(\epsilon_i', \epsilon_{i1}') \\
\theta_2(\ldots) &= \frac{\Gamma(n+2)}{(\Gamma(n))^2}(1 - \epsilon_i - \epsilon_{i1})\,(1 - H(\epsilon_i, \epsilon_{i1}))/I.
\end{aligned} \tag{6.3.11}$$

The expression of $\theta_2(\ldots)$ is suggested by the requirement that the second term in the right-hand side, $(1 - H(\epsilon_i', \epsilon_{i1}'))\theta_2(\ldots)$, must satisfy reciprocity; the definition of I is chosen to get the normalization of $\theta_2(\ldots)$.

Cercignani and Lampis[33] proposed the following kernel θ_0 containing two parameters a and b:

$$\theta_0(\epsilon_i', \epsilon_{i1}' \to \epsilon_i, \epsilon_{i1}; a, b) = \frac{ba}{\pi} \epsilon_i^{-\frac{n-2}{4}} \epsilon_{i1}^{-\frac{n-2}{4}} \epsilon_i'^{-\frac{n-2}{4}} \epsilon_{i1}'^{-\frac{n-2}{4}} \frac{(1-\epsilon_i-\epsilon_{i1})^{\frac{1}{2}}}{(1-\epsilon_i'-\epsilon_{i1}')^{\frac{1}{2}}}$$
$$\times \exp[-a(\epsilon_i - \epsilon_i')^2 - a(\epsilon_{i1} - \epsilon_{i1}')^2]. \qquad (6.3.12)$$

To avoid a singularity in the expression of the kernel (6.3.12) and therefore also in that of $H(\epsilon_i', \epsilon_{i1}')$, it is assumed that $E - E_i' - E_{i1}' = mc_r'^2/4 > \tau$, where τ is a small positive parameter: Then $\theta_0(\ldots)$ and $H(\epsilon_i', \epsilon_{i1}')$ are limited. This approximate kernel $\theta_0(\ldots)$ for $a \to \infty$ tends to $b\theta_{\text{el}}(\epsilon_i', \epsilon_{i1}', \epsilon_i, \epsilon_{i1})$, so that the full kernel given by Eq. (3.6.10) tends to the Larsen–Borgnakke model. For finite values of a, $\theta_0(\ldots)$ describes a collision in which ϵ_i is not exactly equal to ϵ_i' (and similarly for ϵ_{i1}, ϵ_{i1}') but may be different, according to a Gaussian distribution.

By fitting the theoretical expression for viscosity with some experimental data (from standard handbooks), the aforementioned authors[33] have obtained the values of the parameters in the case of N_2 and O_2.

The aforementioned authors, in collaboration with J. Struckmeyer[34] found that a good fitting of some experimental data can be obtained. However, they were unable to obtain a unique determination of the parameters a and b. Some examples of possible choices of a and b have been given in Ref. 32, in the case of N_2 and O_2. In those examples, low values of a (for instance $a = 1, a = 0.1$), and for each of these a value of b close to its maximum value were chosen, but it is also possible to choose much higher values of a.

To obtain more information about the values of the parameters a and b, an obvious way would be to try to fit a second transport coefficient. Unfortunately, the experimental data for η_v versus temperature are very scanty.[35,36] The bulk viscosity of gases can only be measured by the attenuation and dispersion of an ultrasonic, acoustic signal. Moreover, it is a difficult method subject to experimental error.[36] Because of this, it is not possible to draw conclusions on the range of applicability of the model. In the application of the kernel to the calculation of transport coefficients, everything does work also without introducing the cutoff. The situation may be different in other problems, for instance in the application of DSMC. Therefore a similar model that does not present singularities and does not require a cutoff was introduced,[34] based on the following kernel:

$$\theta_0(\epsilon_i', \epsilon_{i1}' \to \epsilon_i, \epsilon_{i1}; a, b)$$
$$= \frac{2ba}{\pi} \epsilon_i^{-\frac{n-2}{4}} \epsilon_{i1}^{-\frac{n-2}{4}} \epsilon_i'^{-\frac{n-2}{4}} \epsilon_{i1}'^{-\frac{n-2}{4}} \frac{(1-\epsilon_i-\epsilon_{i1})}{(1-\epsilon_i-\epsilon_{i1}) + (1-\epsilon_i'-\epsilon_{i1}')}$$
$$\times \exp[-a(\epsilon_i - \epsilon_i')^2 - a(\epsilon_{i1} - \epsilon_{i1}')^2]. \qquad (6.3.13)$$

Using this kernel, they repeated the calculation of heat conductivity, which follows the same procedure as before. The results given in Ref. 32 about heat conductivity versus temperature are identical to those calculated with the new kernel.

We end the review of models for the polyatomic gases by remarking that a model generalizing the ES model (see Chapter 1, Section 1.9) to polyatomic gases has been discussed by Andries et al.,[37] who have also shown that the H-theorem holds for this model for polyatomic gases as well.

Concerning the boundary conditions for the distribution function for polyatomic gases, we remark that there is not much material published on this subject, perhaps because the theory is not much different from that holding in the monatomic case. Concerning specific models, one should mention an extension of the CL model to polyatomic molecules proposed by Lord[38] as a generalization of the Cercignani–Lampis model. We refer to the next chapter (Section 7.2) for results obtained with this model.

6.4. The H-Theorem for Classical Polyatomic Molecules

In this section, following a paper by Lampis and the author,[17] we shall be concerned with the aspects of the Boltzmann equation related to polyatomic gases, with particular reference to the H-theorem, which provoked Lorentz's objections mentioned in the previous section. First, we note that there are no formal difficulties to writing a Boltzmann equation for molecules with $n > 3$ degrees of freedom. Together with the position vector of the center of mass $\mathbf{x}$ and the corresponding velocity $\boldsymbol{\xi}$ we shall also need other variables; we shall denote by $\mathbf{p}$ a vector in a $(2n-3)$-dimensional space that includes all the variables with the exception of the coordinates of the center of mass (while the components of $\boldsymbol{\xi}$ are not excluded). Then the equations of the motion between two subsequent collisions are

$$\dot{\mathbf{x}} = \boldsymbol{\xi}; \qquad \dot{\mathbf{p}} = \mathbf{P}(\mathbf{x}, \mathbf{p}), \tag{6.4.1}$$

where $\mathbf{P}$ is a vector in $2n - 3$ dimensions that describes the partial derivatives (with the appropriate sign) of the Hamiltonian with respect to the coordinates and momenta different from $\boldsymbol{\xi}$ (but $\mathbf{x}$ will now be included). The Boltzmann equation for the distribution function $f(\mathbf{x}, \mathbf{p}, t)$ can be written as follows:

$$\frac{\partial f}{\partial t} + \boldsymbol{\xi} \cdot \frac{\partial f}{\partial \mathbf{x}} + \mathbf{P} \cdot \frac{\partial f}{\partial \mathbf{p}}$$
$$= \int_{\Re^{6n-9}} [f' f'_* W(\mathbf{p}', \mathbf{p}'_* \to \mathbf{p}, \mathbf{p}_*) - f f_* W(\mathbf{p}, \mathbf{p}_* \to \mathbf{p}', \mathbf{p}'_*)] d\mathbf{p}_* d\mathbf{p}' d\mathbf{p}'_*, \tag{6.4.2}$$

where f', f'_*, and f_* denote, as usual, that the function f has (besides $\mathbf{x}$ and t) arguments $\mathbf{p}'$, $\mathbf{p}'_*$, and $\mathbf{p}_*$, respectively, whereas $W(\mathbf{p}', \mathbf{p}'_* \to \mathbf{p}, \mathbf{p}_*)$ (which will include some factors given by Dirac's delta functions, in order to ensure conservation of momentum and energy) is essentially the differential scattering cross section multiplied by the relative speed. In fact one has

$$\iint W(\mathbf{p}', \mathbf{p}'_* \to \mathbf{p}, \mathbf{p}_*) d\mathbf{p}' d\mathbf{p}'_* = |\boldsymbol{\xi} - \boldsymbol{\xi}_*| \Sigma_t, \tag{6.4.3}$$

where Σ_t is the total scattering cross section (assumed to be finite). The latter may, of course, depend upon $\mathbf{p}$ and $\mathbf{p}_*$. One can, however, get rid of this dependence with a trick that makes the proof much simpler. Since the total cross section is assumed to be finite, there will be a maximum distance r_0 beyond which the molecules do not mutually interact. We can then let $\Sigma_t = \pi r_0^2$, provided we introduce fake collisions in which no change occurs for those values of $\mathbf{p}$ and $\mathbf{p}_*$ for which, possibly, there is no interaction, although $r < r_0$. The case in which the collision cross section is infinite can be obtained by first cutting off the interactions occurring for $r > r_0$ and then letting $r_0 \to \infty$ in the final inequality that we are going to prove.

The microscopic motion equations (6.4.1) are, of course, assumed to be time-reversible: This means that there exists a transformation $(\mathbf{x}, \mathbf{p}, t) \to (\mathbf{x}, \mathbf{p}^-, -t)$ (where, typically, a component of $\mathbf{p}^-$ shall be equal to the corresponding one of $\mathbf{p}$ if it has the meaning of a coordinate and opposed to the corresponding one of $\mathbf{p}$ if it has the meaning of a momentum canonically conjugated to a coordinate). Incidentally, we remark that one frequently treats the changes in the coordinates as if they could be ignored in the average and just considers the changes in the momenta; this aspect, however, will not enter in what follows.

It is important to notice that the transformation of variables from the $\mathbf{p}$s to the $\mathbf{p}^-$s, with fixed $\mathbf{x}$, preserves the volume in the space described by the variables $\mathbf{p}$.

The time reversibility of the microscopic equations even during a collision implies that the following relation, called *reciprocity*, holds:

$$W(\mathbf{p}', \mathbf{p}'_* \to \mathbf{p}, \mathbf{p}_*) = W(\mathbf{p}^-, \mathbf{p}^-_* \to \mathbf{p}^{-\prime}, \mathbf{p}^{-\prime}_*). \tag{6.4.4}$$

If the interaction possesses spherical symmetry, as is the case for point masses or perfectly smooth hard spheres, then the following, stronger property holds:

$$W(\mathbf{p}', \mathbf{p}'_* \to \mathbf{p}, \mathbf{p}_*) = W(\mathbf{p}, \mathbf{p}_* \to \mathbf{p}', \mathbf{p}'_*). \tag{6.4.5}$$

this is called *detailed balance*. This property is more or less explicitly used in the proof of the H-theorem for monatomic gases. Whenever it holds, the proof given in Chapter 1 can be transferred without changes to the case of polyatomic gases. Boltzmann's mistake, pointed out by Lorentz, lies in the implicit assumption that (6.4.5) holds for a generic polyatomic gas. If this property fails one can introduce Boltzmann's argument on the "closed cycle of collisions," which, although not so convincing, contains the basic idea of the proof that we shall presently give (i.e., the fact that one must not consider single collisions but subsets of collisions).

The key elements are (6.4.3), whose right-hand side is clearly invariant under time reversal, and the reciprocity property, expressed by (6.4.4). Let us integrate both sides of the latter relation with respect to $\mathbf{p}$ and $\mathbf{p}_*$. We obtain

$$\int\int W(\mathbf{p}', \mathbf{p}'_* \to \mathbf{p}, \mathbf{p}_*) d\mathbf{p} d\mathbf{p}_* = \int\int W(\mathbf{p}^-, \mathbf{p}^-_* \to \mathbf{p}^{-\prime}, \mathbf{p}^{-\prime}_*) d\mathbf{p} d\mathbf{p}_*$$
$$= \int\int W(\mathbf{p}, \mathbf{p}_* \to \mathbf{p}^{-\prime}, \mathbf{p}^{-\prime}_*) d\mathbf{p} d\mathbf{p}_*, \quad (6.4.6)$$

where, in the last step, we changed the integration variables from $\mathbf{p}, \mathbf{p}_*$ to $\mathbf{p}^-$, $\mathbf{p}^-_*$ (using the aforementioned invariance of the volume element) and subsequently abolished the superscript $-$, which is no longer required. But the last integral is that which appears in the left-hand side of (6.4.3), apart from the presence of the superscript $-$ in the second pair of variables; but, because of the invariance of the right-hand side of the same equation with respect to the transformation from variables with superscript $-$ to variables without the same superscript, we can suppress the latter in the last integral of (6.4.6) and get

$$\int\int W(\mathbf{p}', \mathbf{p}'_* \to \mathbf{p}, \mathbf{p}_*) d\mathbf{p} d\mathbf{p}_* = \int\int W(\mathbf{p}, \mathbf{p}_* \to \mathbf{p}', \mathbf{p}'_*) d\mathbf{p} d\mathbf{p}_*. \quad (6.4.7)$$

This is the new relation that we shall use to prove the H-theorem for polyatomic molecules. The importance of a relation of this kind was underlined by Waldmann[9] who, guided by an analogy with the quantum case, wrote it without proof in the particular case of dumbbell-shaped classical molecules, remarking that "one must get the (purely mechanical) normalization property" expressed by (6.4.7). As a possible proof, the same author[9] seems to hint at a complete calculation with the simplifying assumption of "averaging over all possible phase angles [...] before and after collision." This average, albeit useful in some cases to simplify the relations, is not required in the proof of the aforementioned paper,[17] according to which (6.4.7) is a general property, which follows from the time reversibility of the microscopic equations of the molecular motion.

Having shown that (6.4.7) holds, it is now a relatively simple matter to prove the H-theorem, or, more precisely, the Boltzmann lemma (whence the

H-theorem follows). Accordingly, if we let

$$Q(f,f) = \int_{\Re^{6n-9}} [f'f_*'W(\mathbf{p}',\mathbf{p}_*' \to \mathbf{p},\mathbf{p}_*) - ff_*W(\mathbf{p},\mathbf{p}_* \to \mathbf{p}',\mathbf{p}_*')]d\mathbf{p}_*d\mathbf{p}'d\mathbf{p}_*' \tag{6.4.8}$$

we get

$$\int \log f\, Q(f,f)d\mathbf{p} \le 0. \tag{6.4.9}$$

To prove this result, let us multiply (6.4.8) by $\log f$ and integrate with respect to $\mathbf{p}$, thus obtaining

$$\int \log f\, Q(f,f)d\mathbf{p} = \frac{1}{2}\int_{\Re^{8n-12}} ff_* \log\left(\frac{f'f_*'}{ff_*}\right) W(\mathbf{p},\mathbf{p}_* \to \mathbf{p}',\mathbf{p}_*')d\mathbf{p}d\mathbf{p}_*d\mathbf{p}'d\mathbf{p}_*'. \tag{6.4.10}$$

This relation can be easily obtained with the same manipulations discussed in Chapter 1 (i.e., with suitable changes of variables and indices) without using any property of $W(\mathbf{p},\mathbf{p}_* \to \mathbf{p}',\mathbf{p}_*')$. Equation (6.4.10), however, does not permit us to apply the argument used in Chapter 1; to this end, in fact, we should sum (6.4.10) and the relation that can be obtained from it by exchanging the primed and unprimed variables and use the property of detailed balance (6.4.5), which, however, does not generally hold. At this point we must make use of a trick, which apparently was first used by Pauli in an appendix to a paper by Stueckelberg[39] on quantum statistical mechanics. Together with the identity expressed by (6.4.10) let us consider the following one as well:

$$\int_{\Re^{8n-12}} f'f_*'W(\mathbf{p}',\mathbf{p}_*' \to \mathbf{p},\mathbf{p}_*)d\mathbf{p}d\mathbf{p}_*d\mathbf{p}'d\mathbf{p}_*' = \int_{\Re^{8n-12}} ff_*W(\mathbf{p},\mathbf{p}_* \to \mathbf{p}',\mathbf{p}_*')d\mathbf{p}d\mathbf{p}_*d\mathbf{p}'d\mathbf{p}_*'. \tag{6.4.11}$$

This can be obtained with an exchange of the name of the variables and expresses the conservation of the number of molecules in a collision. We can now make use of (6.4.7) in the left-hand side of (6.4.11) and rewrite it as follows:

$$\int_{\Re^{8n-12}} f'f_*'W(\mathbf{p},\mathbf{p}_* \to \mathbf{p}',\mathbf{p}_*')d\mathbf{p}d\mathbf{p}_*d\mathbf{p}'d\mathbf{p}_*' = \int_{\Re^{8n-12}} ff_*W(\mathbf{p},\mathbf{p}_* \to \mathbf{p}',\mathbf{p}_*')d\mathbf{p}d\mathbf{p}_*d\mathbf{p}'d\mathbf{p}_*', \tag{6.4.12}$$

or, equivalently,

$$\frac{1}{2}\int_{\Re^{8n-12}} ff_*\left(\frac{f'f'_*}{ff_*}-1\right)W(\mathbf{p},\mathbf{p}_*\to\mathbf{p}',\mathbf{p}'_*)d\mathbf{p}d\mathbf{p}_*d\mathbf{p}'d\mathbf{p}'_*=0. \tag{6.4.13}$$

We can then subtract the integral appearing in the left-hand side of (6.4.13) from the right-hand side of (6.4.10), without changing anything. We then have

$$\int \log fQ(f,f)\,d\mathbf{p}=\frac{1}{2}\int_{\Re^{8n-12}} ff_*\left[\log\frac{f'f'_*}{ff_*}-\left(\frac{f'f_*'}{ff_*}-1\right)\right] \times W(\mathbf{p},\mathbf{p}_*\to\mathbf{p}',\mathbf{p}'_*)d\mathbf{p}d\mathbf{p}_*d\mathbf{p}'d\mathbf{p}'_*. \tag{6.4.14}$$

Let us now make use of the fact that both f and W are nonnegative and that the following (previously used) elementary inequality holds:

$$\log x-(x-1)\le 0 \qquad (=0 \quad \text{iff}\, x=1)\,. \tag{6.4.15}$$

We can then conclude that (6.4.9) has been proved and that the equality sign holds if and only if (almost everywhere)

$$f'f'_*=ff_*, \tag{6.4.16}$$

that is, if the distribution function describes an equilibrium state.

6.5. Chemical Reactions

Chemical reactions are important in high altitude flight because of the high temperatures that develop near a vehicle flying at hypersonic speed (i.e., with Mach number larger than 5). Up to 2,000 K, the composition of air can be considered to be the same as at standard conditions. Beyond this temperature, N_2 and O_2 begin to react and form NO. At 2,500 K diatomic oxygen begins to dissociate and form atomic oxygen O, till O_2 completely disappears at about 4,000 K. Nitrogen begins to dissociate at a slightly higher temperature (about 4,250 K). NO disappears at about 5,000 K. Ionization phenomena start at about 8,500 K.

As we implicitly remarked when we wrote Eq. (6.2.6), the kinetic theory of gases is an ideal tool for dealing with bimolecular reactions, which can be written schematically as

$$A+B\leftrightarrow C+D, \tag{6.5.1}$$

where A, B, C, and D represent different molecular species. We have already used the term “molecule,” as usual in kinetic theory, to mean also atom (a monatomic molecule); in this section we shall further enlarge the meaning of this term to include ions, electrons, and photons as well, when we have to deal with ionization reactions and interaction with radiation.

As long as the reaction takes place in a single step with the presence of no other species than the reactants, it is a well-known circumstance that the change of concentration of a given species (A, say) in a space-homogeneous mixture can be written as follows:

$$\frac{dn_A}{dt} = k_b(T)n_C n_D - k_f(T)n_A n_B. \tag{6.5.2}$$

Note that, in chemistry, one uses the molar density in place of the number density used here, the two being obviously related through Avogadro's number.

The rate coefficients k_f and k_b for the forward and backward (or reverse) reactions, respectively, are functions of temperature and are usually written by a semiempirical argument, which generalizes the Arrhenius formula, in the form

$$k(T) = \Lambda T^\eta \exp\left(-\frac{E_a}{kT}\right), \tag{6.5.3}$$

where Λ and η ($=0$ in the Arrhenius equation) are constants, and E_a is the *activation energy* of the reaction. It is clear that these equations, though having a flavor of kinetic theory, are essentially macroscopic and can be assumed to hold when the distribution function is essentially Maxwellian.

In fact, the above reaction theory can be obtained by assuming that the distribution functions are Maxwellians, whereas the role of internal degrees of freedom may be ignored and the *reaction cross section* vanishes if the translational energy E_t in the center-of-mass system is less than E_a and equals a constant σ_R if the energy is larger than E_a. A more accurate theory is obtained[40,41,28] by assuming that the ratio of the reaction cross section to the total cross section is zero when the total collision energy E_c (equal to the sum of E_t and the total internal energy of the two colliding molecules E_i) is less than E_a and proportional to the product of a power of $E_c - E_a$ and a power of E_c. The exponents and the proportionality factor are essentially dictated by the number of internal degrees of freedom, the exponent of the temperature in the diffusion coefficient of species A in species B, and the empirical exponent η appearing in Eq. (6.5.3). This theory provides a microscopic reaction model that can reproduce the conventional rate equations (6.5.2–6.5.3) in the continuum limit. The model is, however, as in the case of gas–surface interaction and models for polyatomic gases, largely based on phenomenological considerations and mathematical tractability. The ideal microscopic model would consist of complete tabulations of the differential cross sections as functions of the energy states and **n**. Some microscopic data, coming from extensive quantum-mechanical computations, supported by experiments, are available, but, unfortunately, not very much is known for reactions of engineering interest. When comparisons can be made, the reaction cross section provided by the phenomenological model is of the correct order of magnitude. This provides some measure of optimism about the validity of the results obtained with these models for the highly nonequilibrium rarefied gas flows.

Termolecular reactions provide some difficulty to kinetic theory, because the Boltzmann equation essentially describes the effect of binary collisions. They are, however, of essential importance in high-temperature air, where the reverse (or backward) reaction of a dissociation one is a recombination reaction, which is necessarily termolecular, as we shall presently explain. A typical dissociation–recombination reaction can be represented as

$$AB + X \leftrightarrow A + B + X, \tag{6.5.4}$$

where AB, A, B, and X represent the dissociating molecule, the two molecules produced by the dissociation, and a third molecule (of any species), respectively. The latter molecule, in the forward reaction, collides with AB and causes its dissociation. This process is described by a binary collision and is an endothermic reaction, requiring a certain amount of energy, the dissociation energy E_d.

The recombination process is an exothermic reaction and it might seem that one could dispense with the "third body" X and consider it as a bimolecular reaction

$$AB \leftarrow A + B. \tag{6.5.5}$$

However, one can easily see that the energy balance for this event cannot be satisfied in the presence of energy release. In fact if two molecules form an isolated system and are assumed to interact with a potential energy that is attractive at large distances and repulsive at short distances, they can come close enough to orbit one another but the repulsive part will eventually separate them. In fact, they cannot form a stable molecule; this is seen by writing the energy equation in the center-of-mass system. The final kinetic energy in this reference system should be zero for the molecule whereas it was positive when the two molecules approached each other; note that the change of potential energy is negative. A third molecule X is required to describe the recombination process. In order to keep the binary collision analysis, appropriate for a rarefied gas, we must think of the recombination process as a sequence of two binary collisions.

The first of these forms an (unstable) orbiting pair P, which is stabilized by a second collision of this pair with X, as long as this collision occurs within a sufficiently small elapsed time. One can then extend the previous theory based on a binary collision analysis. If the activation energy is assumed to be zero, then the main change is that the cross section acquires a factor proportional to the number density of the species X.

This simple theory is based on a molecular interaction that is attractive at large distance and repulsive at short distances. One can assume a highly simplified scheme by taking a hard sphere core with diameter σ and a scattering of the square-well type at some larger distance σ_*. The potential energy is, say, $-Q$ ($Q > 0$) between σ and σ_*. Then if m is the mass of a molecule A, for a distance $r > \sigma_*$, the trajectory of one molecule with respect to the other before the interaction begins will be a straight line $r \cos \vartheta = b$ (in polar coordinates), where b is the impact parameter. The condition $b < \sigma_*$ must be satisfied if the molecules actually interact. Then, assuming for simplicity that $A = B$ (as in the case of recombination of oxygen) the conservation of energy and angular momentum give (Problem 6.5.1)

$$\frac{m}{4}\left[\left(\frac{d}{d\vartheta}\frac{1}{r}\right)^2 + \frac{1}{r^2}\right] - \frac{4Q}{mb^2V^2} = \frac{1}{b^2} \quad (\sigma < r < \sigma_*), \tag{6.5.6}$$

where V is, as usual, the relative speed. We easily verify that the trajectory has a corner point at a distance $r = \sigma_*$ and the molecule we are following is deflected toward the other by an angle (Problem 6.5.1)

$$\theta_0 = \cos^{-1}\left(\frac{b}{\sigma_*}\right) - \cos^{-1}\left[\frac{b}{\sigma_*}\left(1 + \frac{4Q}{mV^2}\right)^{-1/2}\right]. \tag{6.5.7}$$

The orbiting pair P has in this case a very simple motion: The molecules A approach each other and then have a hard sphere collision. After that they tend to separate again; and the "molecule" P will disappear and two molecules A emerge again, unless a third molecule X collides with P, which is stabilized into an A_2 molecule. The unstable pair P is endowed with an internal energy

$$E_B = \frac{m}{4}V^2 + Q. \tag{6.5.8}$$

This energy is stored to be, possibly, converted into kinetic energy (and hence, from a macroscopic viewpoint, heat of reaction) through the process (6.5.4).

Even for this simple scheme we must write three Boltzmann equations: one for the species A, one for the species A_2, and one for the unstable species P;

even if A is monatomic, A_2 and P are diatomic and hence have an internal energy. The species X used in the above argument can be any of the three aforementioned species. The species A loses particles when colliding with A (formation of P), and P gains in the same process, but P loses in most collisions with only molecule.

This model can be slightly complicated by assuming that there is a potential barrier between σ_* and σ_{**} ($\sigma_{**} > \sigma_*$). If the potential energy is $E_a > 0$ for these distances between the A molecules, then the formation can occur if, and only if, the relative speed V is larger than $4E_a/m$. Then E_a plays the role of the activation energy.

The resulting system of Boltzmann equations reads as follows:

$$\begin{aligned}\frac{\partial f}{\partial t} + \boldsymbol{\xi} \cdot \frac{\partial f}{\partial \mathbf{x}} = &-2\int_{R^3}\int_{\mathcal{B}_+} ff_*\mathcal{B}^r_{AA}|d\boldsymbol{\xi}_* d\mathbf{n} + \int_{R^3}\int_{\mathcal{B}_+} (f'F'_* - fg_*)\mathcal{B}_{AA_2} d\boldsymbol{\xi}_* d\mathbf{n} \\ &+ \int_{R^3}\int_{\mathcal{B}_+} (f'f'_* - ff_*)\mathcal{B}^e_{AA} d\boldsymbol{\xi}_* d\mathbf{n} \\ &+ \int_{R^3}\int_{\mathcal{B}_+} (f'F'_* - fF_*)\mathcal{B}_{AA_2} d\boldsymbol{\xi}_* d\mathbf{n}, \end{aligned} \qquad (6.5.9)_1$$

$$\begin{aligned}\frac{\partial F}{\partial t} + \boldsymbol{\xi} \cdot \frac{\partial F}{\partial \mathbf{x}} = &\int_{R^3}\int_{\mathcal{B}_+} g'f'_*\mathcal{B}_{BA} d\boldsymbol{\xi}_* d\mathbf{n} + \int_{R^3}\int_{\mathcal{B}_+} g'F'_*\mathcal{B}_{BA_2} d\boldsymbol{\xi}_* d\mathbf{n} \\ &+ \int_{R^3}\int_{\mathcal{B}_+} (F'F'_* - Fg_*)\mathcal{B}_{BA_2}|d\boldsymbol{\xi}_* d\mathbf{n} \\ &+ \int_{R^3}\int_{\mathcal{B}_+} (F'f'_* - Ff_*)\mathcal{B}_{AA_2} d\boldsymbol{\xi}_* d\mathbf{n} \\ &+ \int_{R^3}\int_{\mathcal{B}_+} (F'F'_* - FF_*)\mathcal{B}_{A_2A_2} d\boldsymbol{\xi}_* d\mathbf{n}, \end{aligned} \qquad (6.5.9)_2$$

$$\begin{aligned}\frac{\partial g}{\partial t} + \boldsymbol{\xi} \cdot \frac{\partial g}{\partial \mathbf{x}} = &\int_{R^3}\int_{\mathcal{B}_+} f'f'_*\mathcal{B}^r_{AA} d\boldsymbol{\xi}_* d\mathbf{n} - \int_{R^3}\int_{\mathcal{B}_+} gf_*\mathcal{B}_{BA} d\boldsymbol{\xi}_* d\mathbf{n} \\ &- \int_{R^3}\int_{\mathcal{B}_+} gF_*\mathcal{B}_{BA_2} d\boldsymbol{\xi}_* d\mathbf{n}, \end{aligned} \qquad (6.5.9)_3$$

where t, F, and g denote the distribution functions for species A, A_2, and P, whereas the superscripts r and e are used to discriminate between reactive and elastic collisions when necessary. For simplicity, we have omitted indicating the internal energies of particles A_2 and P. The factors 2 take into account the fact that two particles of a species disappear at the same time. Models for dealing with a chemically reactive gas, akin to the BGK one, have been discussed by Burgers[42] and Yoshizawa.[43]

Problem

6.5.1 Prove that in the case of a square box potential the scattering angle is given by (6.5.7).

6.6. Ionization and Thermal Radiation

Ionization reactions involve the electronic states and it is unlikely that a purely classical theory will be successful in describing them, because of the selection rules. Yet, one can use the phenomenological approach to provide at least an upper bound for the reaction rates.

As mentioned above, one can, in principle, think of interaction with radiation as if it were a reaction involving photons as "molecules." Here spontaneous emission should also be taken into account. It becomes harder to develop phenomenological models, because one should consider as many species as there are excited levels for each molecule.

Photons can be described by means of the so-called radiative transfer (or radiation transport) equation, which looks like a Boltzmann equation. The analogy is, however, in a sense, artificial, because the number of photons is not conserved. The equation reads

$$\frac{\partial f}{\partial t} + c\boldsymbol{\omega} \cdot \frac{\partial f}{\partial \mathbf{x}} = \int_{R^3} K(\mathbf{x}, \mathbf{k}' \to \mathbf{k}) f(\mathbf{x}, \mathbf{k}') d\mathbf{k}' - \nu(\mathbf{x}, \mathbf{k}) f(\mathbf{x}, \mathbf{k}) + s(\mathbf{x}, \mathbf{k}). \tag{6.6.1}$$

Here K is the scattering probability from a wavevector $\mathbf{k}'$ to a wavevector $\mathbf{k}$. The direction of $\mathbf{k}$ is given by $\boldsymbol{\omega}$ and its magnitude is the radiation frequency multiplied by the speed of light c, not to be confused with $\nu(\mathbf{x}, \mathbf{k})$, the total frequency of scattering and absorption events. The term $s(\mathbf{x}, \mathbf{k})$ describes the volume radiation source, due to photon emission. If inelastic scattering effects, such as fluorescence and stimulated emission, are neglected, then there is no interaction between photons of different frequency and we can replace the arguments $\mathbf{k}'$ and $\mathbf{k}$ by the corresponding unit vectors $\boldsymbol{\omega}$ and $\boldsymbol{\omega}'$. The emission term can be expressed as a product of the Planck distribution by the volume emission coefficient $\epsilon_{|\mathbf{k}|,T}$.

Boundary conditions for radiation can be described in a way similar to that used for gas–surface interaction. Of course, emission and absorption of radiation occur, along with reflection.[44,45]

This discussion of ionization and radiation phenomena just touches upon a vast area. The great and important field of rarefied plasmas is left completely outside the realm of this book.

6.7. Concluding Remarks

When looked at in a superficial way, the kinetic theory of polyatomic gases and mixtures seems to be a rather formal extension of the kinetic theory of a single monatomic gas. We took adavantage of this aspect to introduce the topic, but the attentive reader may have already perceived that this is not the case. Conceptual as well as practical problems arise when dealing with polyatomic gases, as shown by the proof of the H-theorem, discussed in Section 6.4 and by the necessity of occupying a vastly larger amount of memory when simulating flows of polyatomic gases by any numerical technique, even when we neglect the spin orientation and just consider the rotational kinetic energy associated with it.

In order to show the unusual kind of phenomena associated with the presence of the spin, we shall briefly deal with the so-called Scott effect.[25] This effect was discovered accidentally and treated by many authors, not always correctly.[25] It amounts to the following phenomenon: In a gas of polyatomic molecules between two coaxial cylinders kept at different temperatures, the presence of a magnetic field in the direction of the axis gives rise to a torque on the inner cylinder, which is hanging on a torsion wire. This phenomenon is essentially related to thermal stresses; the magnetic field is required to couple, via the magnetic moment of the molecules, the shear stress with the thermal stress due to a purely radial change of the temperature.

Concerning mixtures, even in the case of monatomic gases, we met a conceptual problem when defining the stress tensor; although the definition we gave for the partial stress seems to be the most natural and is widely adopted, the choice we have made is not universally used. In fact one could attribute a part of the random motions contributing to the stresses to the diffusive motion with respect to the bulk velocity of the mixture.

Phenomena occurring in mixtures are very important from a practical viewpoint, because they can be exploited to separate isotopes. We just mention thermal diffusion, which had escaped the attention of Maxwell and Boltzmann, because the corresponding coefficient vanishes in the case of Maxwell molecules. Another important phenomenon is the overshoot in the temperature profiles of shock waves. We saw in Chapter 4 that this is a phenomenon that is barely appreciable for a single gas. However, when we deal with binary mixtures, we witness the new fact that the partial temperatures (or, better, thermal energies) of the two species have overshoots that are completely different. The heavier species has a much more strongly marked overshoot.[46–49]

References

[1] A. Kundt and E. Warburg, "Ueber die specifische Wärme des Quecksilbergases," *Ann. der Physik* **157**, 353–369 (1876).
[2] C. Cercignani, *Mathematical Methods in Kinetic Theory*, revised edition, Plenum Press, New York (1990).
[3] C. Cercignani, *The Boltzmann Equation and Its Applications*, Springer-Verlag, New York (1988).
[4] L. Sirovich, "Kinetic modeling of gas mixtures," *Phys. Fluids* **5**, 908–918 (1962).
[5] T. F. Morse, "Kinetic model equations in a fluid," *Phys. Fluids* **7**, 2012–2013 (1964).
[6] F. B. Hanson and T. F. Morse, "Kinetic models for a gas with internal structure," *Phys. Fluids* **10**, 345–353 (1967).
[7] C. S. Wang Chang and G. E. Uhlenbeck, in *Studies in Statistical Mechanics*, J. de Boer and G.E. Uhlenbeck, eds., Vol. II, Part c, North Holland, Amsterdam (1964).
[8] L. Waldmann, "Transporterscheinungen in Gasen von mittlerem Druck," in *Handbuch der Physik*, S. Flügge, ed., Vol. XII, 295–514, Springer-Verlag, Berlin (1958).
[9] L. Waldmann, "On Kinetic equations for particles with internal degrees of freedom," in *The Boltzmann Equation. Theory and Application*, E. G. D. Cohen and W. Thirring, eds., 223–246, Springer-Verlag, Vienna (1973).
[10] R. F. Snider, "Quantum-mechanical modified Boltzmann equation for degenerate internal states," *J. Chem. Phys.* **32**, 1051–1060 (1960).
[11] L. Boltzmann, "Weitere Studien über das Wärmegleichgewicht unter Gasmolekülen," *Sitzungsberichte Akad. Wiss.* II, **66**, 275–370 (1872).
[12] H. A. Lorentz, "Über das Gleichgewicht der lebendingen Kraft unter Gasmolekülen," *Sitzungsberichte Akad. Wiss.* **95**, 115–152 (1887).
[13] L. Boltzmann, "Neuer Beweis zweier Sätze über das Wärmegleichgewicht unter mehratomigen Gasmolekülen," *Sitzungsberichte Akad. Wiss.* **95**, 153–164 (1887).
[14] L. Boltzmann, *Vorlesungen über Gastheorie*, 2 vols., J. A. Barth, Leipzig (1895–1898).
[15] R. C. Tolman, *The Principles of Statistical Mechanics*, Clarendon Press, Oxford, (1938).
[16] G. E. Uhlenbeck, "The validity and the limitations of the Boltzmann equation," in *The Boltzmann Equation. Theory and Application*, E. G. D. Cohen and W. Thirring, eds., 107–119, Springer-Verlag, Vienna (1973).
[17] C. Cercignani and M. Lampis, "On the H-theorem for polyatomic gases," *J. Stat. Phys.* **26**, 795–801 (1981).
[18] J. A. Lordi and R. E. Mates, "Rotational relaxation in nonpolar diatomic gases," *Phys. Fluids* **13**, 291–308 (1970).
[19] C. Muckenfuss and C. F. Curtiss, "Kinetic theory of nonspherical molecules. III," *J. Chem. Phys.* **29**, 1257–1277 (1958).
[20] J. H. Jeans, "On the partition of energy in a system of loaded spheres," *Q. J. Pure Appl. Math* **35**, 224–238 (1904).
[21] J. S. Dahler and N. F. Sather, "Kinetic theory of loaded spheres. I," *J. Chem. Phys.* **38**, 2363–2382 (1962).

[22] S. I. Sandler and N. F. Dahler, "Kinetic theory of loaded spheres. IV. Thermal diffusion in a dilute gas mixture of D_2 and HT," *J. Chem. Phys.* **47**, 2621–2630 (1967).
[23] G. H. Bryan, *Rep. Brit. Assoc. Advmt. Sci.*, 83 (1894).
[24] F. R. W. Mc Court, J. J. M. Beenakker, W. E. Köhler, and I. Kuščer, *Nonequilibrium Phenomena in Polyatomic Gases*, Clarendon Press, Oxford (1990).
[25] C. Cercignani and M. Lampis, "Kinetic theory analysis of the Scott effect," in *Rarefied Gas Dynamics*, V. Boffi and C. Cercignani, eds., Vol. I, 336–344, B. G. Teubner, Stuttgart (1986).
[26] I. Kuščer, "Models of energy exchange in polyatomic gases," in *Operator Theory: Advances and Applications*, Vol. 51, 180–188, Birkhäuser Verlag, Basel (1991).
[27] C. Borgnakke and P. S. Larsen, "Statistical collision model for Monte Carlo simulation of polyatomic gas mixture," *J. Comput. Phys.* **18**, 405–420 (1975).
[28] G. A. Bird, *Molecular Gas Dynamics and the Direct Simulation of Gas Flows*, Clarendon Press, Oxford (1994).
[29] G. Benettin, L. Galgani, and A. Giorgilli, "Realization of holonomic constraints and freezing of high frequency degrees of freedom in the light of classical perturbation theory. Part I," *Comm. Math. Phys.* **113**, 87–103 (1987).
[30] G. Benettin, L. Galgani, and A. Giorgilli, "Realization of holonomic constraints and freezing of high frequency degrees of freedom in the light of classical perturbation theory. Part II," *Comm. Math. Phys.* **121**, 557–601 (1989).
[31] L. Boltzmann, "On certain questions of the theory of gases," *Nature* **51**, 413–415 (1895).
[32] C. Cercignani, *Ludwig Boltzmann. The Man Who Trusted Atoms*, Oxford University Press, Oxford (1998).
[33] C. Cercignani and M. Lampis, "A new model for the differential cross-section of a polyatomic gas," in *Rarefied Gas Dynamics 19*, Ching Shen, ed., 731–736, Peking University Press, Beijing (1997).
[34] C. Cercignani, M. Lampis, and J. Struckmeier, "New models for the differential cross section of a polyatomic gas in the frame of the scattering kernel theory," *Mech. Res. Comm.* **25**, 231–236 (1998).
[35] G. J. Prangsma, A. H. Alberga, and J. J. M. Beenakker, "Ultrasonic determination of the volume viscosity of N_2, CO, CH_4 and CD_4 between 77 and 300 K," *Physica* **64**, 278–288 (1973).
[36] G. Emanuel, "Bulk viscosity of a dilute polyatomic gas," *Phys. Fluids A* **2**, 2252–2254 (1990).
[37] P. Andries, P. Le Tallec, J. P. Perlat, and B. Perthame, "The Gaussian–BGK model of Boltzmann equation with small Prandtl number," to appear in *Euro. J. Mecha. B* (1998).
[38] R. G. Lord, "Some extensions to the Cercignani–Lampis gas scattering kernel," *Phys. Fluids A* **3**, 706–710 (1991).
[39] E. C. G. Stueckelberg, "Théorème *H* et unitarité de *S*," *Helv. Phys. Acta* **25**, 577–580 (1952).
[40] G. A. Bird, "Simulation of multi-dimensional and chemically reacting flows," in *Rarefied Gas Dynamics*, R. Campargue, ed., 365–388, CEA, Paris (1979).
[41] G. A. Bird, "Monte-Carlo simulation in an engineering context," in *Rarefied Gas Dynamics*, Part I, 239–255, AIAA, Washington (1981).
[42] J. M. Burgers, *Flow Equations for Composite Gases*, Academic Press, New York (1969).

[43] Y. Yoshizawa, "Wave structures of a chemically reacting gas by the kinetic theory of gases," in *Rarefied Gas Dynamics*, J. L. Potter, ed., Part I, 501–517, AIAA, New York (1977).
[44] R. Siegel and J. R. Howell, *Thermal Radiation Heat Transfer*, Hemisphere Publishing, Washington (1992).
[45] A. Kersch, W. J. Morokoff, *Transport Simulation in Microelectronics*, Birkhäuser, Basel (1995).
[46] B. B. Hamel, "Disparate mass mixture flows," in *Rarefied Gas Dynamics*, J. L. Potter, ed., Part I, 171–195, AIAA, New York (1977).
[47] G. A. Bird, "The structure of normal shock waves in a binary gas mixture," *J. Fluid Mech.* **31**, 657–668 (1968).
[48] G. A. Bird, "Shock wave structure in gas mixtures," in *Rarefied Gas Dynamics*, H. Oguchi, ed., Vol. I, 174–182, University of Tokyo Press, Tokyo (1984).

7

Solving the Boltzmann Equation by Numerical Techniques

7.1. Introduction

Kinetic models and perturbation methods are very useful in obtaining approximate solutions and forming qualitative ideas on the solutions of practical problems, but in general, as already remarked in Chapter 2, they are not sufficient to provide detailed and precise answers for practical problems. Various numerical procedures exist that either attempt to solve for f by conventional techniques of numerical analysis or efficiently bypass the formalism of the integro-differential equation and simulate the physical situation that the equation describes (Monte Carlo simulation). Only recently have proofs been given that these partly deterministic, partly stochastic games provide solutions that converge (in a suitable sense) to solutions of the Boltzmann equation. Numerical solutions of the Boltzmann equation based on finite difference methods meet with severe computational requirements owing to the large number of independent variables. Recent ideas based on the use of discrete velocity models will be discussed at the end of this chapter (Section 7.6).

The only method that has been used for space-inhomogeneous problems in more than one space dimension is the technique of Hicks–Yen–Nordsiek,[1,2] which is based on a Monte Carlo quadrature method to evaluate the collision integral. This method was further developed by Aristov and Tcheremissine[3,4] and has been applied with some success to a few two-dimensional flows.[5,6]

An additional difficulty for traditional numerical methods is the fact that chemically reacting and thermally radiating flows (and even simpler flows of polyatomic gases) are hard to describe with theoretical models having the same degree of accurateness as the Boltzmann equation for monatomic nonreacting and nonradiating gases. These considerations paved the way to the development of simulation schemes, which started with the work of Bird on the so-called Direct Simulation Monte Carlo (DSMC) method,[7] and have become a powerful tool for practical calculations. There appear to be very few limitations to the complexity of the flow fields that can be handled with this approach. Chemically reacting and ionized flows can be and have been analyzed by these methods.

A problem that arises in the applications of the DSMC method is the choice of a model for the molecular collisions. The issue is to concoct simple computing rules by discarding what is physically insignificant. In the case of monatomic gases, the relation of the deflection angle to the impact parameter and the relative speed appears to be the most important piece of physics. It turns out, however, that the scattering law has little effect and that the observable effects are strongly correlated with the cross-section change with relative speed. This realization led to the idea of the variable hard-sphere (VHS) model,[8] which combines the scattering simplicity of the hard-sphere model with a variable cross section based on a molecular diameter proportional to some power (ω minus one half) of the relative speed (ω being the power of absolute temperature ruling the change of the viscosity coefficient). This does not produce problems in a single gas, because the heat conductivity varies with approximately the same law as the viscosity coefficient, but problems arise for mixtures. If one wants the correct diffusion coefficient, another modification is needed. Koura and Matsumoto[9] developed the variable soft sphere (VSS) model, which introduces an additional power-law parameter and gives the necessary flexibility for mixtures. More complicated models can be devised when the attractive part of the intermolecular force is taken into account.[10] A simpler method has also been proposed.[11]

Before discussing the DSMC method and some of its applications in some detail, we remark that, although the DSMC has no rivals for practical computations, some other methods may turn out to be of interest in the future as much more powerful computers become available. In Section 7.6 we shall discuss the discrete velocity models that were intensely studied for many years before evolving into a systematic method of approximating the Boltzmann equation.

7.2. The DSMC Method

As we discussed already in Chapter 3, in the DSMC method, the molecular collisions are considered on a probabilistic rather than a deterministic basis. The main aim is to calculate practical flows through the use of the collision mechanics of model molecules. In fact, the real gas is modeled by some hundreds of thousands or millions of simulated molecules on a computer. For each of them the space coordinates and velocity components (as well as the variables describing the internal state, if we deal with polyatomic molecules) are stored in memory and are modified with time as the molecules are simultaneously followed through representative collisions and boundary interactions in the simulated region of space. In most applications, the number of simulated molecules is extremely small in comparison with the number of molecules that would be present in a real gas flow. Thus, in the simulation, each model molecule is representing the appropriate number of real molecules. The calculation is unsteady and the steady solutions are obtained as asymptotic limits of unsteady solutions. The flow field is subdivided into cells, which are taken to be small enough for the solution to be approximately constant throughout the cell. The time variable is advanced in discrete steps of size Δt, small with respect to the mean free time (i.e., the time between two subsequent collisions of a molecule). This permits a separation of the inertial motion of the molecules from the collision process: One first moves the molecules according to collision-free dynamics and subsequently the velocities are modified according to the collisions occurring in each cell. The rate of occurrence of collisions is given by (hard spheres):

$$r = \sum_{i,j<i} r_{ij}, \tag{7.2.1}$$

$$r_{ij} = \frac{\rho}{Nm} \int_{\mathcal{B}_+} \mathbf{n} \cdot (\boldsymbol{\xi}_i - \boldsymbol{\xi}_j) d\mathbf{n}, \tag{7.2.2}$$

where N is the number of molecules in the sample and $\boldsymbol{\xi}_i$ is the velocity of the ith molecule ($i = 1, \ldots, N$). Each time a collision occurs, the velocities of a collision pair are modified (and, as a consequence, r also varies). Let T_k be the length of the time interval between the $(k-1)$-th and the kth collision in the time interval $[0, \Delta t]$ ($t = 0$ is the time of the zeroth collision by definition). The time intervals T_k are chosen in this way: After the $(k-1)$-th collision, one samples a pair (i, j) on the basis of the probability distribution $p_{ij} = r_{ij}/r$ (with a fixed velocity for each pair); then one takes $T_k = 2N/[(N-1)r_{ij}]$. Then one samples a direction ω of the postcollisional velocity $\boldsymbol{\xi}_i - \boldsymbol{\xi}_j$ with the probability distribution $p(\omega) = r_{ij}/r$ (fixed i, j and $|\boldsymbol{\xi}_i - \boldsymbol{\xi}_j|$) and replaces $\boldsymbol{\xi}_i$ and $\boldsymbol{\xi}_j$ by $\boldsymbol{\xi}_i^+$ and $\boldsymbol{\xi}_j^+$, given by

$$\boldsymbol{\xi}_i^+ = \frac{1}{2}(\boldsymbol{\xi}_i + \boldsymbol{\xi}_j + \omega\, |\boldsymbol{\xi}_i - \boldsymbol{\xi}_j|), \tag{7.2.3}$$

$$\boldsymbol{\xi}_i^+ = \frac{1}{2}(\boldsymbol{\xi}_i + \boldsymbol{\xi}_j - \omega\, |\boldsymbol{\xi}_i - \boldsymbol{\xi}_j|). \tag{7.2.4}$$

The operation is repeated until $\sum_k T_k$ exceeds ΔT.

Some variations of Bird's method have appeared, due to Koura,[12] Belotserkovskii and Yanitskii,[13] and Deshpande.[14] They differ from Bird's method because of the procedure used to sample the time interval between two subsequent collisions. In particular Koura[12] introduced the "null collision technique," which uses, for models different from hard spheres, the maximum of the cross section, when estimating the possible collisions. Then for values of the cross section less than the maximum, some collisions produce a null effect.

A different method was proposed by Nanbu.[15] One does not subdivide Δt and works with the probability P_i that the ith molecule collides in $[0, \Delta t]$. One has

$$P_i = \Delta t \sum_{j=1}^{N} r_{ij}. \tag{7.2.5}$$

One starts with r_{ij} evaluated at $t = 0$ and samples a random number ω in [0,1]. According to whether ω is smaller or larger than P_i, the ith molecule collides or does not collide in $[0, \Delta t]$. If a collision occurs, one samples a collision partner of the ith molecule from the probability distribution

$$P_j^{(i)} = r_{ij} \Big/ \sum_{k=1}^{N} r_{ik}. \tag{7.2.6}$$

Then one samples a direction $\boldsymbol{\omega}$ of the postcollisional velocity $\boldsymbol{\xi}_i^+ - \boldsymbol{\xi}_j^+$ with the probability distribution $p(\boldsymbol{\omega}) = r_{ij}/r$ (for fixed values of i, j and $|\boldsymbol{\xi}_i^+ - \boldsymbol{\xi}_j^+|$) and replaces $\boldsymbol{\xi}_i$ by $\boldsymbol{\xi}_i^+$ given by Eq. (7.2.4). Then the procedure is repeated for all the values of i.

The method of Nanbu was criticized by Koura,[16] who asserted that momentum and energy are not conserved by collisions, because one does not change the velocity of the jth molecule when one changes the velocity of the ith molecule. The criticism is not well-founded, however, because, as Nanbu pointed out,[17] the Boltzmann equation requires the overall conservation of momentum and energy of the system at each space point, not the conservation of the same quantities at each single simulated event. What is new in Nanbu's method is precisely the circumstance that it does not try to simulate the N-body dynamics, but rather offers a description of the system given by the Boltzmann equation. Nanbu's method is now well understood, from both a physical and mathematical standpoint and has been rigorously proven to yield approximations to solutions of the Boltzmann equation, provided the number of test molecules is sufficiently large. The relevant theorem reads as follows:[18,19]

Theorem. *If the Boltzmann equation with initial data f_0 has a smooth, nonnegative solution $f(\mathbf{x}, \boldsymbol{\xi}) \in L^1$, then the solution $\hat{f}$ of Nanbu's method converges weakly in L^1 to f, in the sense that, for any test function $\phi(\mathbf{x}, \boldsymbol{\xi}) \in L^\infty$*

$$\int \phi \hat{f} \, d\mathbf{x} d\boldsymbol{\xi} \to \int \phi f \, d\mathbf{x} d\boldsymbol{\xi} \tag{7.2.7}$$

as $N \to \infty$, $\Delta x, \Delta t \to 0$.

Subsequently, a similar result was proved for Bird's method.[20,21]

The fluctuations inevitably occurring in a DSMC calculation can cause problems if the number of molecules is not large enough to representatively sample each area of the flow domain. They can, however, have a physical significance and contain information on those occurring in a real gas. The comparisons between the fluctuations in DSMC and the predictions of fluctuating hydrodynamic theory have been reviewed by Garcia[22] and appear to be consistent with the fluctuations in a real gas.

The computing task of a simulation method varies with the molecular model. For models other than Maxwell's it is proportional to N for Bird's method, while it is proportional to N^2 for Nanbu's method. As mentioned in Chapter 2, Babovsky[18] found a procedure to reduce the computing task of Nanbu's method and make it proportional to N. His modification is based on the idea of subdividing the interval $[0,1]$ into N equal subintervals. If the random number ω lies in the jth segment, one calculates only $P_{ij} = r_{ij}\Delta t$ and there is no collision if $\omega < (j/n) - P_{ij}$, while there is a collision with the jth molecule if $\omega \geq (j/n) - P_{ij}$. There is also a condition that P_{ij} must satisfy (i.e., $P_{ij} < 1/N$), but this is usually automatically verified, given the size of Δt. Application of Nanbu's method in the form modified by Babovsky shows that the computing task is not only theoretically but also practically comparable to that of Bird's method.[23] This modification eventually evolved into what is called the "Finite-Pointset" method (see Section 7.4).

We remark that one may take advantage of flow symmetries in physical space, but all collisions are calculated as three-dimensional events.

As for the boundary conditions, Maxwell's model of diffuse reflection (see Chapter 1, Section 1.11) is adequate for many problems. There are many cases, however, in which it is far from adequate, as mentioned in Chapter 1. The CL model[24] has been adapted and extended by Lord[25,26] for application in DSMC studies. The resulting CLL model has been shown[27] to provide a realistic boundary condition with incomplete accommodation. More complicated models would be required to describe chemical reactions that can occur at the surface for high impact energies.

7.3. Applications of the DSMC Method to Rarefied Flows

Monte Carlo simulations started in the early 1960s but the first approaches were of a tentative nature, given the fact that the available computer power was very small. Of these early approaches we just quote an attempt to determine the thickness of shock waves.[28]

As we saw in Chapter 4, the first significant application of the DSMC method dealt with the structure of a normal shock wave,[29] but only a few years later Bird was able to calculate shock profiles[30] that allowed meaningful comparisons with the experimental results then available[31] and with subsequent experiments.[32] This long time span is understandable: The method is very demanding of computer resources. In 1964, even with the fastest computers, the restriction on the number of molecules that could be used was such that large random fluctuations had to be expected in the results, and it was difficult to arrive at definite conclusions. Thus the number of simulated molecules and the sample sizes in the computations that could be performed in those years were extremely small in comparison with those that have been routinely employed by an increasing number of workers. The problem of the shock wave structure has continued to be an important test case. Later studies have included comparisons of measured and computed velocity distribution functions within strong shock waves in helium.[33]

Early DSMC studies were also devoted to the problem of a hypersonic leading edge. This arises in connection with the flow of a gas past a very sharp plate, parallel to the oncoming stream. When the Reynolds number $\mathrm{Re} = \rho_\infty V_\infty L/\mu_\infty$, based on the plate length, is very large, the picture, familiar from continuum mechanics – of a potential flow plus a viscous boundary layer – is valid everywhere except near the leading and the trailing edges. Estimates obtained already in the late 1960s by Stewartson[34] and Messiter[35] showed that the Knudsen number at the trailing edge is of order $\mathrm{Ma}_\infty \mathrm{Re}^{-3/4}$, where Ma_∞ is the upstream Mach number. As a consequence, kinetic theory is not needed (for large values of Re) at the trailing edge. For the leading edge, the Knudsen number is of order Ma_∞; hence in supersonic, or, even more, hypersonic flow ($\mathrm{Ma}_\infty \geq 5$), the flow in the region about the leading edge must be considered as a typical problem in kinetic theory.

In particular, the viscous boundary layer and the outer flow are no longer distinct from each other, although a shocklike structure may still be identified.[36–38] It is in this connection that the name of merged-layer regime, mentioned in Chapter 2, arose. There are several methods based on simplified continuum models, represented by the papers of Oguchi,[39] Shorenstein and Probstein,[40] Chow,[41,42] Rudman and Rubin,[43] Cheng et al.,[44] and Kot and Turcotte,[45] which usefully predict surface and other gross properties in this regime. The good agreement between these approaches and experiment gave new evidence for the importance of the Navier–Stokes equations. Nevertheless, if we go sufficiently close to the leading edge, the Navier–Stokes equations must be given up in favor of the Boltzmann equation. Huang and coworkers[46–48] carried out extensive computations based on discrete ordinate methods for the BGK model and were able to show the process of building the flow picture assumed in the simplified continuum models mentioned above.

The first DSMC is due to Vogenitz et al.[49] and exhibits a flow structure qualitatively different from the predictions of earlier studies. Their results are supported by the experiments of Metcalf et al.[50] Validation studies of the DSMC method were also conducted at the Imperial College.[51]

Hypersonic flows past blunt bodies were also the object of many simulations; most of the calculations were those made for the Shuttle Orbiter reentry, for which useful comparisons with measured data were possible.[52] This comparison was concerned with the windward centerline heating and employed an axially symmetric equivalent body. Later comparisons[53] with Shuttle data were for the aerodynamic characteristics of the full three-dimensional shape.

Another interesting problem that has been simulated by Ivanov and his coworkers is the reflection on a plane wall of an oblique shock wave generated by a wedge.[54,55] As is well-known from continuum mechanics based on the compressible Euler equations[56,57] two types of reflection are possible: 1. the regular reflection in which the reflected shock wave originates at a point of the plane and 2. the Mach reflection. In fact, if the Mach number after the incident shock is lower than the detachment Mach number corresponding to the angle of deflection of the flow toward the plane wall, a normal, or nearly normal, shock (the Mach stem) appears near the wall and forms a triple intersection point (a line in three dimensions), from which the reflected shock starts (Fig. 7.1). A further discontinuity surface, a weak slipstream (or contact discontinuity) separates the gas behind the reflected shock from that behind the Mach stem (which has a different entropy). If α is the angle between the incident shock and the wall, there is, theoretically, a range of values of α, (α_N, α_D), for which both reflections are possible. The difference between α_D (the detachment angle) and α_N (the von Neumann angle) is about 10^0 for strong shock waves. Experiments show[58] that the transition invariably occurs for $\alpha = \alpha_N$. The first DSMC calculation[54] showed a hysteresis effect, which was first predicted by Hornung et al.[59] and at variance with the experiments.[58]

The second set of calculations[55] confirmed the results of the first one,[54] but it was accompanied by a study of the stability of the solution with respect to perturbations. The conclusions will be presently summarized.

First, an upstream uniform flow generates either a regular reflection (for $\alpha_* < \alpha_D$; Fig. 7.2) or a Mach reflection ($\alpha_* > \alpha_D$; Fig. 7.3). Starting from the latter as initial data and slightly decreasing the angle α, one observes the hysteresis effect and produces a steady-state solution with a Mach reflection for angles lower than α_D (Fig. 7.4).

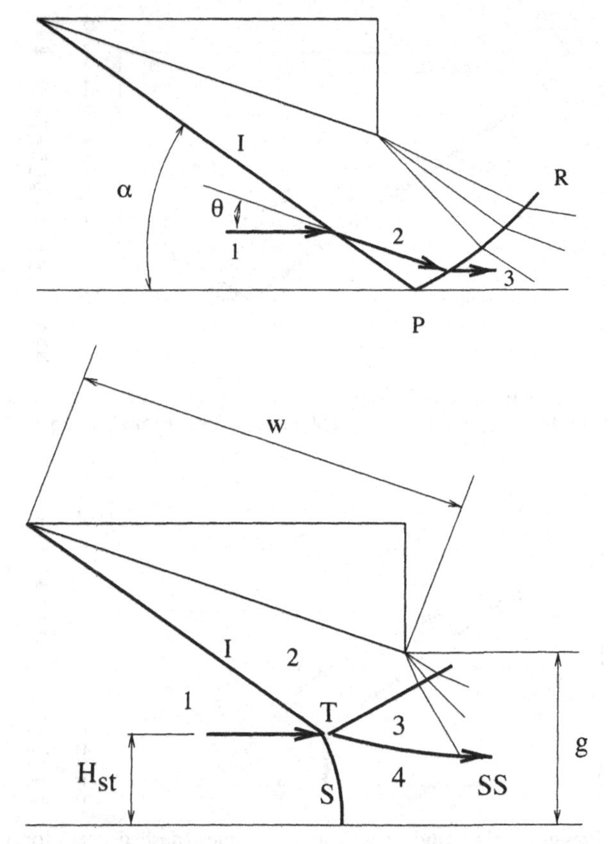

Figure 7.1. When a steady supersonic flow meets a wedge, an oblique shock wave is formed. If the latter impinges on a wall two kinds of reflection may occur, according to the circumstances detailed in the text: the regular reflection (upper picture) and the Mach reflection (lower picture). R denotes the regularly reflected shock wave, S the Mach stem, and T the quadruple point where the impinging wave, the Mach stem, the reflected stem, and a slipstream (SS) meet (From Ref. 55.)

There is an angle $\alpha_* > \alpha_N$ such that the Mach configuration exists down to α_*. Below this critical value α_* a regular reflection always occurs. The latter angle, however, depends on the Knudsen number and approaches α_N when the Knudsen number decreases (Fig. 7.5). This fact, together with the circumstance that the height of the Mach stem decreases and becomes comparable with the shock thickness for angles close to α_N can explain the difficulty of observing the effects in the experiments. It is remarkable, however, that the lack of uniqueness in the steady-state Euler equations is also present in the Boltzmann equation, enhanced by rarefaction effects.

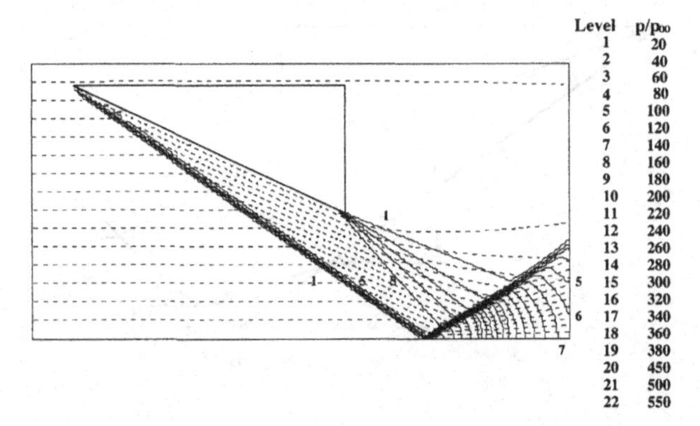

Figure 7.2. Pressure field (solid lines) and streamlines (dashed lines) for a simulated flow, initially uniform at $M = 16$, $Kn = 0.0025$, $\alpha = 35°$. (From Ref. 55.)

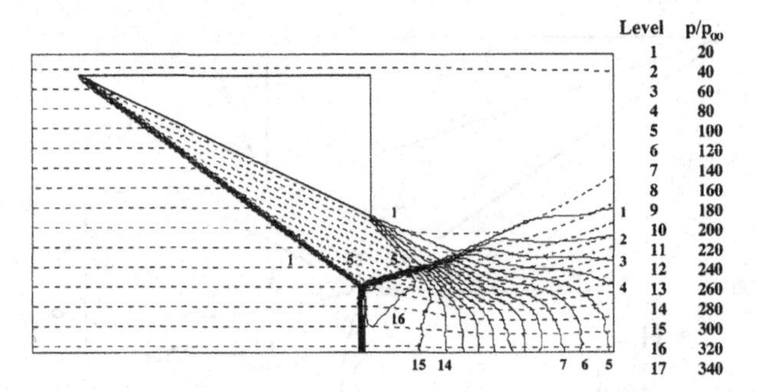

Figure 7.3. Pressure field (solid lines) and streamlines (dashed lines) for a simulated flow, initially uniform at $M = 16$, $Kn = 0.0025$, $\alpha = 36°$. (From Ref. 55.)

Three-dimensional DSMC calculations have also been made for the flow past a delta wing.[60] The results compare well with wind tunnel measurements[61] of the flow field under the same conditions.

Other important problems are related to separated flows, especially wake flows and flows involving viscous boundary layer separation and reattachment. The first calculations referred to the two-dimensional flow over a sharp flat plate followed by an angled ramp.[62] The results were in a reasonably good agreement with wind tunnel studies, which are not truly two-dimensional because of inevitable sidewall effects. Similar experiments were therefore performed[63] for the corresponding axially symmetric flow, which is less subject to the aforementioned nonuniformity. The DSMC calculations for these cases[64] show excellent agreement with experimental results. In particular, separation and reattachment of a viscous boundary layer in the laminar regime are correctly predicted.

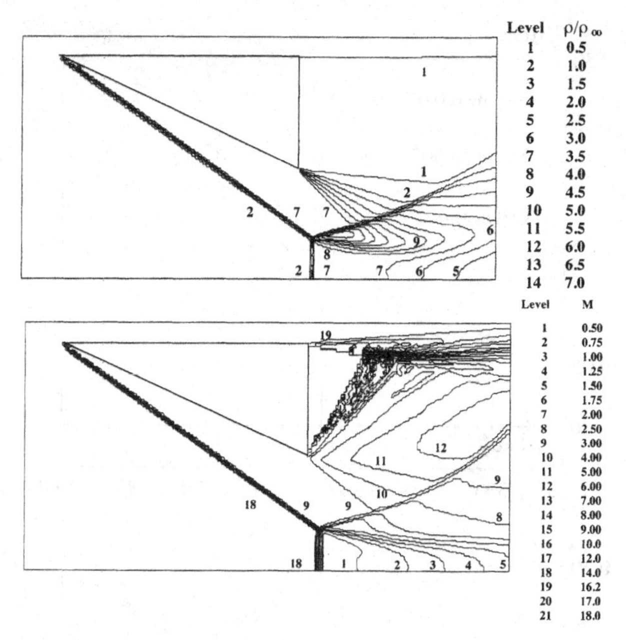

Figure 7.4. Density field (upper picture) and Mach number field (lower picture) for a simulated flow at $M = 16$, $Kn = 0.0025$, $\alpha = 35°$, and initially uniform. At variance with the case of Figure 7.1, the initial data are taken to be those of the steady solution of Figure 7.3. The Mach stem is now clearly formed. (From Ref. 55.)

The most remarkable wake flow simulation was for a 70° spherically blunted cone model that had been tested in several wind tunnels.[65,66] The calculations[67] of the lee side flow that contains the vortex are in good agreement with the experiments and with Computational Fluid Dynamics (CFD) studies of the flow based on the Navier–Stokes equations.

In the case of polyatomic gases one has several cross sections, such as elastic, rotational, vibrational, and also reactive, if chemical reactions occur. Koura[68] has extended his null collision technique [12] to these cases and improved it later.[69] He applied this method to simulate the hypersonic rarefied nitrogen flow past a circular cylinder,[69] with particular attention to the simulation of the vibrational relaxation of the gas; he also investigated the effect of changing the number of molecules in each (adaptive) cell and the truncation in the molecular levels. Some results are shown in Figs. 7.6–7.9.

The Knudsen number based on the diameter and the mean free path at infinity is 0.1, the Mach number 20, and the temperature at infinity 300 K. The cylinder surface is assumed to scatter the molecules with complete diffusion at the same temperature as the upstream one. One can easily see the formation of a bow

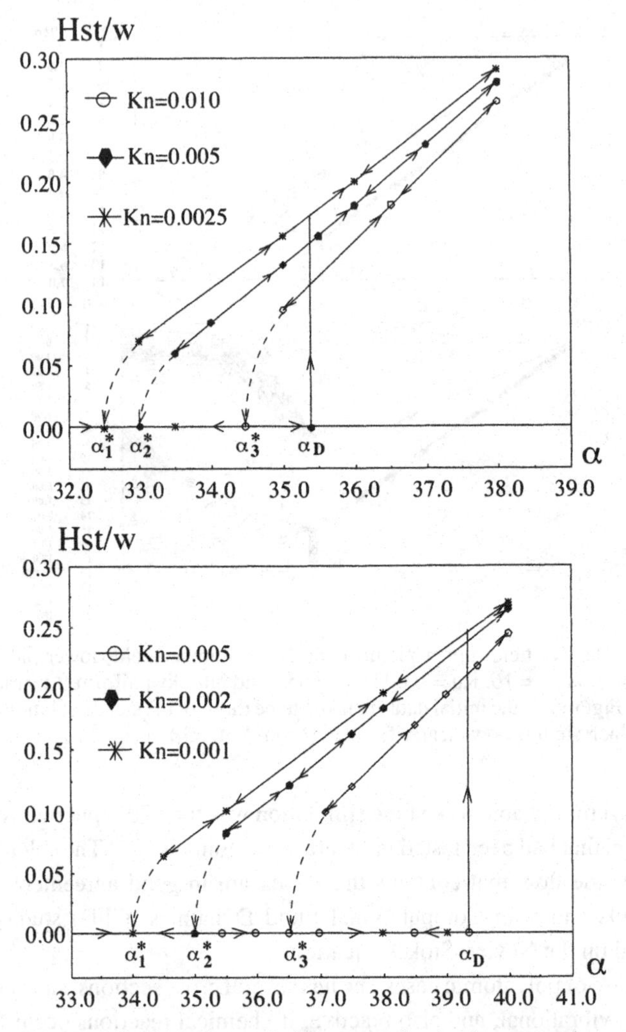

Figure 7.5. The normalized Mach stem height versus the shock wave angle α, as obtained from DMS. H_{st} denotes the height of the stem and w the hypothenuse of the wedge (see Figure 7.1). The upper picture refers to argon at M = 16, the lower one to nitrogen at M = 4.96. (From Ref. 55.)

shock. The peak density and translational, rotational, and vibrational temperatures on the stagnation line are $\rho/\rho_\infty = 35.5$, $T_{tr}/T_\infty = 94.2$, $T_r/T_\infty = 28.3$, and $T_v/T_\infty = 4.0$. The positions of the peaks of internal temperatures in front of the cylinders are closer to the cylinder than that of the translational temperatures. This reflects the physical circumstance that the former relax more slowly than the latter.

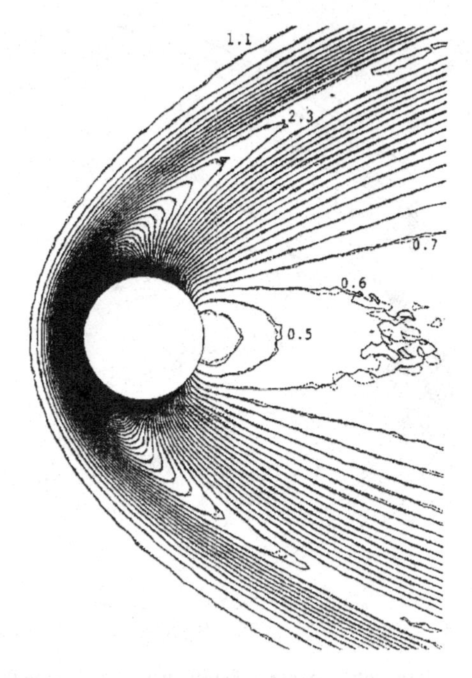

Figure 7.6. Simulation of the vibrational relaxation in nitrogen in the flow past a cylinder at $M = 20$ and $Kn = 0.1$. Here the number density contour lines obtained by the Improved Null Collision (INC) DSMC method are shown. The solid and dashed lines refer to 100 and 10 molecules in each adaptive collision cell. (From Ref. 69.)

7.4. Vortices and Turbulence in a Rarefied Gas

The Direct Simulation Monte Carlo method discussed at the end of the previous section is not only a practical tool for engineers but also a good method for probing into uncovered areas of the theory of the Boltzmann equation, such as stability of the solutions of this equation. An example has been mentioned in Section 7.3, when discussing the transition from regular to Mach reflection.

While the problems related to the instability of laminar flows and their transition to turbulence have been studied for a long time in classical hydrodynamics, the corresponding problems in kinetic theory have received little attention until recently. This circumstance is clearly related to the extremely complex character of the matter.

It is also clear, however, that the study of such problems might be of great importance for the purpose of understanding fundamental phenomena of instability and self-organization in molecular dynamics as well as computing hypersonic rarefied flows. Typical examples in this area are the Rayleigh–Bénard's instability,[70–73] the Taylor vortices,[74,75] and channel flow.[76]

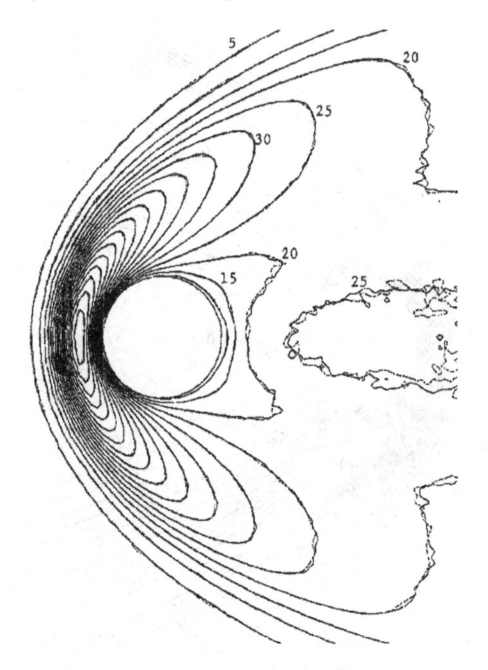

Figure 7.7. Simulation of the vibrational relaxation in nitrogen in the flow past a cylinder at $M = 20$ and $Kn = 0.1$. Here the translational temperature contour lines obtained by the INC-DSMC method are shown. The solid and dashed lines refer to 100 and 10 molecules in each adaptive collision cell. (From Ref. 69.)

Objections have occasionally been raised against the ability of the DSMC technique to faithfully resolve vortex motion, owing to the lack of accurate conservation of angular momentum in collisions. It seems that the first paper to cast doubts on this ability is by Meiburg,[77] who applied both molecular dynamics and Direct Simulation Monte Carlo to the case of a rarefied flow past an impulsively started inclined flat plate. A clear vortex structure was obtained in the wake region of the molecular dynamics calculation, but the wake in the Direct Simulation Monte Carlo was relatively devoid of structure. As pointed out by Bird,[78] the approximations associated with either method cannot be tested for the unsteady flow past a plate because this problem places excessive demands on computer time. He therefore introduced the forced vortex flow produced by a moving wall in a two-dimensional cavity as an alternative test case and came to the conclusion that as long as the cell size requirements are met (see Section 7.7), the lack of conservation of angular momentum in collisions does not appear to have a significant effect on the results of Direct Simulation Monte Carlo calculations. Bird[78] also examined the values of the parameters in

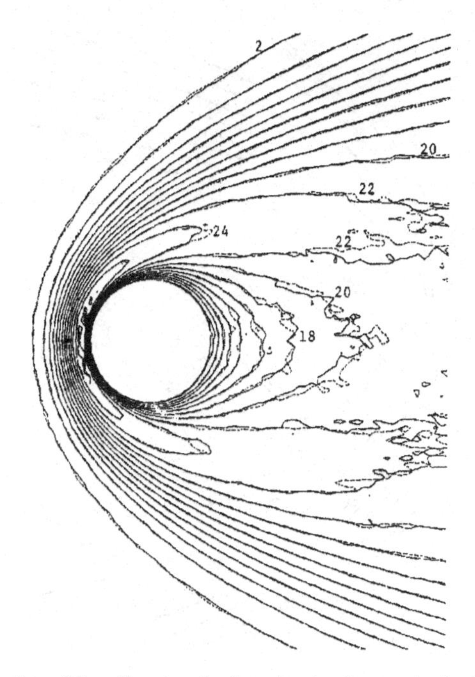

Figure 7.8. Simulation of the vibrational relaxation in nitrogen in the flow past a cylinder at $M = 20$ and $Kn = 0.1$. Here the rotational temperature contour lines obtained by the INC-DSMC method are shown. The solid and dashed lines refer to 100 and 10 molecules in each adaptive collision cell. (From Ref. 69.)

Meiburg's calculations[77] and showed that the density of the gas was too high to employ the DSMC method, because the mean free path was of the order of the molecular diameter and the size of the cell is too large to analyze the vortical wake structure. The issue of lack of conservation of angular momentum was addressed by Nanbu et al.,[79] who showed that if the cell size is sufficiently small, the total angular momentum of the molecules in a cell is almost conserved in the Direct Simulation Monte Carlo calculations.

The results obtained by Stefanov and Cercignani[70–74] (see also the book by Bird[80] and his paper[81], where the formation of Taylor vortices is discussed in some detail) have a preliminary character, but they clearly indicate that the formation of such vortex patterns arising from the aforementioned instabilities are possible. The calculations refer to values of the Knudsen number of order 10^{-2} and different choices of the other parameters (including high values of the Mach number).

Reichelman and Nanbu[82] have also applied the DSMC method to the Taylor–Couette flow in air. They chose the speed ratio and the range of values for

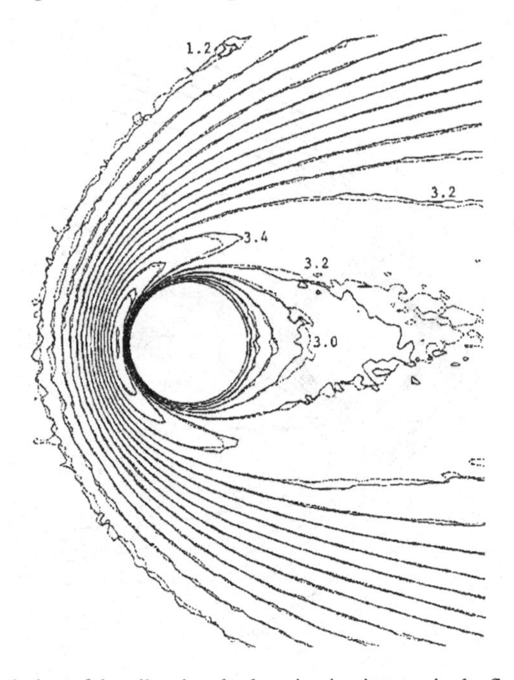

Figure 7.9. Simulation of the vibrational relaxation in nitrogen in the flow past a cylinder at M = 20 and Kn = 0.1. Here the vibrational temperature contour lines obtained by the INC-DSMC method are shown. The solid and dashed lines refer to 100 and 10 molecules in each adaptive collision cell. (From Ref. 69.)

the Knudsen number to match the experiments by Kulthau[83] and detected the formation of vortices through the behavior of the torque coefficient on the fixed cylinder. Their results were in excellent agreement with experiment.

Encouraged by the positive result of these investigations, Cercignani and Stefanov also considered the fluctuations of the macroscopic quantities in a rarefied gas flowing in a channel under the action of a constant external body force in a direction parallel to the walls (which are assumed to be at rest with the same temperature). Their ultimate aim was the study of the transition to turbulence by means of the Boltzmann equation, but they remained far from achieving this goal, since their calculations are restricted to a two-dimensional geometry.[76] The main aim of their calculations was to investigate the time evolution of both the macroscopic quantities and their fluctuations. The numerical experiments refer to values of the Knudsen number between .005 and .1 and three different values of the body force. The analysis of the results indicates an increase in the macroscopic fluctuation for Kn $\leq$ .05 and certain magnitudes of the force. In order to recognize the possible formation of vortex patterns and to estimate the macroscopic fluctuations, a data analysis was performed. If one takes into account the results of this analysis, one is tempted to conclude that one observes a transition from laminar flow to two-dimensional small-scale turbulence.

A common statement in the books and papers dealing with turbulence is that the scales of turbulence and molecular motions are widely separated and thus one may well ask what is the purpose of calculations of the kind discussed above. To answer this objection, one may remark that the common statement alluded to in the previous sentence may always be true for a liquid; but what about a gas? Let us consider the ratio between the mean free path ℓ and the dissipative scale of turbulence ℓ_D:

$$\ell/\ell_D = (\ell/L)(\mathrm{Re})^{3/4} = \mathrm{Kn}(\mathrm{Re})^{3/4} = (\mathrm{Ma})^{3/4}(\mathrm{Kn})^{1/4}, \tag{7.4.1}$$

where L is a macroscopic scale and the relation $\mathrm{Re} = \mathrm{Ma}/\mathrm{Kn}$ has been taken into account. If we take $\mathrm{Re} = 10^4$ for fully developed turbulence, we obtain that ℓ and ℓ_D are of the same order for $\mathrm{Kn} = 10^{-3}$ and $\mathrm{Ma} = 10$. Thus, in hypersonic flow, for moderate rarefaction, turbulence scale and mean free path are of the same order of magnitude!

These estimates are, of course, to be taken with caution, since there might be sizable factors "of order unity." These considerations should, however, not be ignored when considering high Mach number flows. This field is completely unexplored and is certainly worth more attention.

Recently, Stefanov et al.[84] have taken up again the problem of the Rayleigh–Bénard instability to investigate the occurrence of a secondary instability and the transition to a chaotic behavior for low values of the Knudsen number. They used both the DSMC method and Navier–Stokes finite difference calculations. Within a given interval of values of the gravity force both methods exhibit a secondary instability with a final flow regime of a permanently chaotic vortex formation. In nonlinear dynamics a solution of this kind is known under the name of "strange attractor." The existence of a "strange attractor," obtained by methods having rather different bases, opens a new path for the investigation of the transition to turbulence at the edge between molecular and continuum models. To clarify the idea let us give an example: In spite of the numerous results of Navier–Stokes calculations for various incompressible viscous flows showing chaotic behavior for long times, it turns out that the question is frequently raised whether the chaotic solutions of the Navier–Stokes equations produce a turbulent behavior adequately representing reality. The qualitative similarity of the results[84] obtained by using both molecular and continuum calculations gives an important argument for a positive answer: Both approaches exhibit the basic properties of the transition to a chaotic fluid motion.

The paper under discussion[84] contains a comparative analysis of molecule simulation and Navier–Stokes finite difference computations (with slip boundary conditions) of the two-dimensional Rayleigh–Bénard instability problem for $\mathrm{Kn} = 0.001$ and a temperature ratio $T_c/T_h = 0.1$. The Navier–Stokes computations have been completed for a set of Froude numbers $\mathrm{Fr} = [0.8, 1.0, 2.0]$. Due to the notable computational effort, the particle simulations has been run only for the case $\mathrm{Fr} = 1.0$ by using the "Finite-Pointset method,"[85] one of the variations of the DSMC, developed at Kaiserslautern.

The numerical calculations employed a hard-sphere molecular model. A monatomic simple gas with average number density n_0 was studied in a rectangular computational domain $D\left\{(y, z) \in (L' \times L)\right\}$ (i.e., with an aspect ratio $A = L'/L$). The domain is confined in the vertical direction by two diffusively reflecting horizontal walls (the lower wall temperature is greater than the upper one, $T_h > T_c$) and a periodicity condition is implemented at two vertical planes at a mutual distance L'. A constant acceleration $\mathbf{g} = (0, g)$ acts on the gas in each point of the computational domain. As mentioned above, for certain sets of the governing parameters, the final state of the flow is a strange attractor and consequently should be calculated by using an unsteady numerical approach.

To avoid the long transient period of gas stratification to a state when the solution loses its stability and a secondary convective motion begins developing (the first bifurcation of the solution) the time evolution of the solution was computed in two stages. In the first stage, the flow is calculated as one dimensional with macroscopic gradients in the z direction only. The computations start from a uniform gas state with temperature equal to T_h. The steady-state solution obtained for a large time $t_0 \gg 0$ serves as an initial state for the second stage (this procedure was first used by Bird for his DSMC simulation of the Taylor–Couette flow[81]). In the second stage the solution is sought in a two-dimensional domain: The flow field is extended and reorganized as a two-dimensional flow for $(y, z) \in (L' \times L)$, uniform in the y direction. Unlike the DSMC simulated state, which contains macroscopic fluctuations naturally, the Navier–Stokes solution require that the disturbances be imposed artificially and in the case under consideration they are included in the temperature field.

The governing nondimensional parameters are the Knudsen number Kn, the Froude number Fr and the temperature ratio T_c/T_h.

Low Knudsen number calculations are very complicated and time consuming. Thus the authors of the paper under discussion[84] reported only the results concerning $\mathrm{Kn} = 0.001$ and $T_c/T_h = 0.1$.

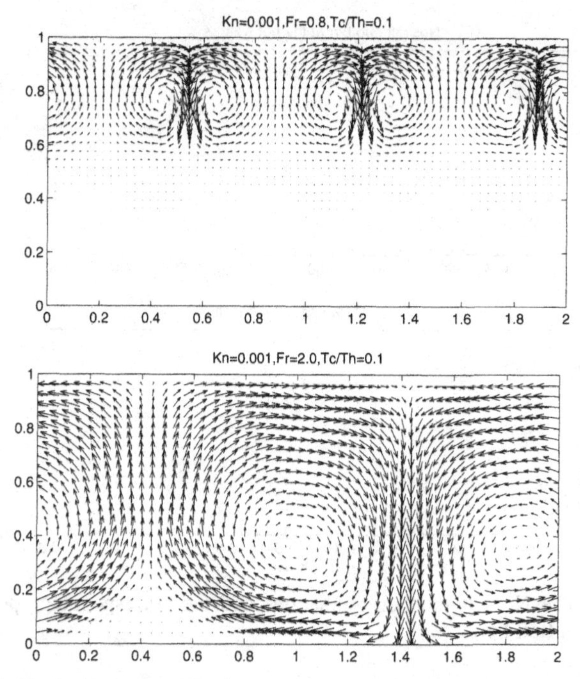

Figure 7.10. Vortical structures for Fr = 0.8 (upper picture) and Fr = 2.0 (lower picture). In the first case the flow near the lower plate is so low that the arrows are barely visible. (From Ref. 84.)

The Navier–Stokes computation was tested on rectangular uniform grids with 100×50, 200×100, and 400×200 nodes respectively, covering a computational domain $L' \times L$ with an aspect ratio $a = L'/L = 2$. A grid with 200×100 nodes turned out to give stable results with good accuracy for the purpose of obtaining a solution sufficiently accurate for a meaningful comparison.

The particle simulation where performed on a fixed domain with size 300×100 but using a super-fine grid of 1536×512 cells with a total number of particles of about 7.86×10^6. The total CPU time to run the simulations over 32.000 discrete time steps was about 250 hours using 64 nodes of the parallel system nCUBE 2S.

It is worth noting that for low Knudsen number flows both the microscopic and macroscopic space scales are clearly separated and this leads naturally to the use of two computational grids. The same argument is valid for the time scales. Thus, the use of two grids allows an additional time averaging over an interval short with respect to the macroscopic time scale but possibly very large with respect to the kinetic scale. In Fig. 7.10 two limiting

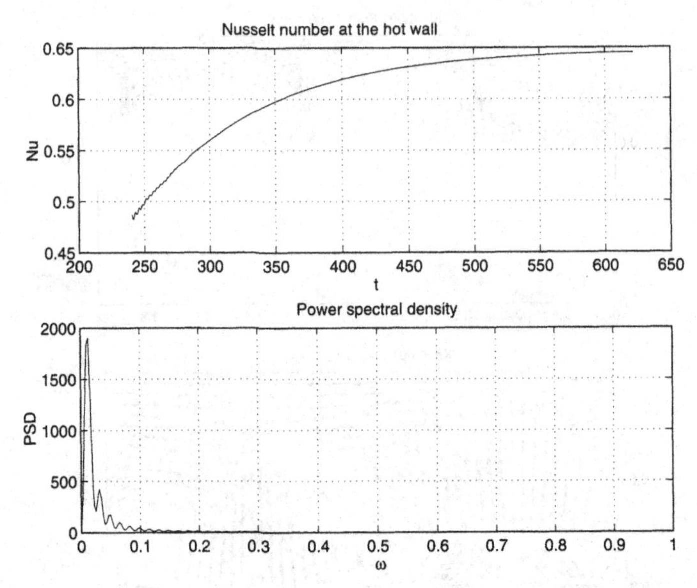

Figure 7.11. The Nusselt number oscillations and the corresponding spectrum for Fr = 0.8. (From Ref. 84.)

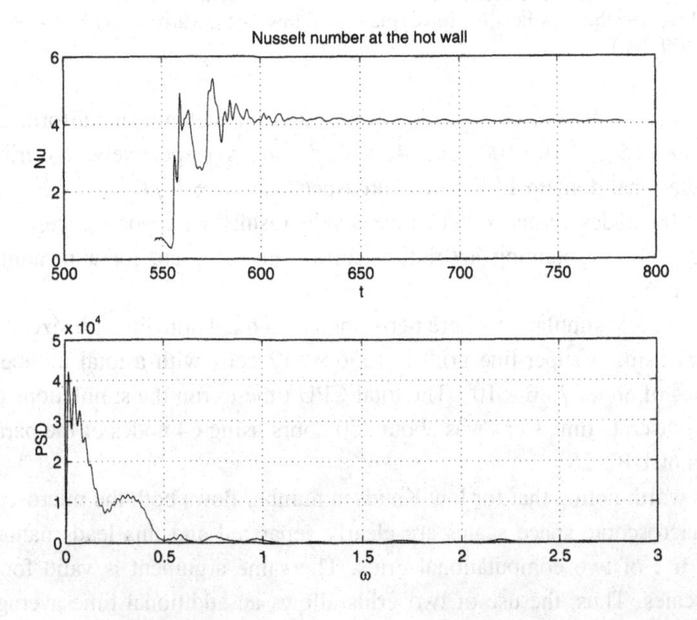

Figure 7.12. The Nusselt number oscillations and the corresponding spectrum for Fr = 2.0. (From Ref. 84.)

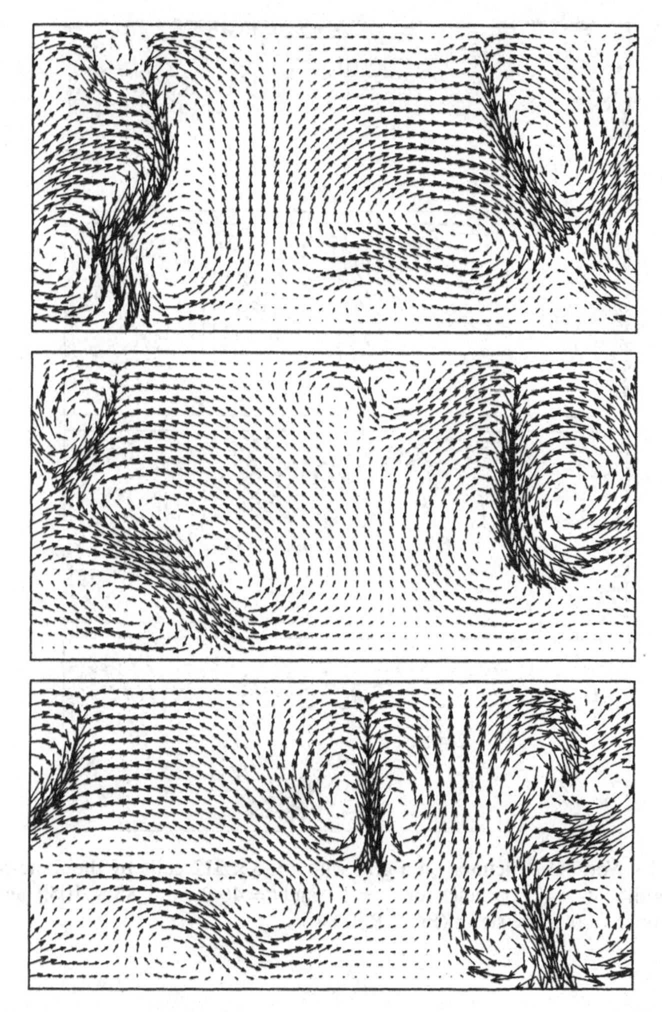

Figure 7.13. Evolving vortical structures on a grid with 200 × 100 cells by NS-FDS for Fr = 1.0 at $t = 1414.0, 1427.9, 1434.0$. (From Ref. 84.)

cases Fr = 0.8 and Fr = 2.0 having a stable vortex structure are presented. The calculations have been performed by using the Navier–Stokes equations but the simulations performed for the same Froude numbers by using Bird's DSMC method[84] showed the same stable vortex structures. Figures 7.11 and 7.12 present the corresponding time evolution of the Nusselt number ($\mathrm{Nu} = q_w L/(\lambda_h T_h)$, where $q_w = \lambda(dT/dz)$), and their spectra for two different

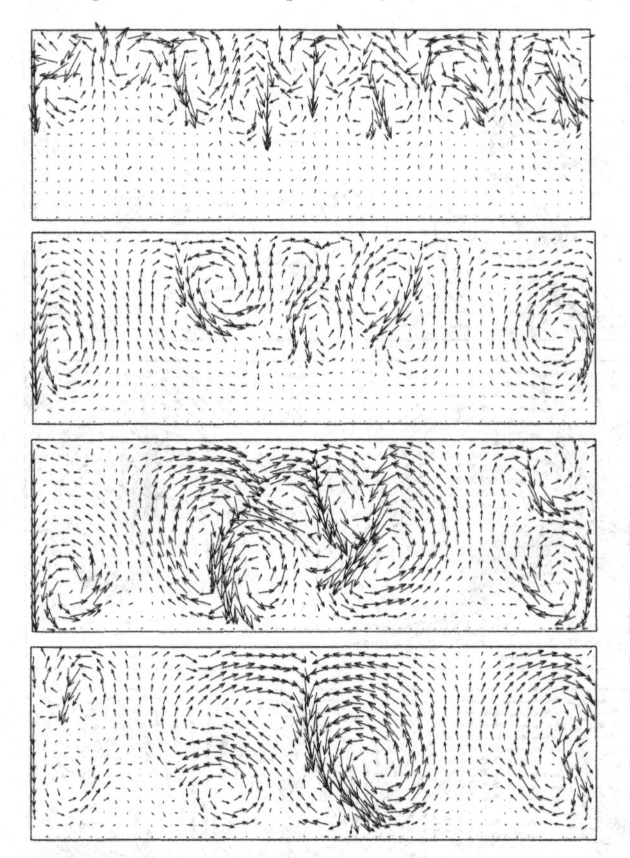

Figure 7.14. Vortical structures on a grid with 1536×512 cells by the finite-pointset method with 7.86×10^6 particles for $\mathrm{Fr} = 1.0$ and $t = 46.2, 66.0, 85.8, 105.6$ and mean averages over 1,000 time steps. (From Ref. 84.)

values of the Froude number. Both methods demonstrate a completely different behavior of the solution for $\mathrm{Fr} = 1.0$, shown in Figs. 7.13 and 7.14.

The two methods start from different initial states; thus it is difficult to make a quantitative comparison of the time evolution of the solutions, but the main remark is that the evolutions of the vortex structure obtained from either calculation have some similarity although they are very complicated and not repeatable.

The influence of the chaotic vortex formation on the nondimensional heat transfer (the Nusselt number) at the hot (respectively, cold) wall is shown in Figs. 7.15 and 7.16. One can observe that the oscillations at the walls have different spectra, which almost fill the whole interval of circular frequencies ω.

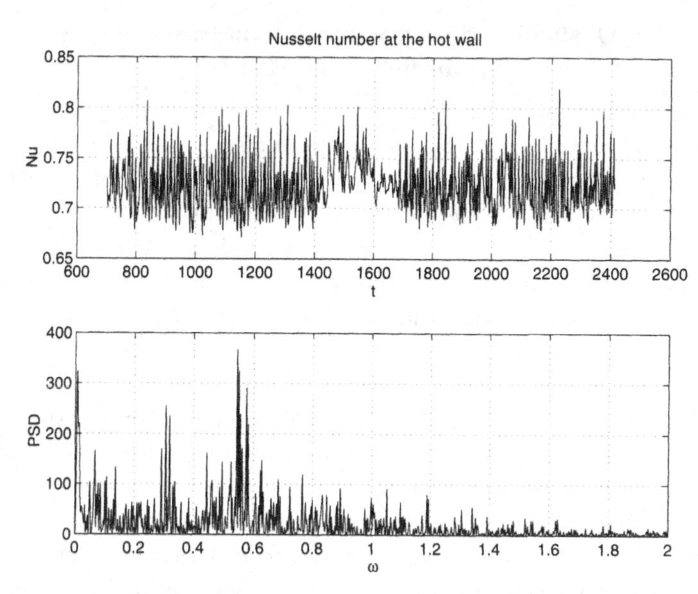

Figure 7.15. The Nusselt number oscillations at the hot wall and the corresponding spectrum for Fr = 1.0, obtained by using NS-FDS. (From Ref. 84.)

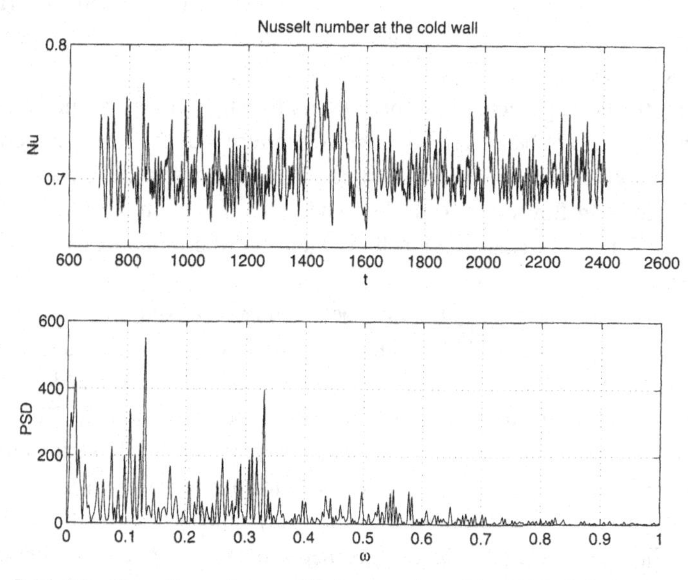

Figure 7.16. The Nusselt number oscillations at the cold wall and the corresponding spectrum for Fr = 1.0, obtained by using NS-FDS. (From Ref. 84.)

7.5. Qualitative Differences between the Navier–Stokes and the Boltzmann Models

The results discussed in the previous section show that there is a qualitative agreement between the Boltzmann equation and the Navier–Stokes equations for sufficiently low values of the Knudsen number. This agreement is also borne out by the computations in unsteady flows presented by Stefanov et al.[86]

There are however flows where this agreement does not occur. They have been especially studied by Sone and his coworkers.[87] In addition to the thermal creep induced along a boundary with a nonuniform temperature (Chapter 5), two new kinds of flow are induced over boundaries kept at uniform temperatures. They are related to the presence of thermal stresses in the gas.

The first effect[87–89] is present even for small Mach numbers and small temperature differences and follows from the fact that there are stresses related to the second derivatives of the temperature. Although these stresses do not change the Navier–Stokes equations, they change the boundary conditions; the gas slips on the wall, and thus a movement occurs even if the wall is at rest. This effect is particularly important in small systems, such as micromachines, since the temperature differences are small but the temperature itself may have relatively large second derivatives; it is usually called the thermal stress slip flow.[87–89]

The second effect is nonlinear[90,91] and occurs when two isothermal surfaces do not have a constant distance (thus in any situation with large temperature gradients, in the absence of particular symmetries).

These effects may occur even for sufficiently large values of the Knudsen number; they cannot be described, however, in terms of the local temperature field. They rather depend on the configuration of the system. They should not be confused with flows due to the presence of a temperature gradient along the wall, such as the transpiration flow[92] and the thermophoresis of aerosol particles.[93]

An example of this flow is the flow between two coaxial elliptic cylinders, which has been studied by Aoki et al.[94] They simulated a rarefied gas confined in the gap between two such cylinders, with axes $(a_0 L, b_0 L)$ and $(a_1 L, L)$ $(a_0 < a_1,\ b_0 < 1)$ with different uniform temperatures T_0 and T_1. They took $T_1/T_0 = 5$, $a_0 = 0.3$, $b_0 = 0.7$, and $a_1 = 1.5$ and performed calculations for Kn ranging between 0.1 and 5. Some velocity fields are reproduced in Fig. 7.17. To see the effect of the temperature ratio, they considered, for Kn = .5, the ratio $T_1/T_0 = 2$. To see the effect of geometry, for the same Knudsen number, we show in Fig. 7.18 the case when the inner cylinder is circular, with radius $a_0 = b_0 = 0.6$.

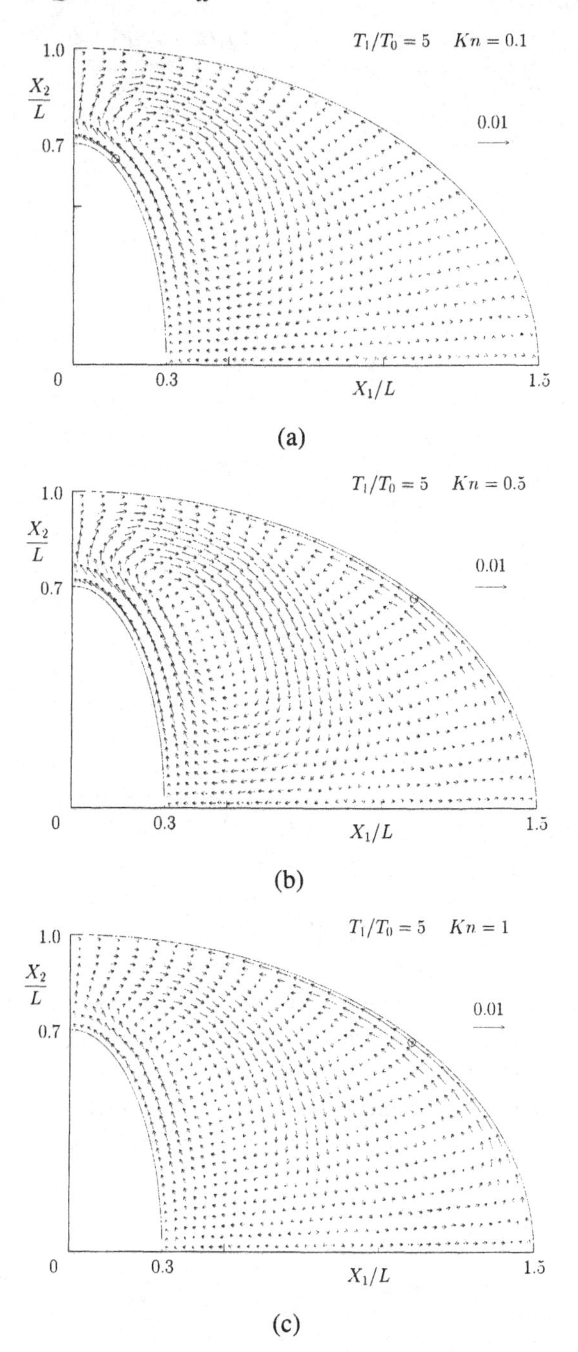

Figure 7.17. Flow between elliptic cylinders induced by a temperature difference. The flow velocity field for $a_0 = 0.3$, $b_0 = 0.7$, $a_1 = 1.5$, $T_1/T_0 = 5$, and (a) Kn = 0.1; (b) Kn = 0.5; (c) Kn = 1; (d) Kn = 2; (e) Kn = 5 is shown. (From Ref. 94.)

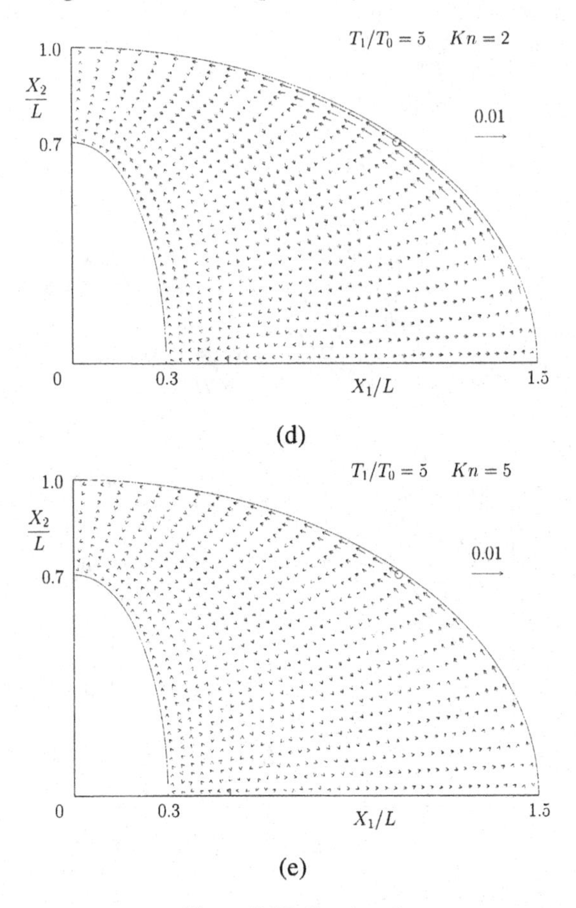

Figure 7.17. *Continued*

Another example where similar effects occur is the flow induced in a rectangular cavity containing a flat plate with sharp edges, kept at a temperature different from that of the boundary of the cavity.[95] A striking difference is that, in this case, a very steep temperature gradient in the direction parallel to the plate arises in the gas near the edges of the plate, and thus a localized but intense flow arises there (Fig. 7.19).

The fact that the solutions of the Boltzmann equation frequently provide macroscopic flow fields practically indistinguishable from those obtained with Navier–Stokes equations suggests that one may reduce the problems of memory and computing time in numerical calculations, if the Boltzmann equation is replaced by the Navier–Stokes equation, or even the Euler equations, in the

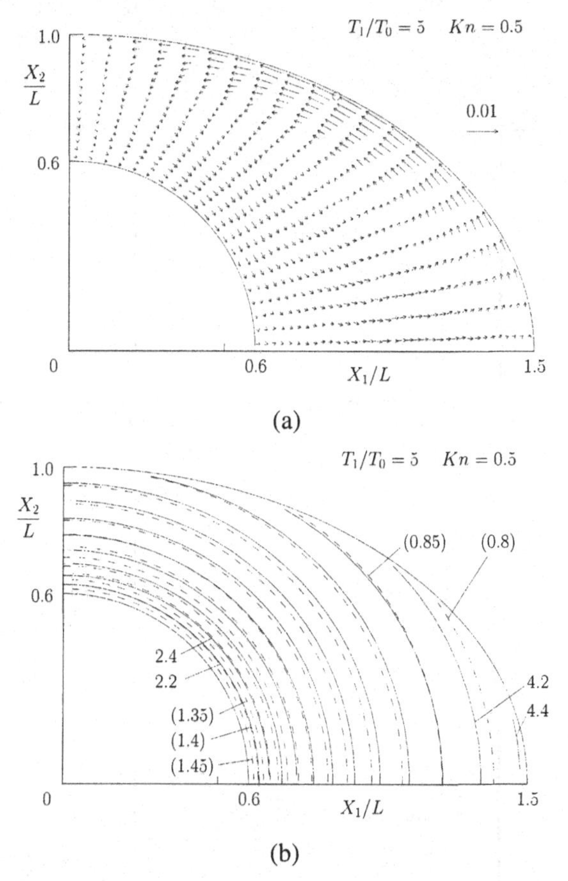

Figure 7.18. Flow between an elliptic and a circular cylinder induced by a temperature difference. (a) The flow velocity field for $a_0 = b_0 = 0.6$, $a_1 = 1.5$, $T_1/T_0 = 2$, and Kn = 0.5. (b) The isodensity lines for $a_0 = b_0 = 0.6$, $a_1 = 1.5$, $T_1/T_0 = 2$, and Kn = 0.5. (From Ref. 94.)

regions far away from shock waves and solid boundaries where the most remarkable kinetic effects occur.

This strategy of *domain decomposition* requires suitable interface conditions where the two kinds of regions (Boltzmann and continuum) come into contact. Usually the coupling conditions are determined by equating the flows of mass, momentum, and energy at the interfaces.

More accurate results are obtained by studying a connection layer[96,97] similar to the Kramers' problem discussed in Chapter 3. The main difference is the fact that the Boltzmann equation is linearized about a drifting Maxwellian

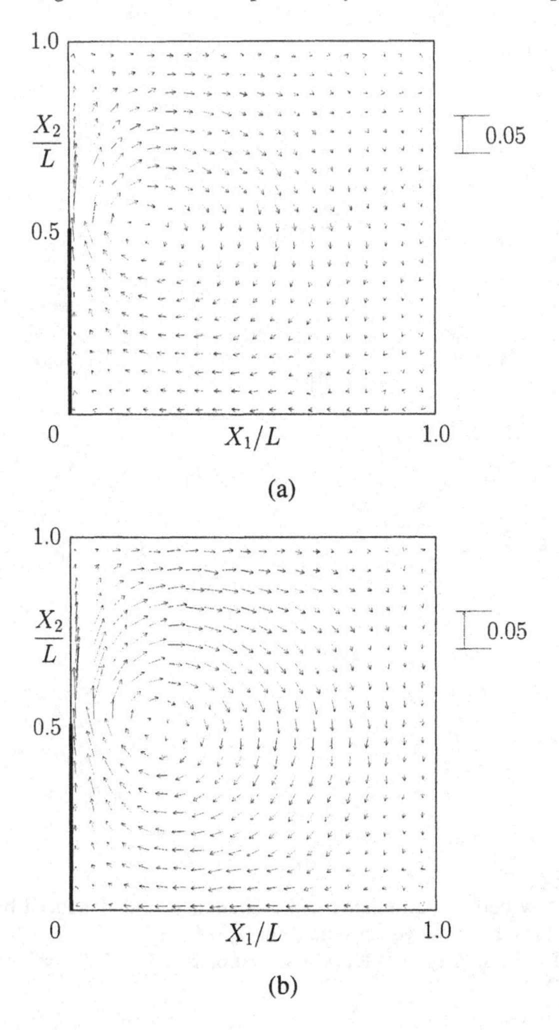

Figure 7.19. Flow velocity near a sharp edge. Arrows indicate the nondimensional velocity $\mathbf{v}(2RT_0)^{-1/2}$. One quarter of the wave field is shown. $T_1/T_0 = 5$. (a) Kn = 0.05; (b) Kn = 0.1; (c) Kn = 0.2; (d) Kn = 0.5; (e) Kn = 2; (f) Kn = 5. (From Ref. 95.)

distribution, thus originating a problem that first showed up in the study of Knudsen layers near an evaporating surface[98,99] (see also Chapter 8).

7.6. Discrete Velocity Models

According to standard definitions, a discrete velocity model[100,101] of a gas is a system of partial differential equations of hyperbolic type (discrete Boltzmann

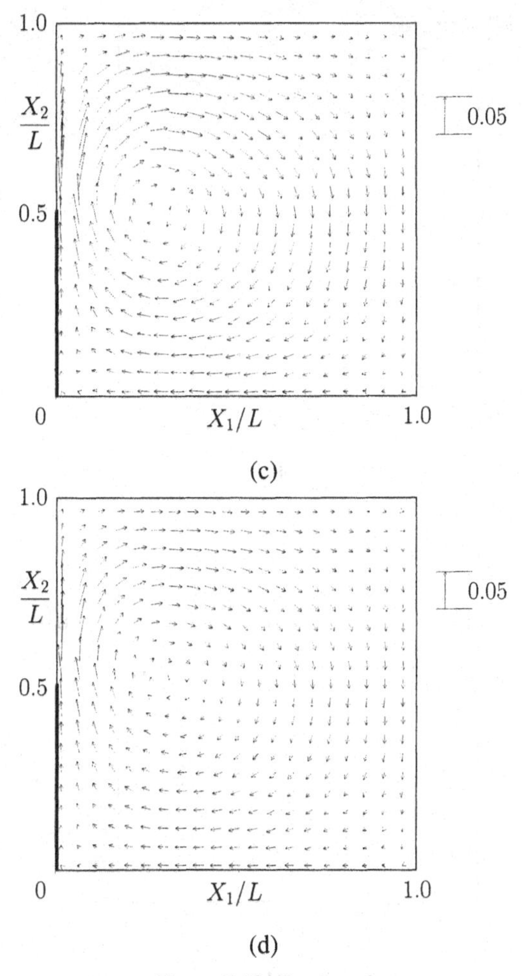

Figure 7.19. *Continued*

equation), having the following form

$$\partial f_{\mathbf{n}}/\partial t + \mathbf{v}_{\mathbf{n}} \cdot \partial f_{\mathbf{n}}/\partial \mathbf{x} = Q_{\mathbf{n}}(f, f), \tag{7.6.1a}$$

$$Q_{\mathbf{n}}(f, f) = \sum_{\mathbf{lk}} c_{\mathbf{nlk}} f_{\mathbf{l}} f_{\mathbf{k}} - \sum_{\mathbf{m}} k_{\mathbf{nm}} f_{\mathbf{n}} f_{\mathbf{m}}, \tag{7.6.1b}$$

where $\mathbf{v}_{\mathbf{n}}$ are the discrete velocities (vectors of R^3) belonging to a prearranged discrete set, $c_{\mathbf{nlk}}$ and $k_{\mathbf{nl}}$ are positive constants, and the vector indices run over a discrete set of vectors with integer components, symmetrical with respect to the origin, whereas $f_{\mathbf{n}}$ are the probabilities (per unit volume) of finding a molecule

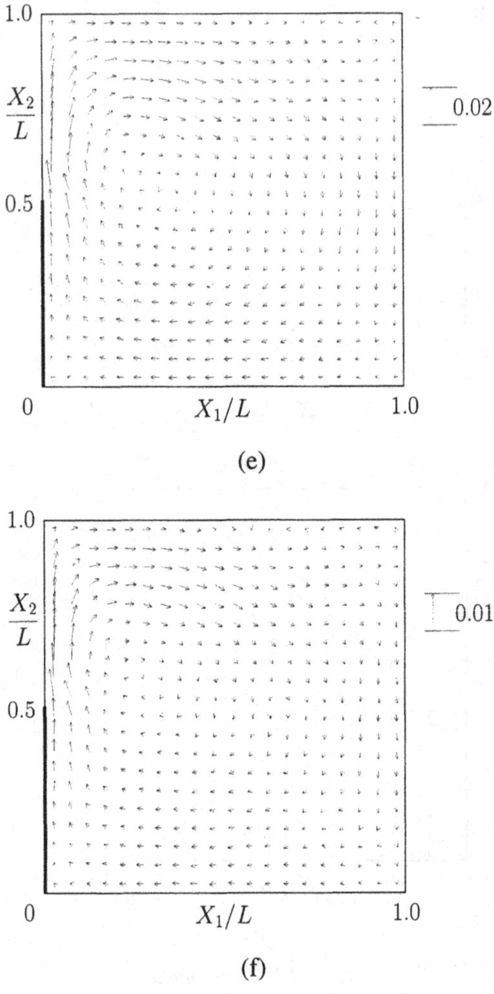

Figure 7.19. *Continued*

at time t at position $\mathbf{x}$ with velocity $\mathbf{v_n}$. We shall occasionally write f for the collection $\{f_\mathbf{n}\}$, as done in the collision term of (7.6.1a) and (7.6.1b).

We certainly must assume that the collision term $Q_\mathbf{n}(f, f)$ satisfies the restrictions needed to guarantee the conservation of mass, momentum, and energy and the entropy inequality. We remark that this seems to be a generalization with respect to the traditional concept of discrete velocity models, where it is assumed that each single collision satisfies momentum and energy conservation.

However, if we only assume that

$$\sum_{\mathbf{n}} Q_{\mathbf{n}}(f, f)\psi_{\alpha\mathbf{n}} = 0 \quad (\alpha = 0, 1, 2, 3, 4), \tag{7.6.2}$$

where ψ_α ($\alpha = 0, 1, 2, 3, 4$) are the five collision invariants $(1, v_1, v_2, v_3, |\mathbf{v}|^2)$ and v_j ($j = 1, 2, 3$) are the Cartesian components of $\mathbf{v}$, we lose several important properties of the Boltzmann equation.

We pause a moment to discuss what the physical interpretation of these extended discrete velocity models could be. If we accept Eq. (7.6.2) as the only restriction on $Q_{\mathbf{n}}(f, f)$, a collision is a more complicated process than in the continuous velocity model. When two molecules meet they undergo a not completely deterministic process, in the sense that we cannot guarantee that another pair will emerge from the collision with certain velocities but only that the precollision momentum and energy will be distributed with a certain probability to a number of pairs; this is true even in the continuous Boltzmann equation because the collision parameters also determine the postcollisional velocities. Here we would go a step further because Eq. (7.6.2) does not require that we exhibit possible pairs of these velocities; thus there are no elementary processes where mass, momentum, and energy are conserved, but we only ensure that momentum and energy are conserved globally.

We need, however, also to assume the validity of an H-theorem and this can be shown to require the conservation of momentum and energy in each single collision, thus ruling out the possibility that we have just discussed. As a consequence of the new assumption, however, we can also show the existence of a (discrete) Maxwellian distribution $\mathcal{M}_{\mathbf{n}}$, such that $Q(\mathcal{M}, \mathcal{M}) = 0$ and $\log \mathcal{M}$ is a linear combination of the collision invariants.

We shall henceforth assume that[100]

$$c_{\mathbf{nlk}} = \sum_{\mathbf{m}} A^{\mathbf{k},\mathbf{l}}_{\mathbf{n},\mathbf{m}}, \tag{7.6.3}$$

$$k_{\mathbf{nm}} = \sum_{\mathbf{k},\mathbf{l}} A^{\mathbf{k},\mathbf{l}}_{\mathbf{n},\mathbf{m}}, \tag{7.6.4}$$

with

$$A^{\mathbf{k},\mathbf{l}}_{\mathbf{n},\mathbf{m}} = A^{\mathbf{n},\mathbf{m}}_{\mathbf{k},\mathbf{l}} = A^{\mathbf{m},\mathbf{n}}_{\mathbf{l},\mathbf{k}} \geq 0. \tag{7.6.5}$$

These assumptions make it easy to satisfy the conservation equations and the H-theorem and might be slightly relaxed.

The usual macroscopic quantities are computed, as usual, by taking averages, weighted with f_i, of the microscopic quantities. Thus the gas density is obtained from

$$\rho(x,t) = \sum_i f_i(x, v_i, t).$$

Starting with the mid 1970s, a considerable amount of research was devoted to discrete velocity models. By the early 1990s discrete velocity models (DVMs) were also becoming a tool to approximate the solutions of the Boltzmann equation, at a theoretical if not at a practical level.[102–105]

A breakthrough came in 1995 when Bobylev, Palczewski, and Schneider proved a consistency result for the DVM as an approximation of the Boltzmann equation.[106]

The extension of DVMs to mixtures poses no special problems for the case of a rational ratio between masses (this limitation is, of course, irrelevant in practice). Yet, only trivial models with a low number of velocities have appeared till 1998, when a thorough discussion of the matter was given and a nontrivial model was proposed.[107] Yet, this model turns out to have spurious collision invariants. It is hard but not impossible to produce models without this inconvenience.[108]

We now introduce a regular lattice as a grid in velocity space with step h such that the grid points will have position vectors

$$\mathbf{v}_{\mathbf{n}} = \mathbf{n}h. \tag{7.6.6}$$

Henceforth we shall consider this set of velocity vectors or some subset of the said set. When we consider the entire set, Galilean invariance is lost but can be replaced by a discrete symmetry (invariance with respect to translation rectilinear motions having velocity $\mathbf{n}_0 h$, where $\mathbf{n}_0$ is any chosen vector with integer components). Models for gases (even polyatomic) on a regular grid have been discussed and used by Goldstein.[109]

The coefficients $A_{\mathbf{l},\mathbf{k}}^{\mathbf{m},\mathbf{n}}$ must vanish if the following conservation equations are not satisfied:

$$\mathbf{l} + \mathbf{k} = \mathbf{m} + \mathbf{n}, \tag{7.6.7}$$

$$|\mathbf{l}|^2 + |\mathbf{k}|^2 = |\mathbf{m}|^2 + |\mathbf{n}|^2. \tag{7.6.8}$$

The most natural and popular model was first proposed by Goldstein, Sturtevant, and Broadwell in 1989.[102] The proof of consistency for this model was provided in Ref. 106 (see also some subsequent papers[110,111]).

We can generalize this kind of model (and the proof of consistency) to the case of mixtures, following the aforementioned paper.[107]

To this end, we consider a mixture with rational masses $m_1, m_2, \ldots, m_s$, where s is the number of species. If we exclude irrational ratios for the masses, without loss of generality we can assume the masses to be given by integers, by a suitable choice of the mass unit.

Let us consider any pair of molecules with masses m_i, m_j and let us put $m_i = m$, $m_j = M$ $(m < M)$. The usual Boltzmann equation for mixtures (with continuous velocities) has the following form:

$$\partial f_i/\partial t + \mathbf{v}\cdot\partial f_i/\partial\mathbf{x} = \sum_j Q_{ij}, \tag{7.6.9}$$

where

$$Q_{ij} = \int d\mathbf{v}d\omega|\mathbf{u}|\sigma(\mu|\mathbf{u}|^2/2, \mathbf{u}\cdot\omega/|\mathbf{u}|)[f(\mathbf{v}')F(\mathbf{w}') - f(\mathbf{v})F(\mathbf{w})]. \tag{7.6.10}$$

Here $f(\mathbf{v}) = f_i(\mathbf{v})$, $F(\mathbf{v}) = f_j(\mathbf{w})$, and

$$\mathbf{u} = \mathbf{v} - \mathbf{w}, \qquad \mu = \frac{mM}{m+M}, \tag{7.6.11}$$

$$\begin{aligned}
\mathbf{v}' &= \frac{m\mathbf{v} + M\mathbf{w}}{m+M} + \frac{|\mathbf{u}|\omega}{2}\frac{\mu}{m},\\
\mathbf{w}' &= \frac{m\mathbf{v} + M\mathbf{w}}{m+M} - \frac{|\mathbf{u}|\omega}{2}\frac{\mu}{M}.
\end{aligned} \tag{7.6.12}$$

The first step toward obtaining a form of the collision term suitable for arriving at discrete velocity models is to adopt $\mathbf{u} = \mathbf{v} - \mathbf{w}$ as an integration variable. We obtain

$$Q_{ij} = \int d\mathbf{u}d\omega|\mathbf{u}|\sigma(\mu|\mathbf{u}|^2/2, \mathbf{u}\cdot\omega/|\mathbf{u}|)[f(\mathbf{v}')F(\mathbf{w}') - f(\mathbf{v})F(\mathbf{v}-\mathbf{u})], \tag{7.6.13}$$

where the primed variables must be expressed according to

$$\begin{aligned}
\mathbf{v}' &= \mathbf{v} + \frac{M}{m+M}(|\mathbf{u}|\omega - \mathbf{u}),\\
\mathbf{w}' &= \mathbf{v} - \mathbf{u} - \frac{m}{m+M}(|\mathbf{u}|\omega - \mathbf{u}).
\end{aligned} \tag{7.6.14}$$

We again change the variables by letting $\mathbf{u} = (m+M)\tilde{\mathbf{u}}$ and then omitting the tilde. The result reads as follows:

$$\begin{aligned}
Q_{ij} &= (m+M)^{1/2}\int d\mathbf{u}d\omega|\mathbf{u}|\sigma(mM(m+M)|\mathbf{u}|^2/2, \mathbf{u}\cdot\omega/|\mathbf{u}|)\\
&\quad\times[f(\mathbf{v} - M\mathbf{u} + M|\mathbf{u}|\omega)F(\mathbf{v} - M\mathbf{u} - m|\mathbf{u}|\omega)\\
&\quad - f(\mathbf{v})F(\mathbf{v} - (M+m)\mathbf{u})].
\end{aligned} \tag{7.6.15}$$

In the following it will be useful, for any vector $\mathbf{u}'$, to use the following notation:

$$\Psi(\mathbf{u},\mathbf{u}') = |\mathbf{u}|\sigma(mM(m+M)|\mathbf{u}|^2/2, \mathbf{u}\cdot\mathbf{u}'/|\mathbf{u}|^2) \times [f(\mathbf{v} - M\mathbf{u} + M\mathbf{u}') \times F(\mathbf{v} - M\mathbf{u} - m\mathbf{u}') - f(\mathbf{v})F(\mathbf{v} - (M+m)\mathbf{u})]. \tag{7.6.16}$$

We have the following lemma of discrete approximation:[107]

Lemma. *Let $\Psi(\mathbf{u},\mathbf{u}') : \Re^3 \times \Re^3 \Rightarrow \Re$ be continuous and have a compact support. Let us introduce*

$$S_h(\Psi) = \sum_{\mathbf{n}\in\mathbf{Z}^3} h^3 \frac{|\Omega_3|}{r_3(|\mathbf{n}|^2)} \sum_{|\mathbf{m}|^2 = |\mathbf{n}|^2} \Psi(\mathbf{n}h, \mathbf{m}h), \tag{7.6.17}$$

where $|\Omega_3|$ is the area of the unit sphere in three dimensions and $r_3(|\mathbf{n}|^2)$ denotes the number of the roots of the equation $|\mathbf{m}|^2 = |\mathbf{n}|^2$, where the vectors with integer components **m** *and* **n** *denote, respectively an unknown and a given vector. Then*

$$S_h(\Psi) \to \int d\mathbf{u}d\omega\Psi(\mathbf{u}, |\mathbf{u}\omega|) \quad as \quad h \to 0. \tag{7.6.18}$$

This lemma provides the desired approximation and hence a rule to construct discrete velocity models for mixtures with arbitrarily many velocities.

We remark that the previous lemma is a slight generalization of a statement first made in another paper[106] and is related to classical problems of number theory.[112,113] The problem is full of subtleties and one can also provide an estimate of the error in the quadrature formula given above.[106] The starting point is that any number which is not congruent to 7 (mod. 8) (i.e., which does not give 7 as a remainder when divided by 8) can be represented as a sum of three squares and the number of possible representations grows sufficiently fast with the size of the number. The main problems do not arise from numbers congruent to 7 (mod. 8) (in fact they never arise since we equate the sum of three squares to a number which is known already to be the sum of three squares) but from special sequences of the form $\{4^n m_0\}$, where m_0 is a fixed number prime with 4 and n grows; in fact for these sequences the number of roots $r_3(4^n m_0)$ grows rather slowly with n. The difficulty can be easily overcome from a theoretical viewpoint. A key tool of the proof is the strong number-theoretical result (uniform distribution of integer points on the surface of a sphere) obtained by Iwaniec in 1987 (details are given in two papers[112,113] mentioned above).

If we assume that the velocities lie in a plane, matters are more complicated (there are considerably fewer integers that are sums of two, rather than three, squares), and the proof of the above lemma does not hold.

As we hinted at above, the main difficulty for the practical use of DVMs is the small number of pairs of discrete postcollisional velocities for a given pair of precollisional velocities. At a theoretical level, this difficulty was brilliantly overcome with the lemma described above, but in practice one needs a very refined grid and the cost of computation becomes prohibitive on present computers. This motivated a paper by Buet et al.,[114] in which two modifications of the DVMs are proposed.

Note that the number of discrete velocities is sufficiently small, it is possible to find exact solutions of DVMs. This area was systematically investigated by Cornille[115–117] and leads, in particular, to exact solutions for the shock wave structure problem.

7.7. Concluding Remarks

The area of numerical simulation of rarefied flows is rapidly expanding. On one hand, industrial processes (micromachines, deposition techniques, etc.) require more and more computer simulations. On the other hand, as we have seen, the methods may become competitive with the numerical solution of the Navier–Stokes equations even for fairly low values of the Knudsen number.

We must add a word of caution here. We have not discussed in detail how to choose the size of the cells, although it was clear that when we compute the probability of a collision, the cells must be smaller than a mean free path (say, one half of it). Many simulations at small Knudsen numbers are performed with larger sizes by putting a large number of simulated molecules in each cell. This problem came out in a more or less clear fashion when we discussed the objections of Meiburg[77] to the DSMC method in Section 7.4. The problem investigated by Meiburg[77] was carefully studied by Koura,[118] who showed that the DSMC method was able to generate large-scale vortices even for collision-cell sizes considerably larger than the mean free path, provided that the average number of molecules per cell was sufficiently large. However, there is a cell-size effect: The vortex structure becomes less clear and the decay length of large-scale vortices becomes shorter when increasing the size of the collision cells. The problem was put under focus in a paper by Alexander et al.,[119] who proved that there is a strong dependence on the cell size of the viscosity coefficient and the heat conductivity in a DSMC simulation for hard spheres. With cells of the size of about a mean free path the error is close to 8%, which is already a nonnegligible effect, but it would reach 20% when using cells of the size of one mean free path.

This unpleasant situation has been analyzed by J. Fan and I. D. Boyd.[120] They pointed out that the requirement, commonly adopted, of having a cell size smaller than the typical length associated with the macroscopic flow gradients should be sufficient. There is also another restriction, which arises from the process of choosing potential pairs: For this purpose the cell must be less than the mean free path. This restriction becomes very tight when the Knudsen number is small. The way out of this difficulty had already been pointed out by Bird[78] when discussing the objection raised by Meiburg.[77] It is based on using cells satisfying the first restriction and on introducing subcells smaller than a mean free path, which are used to select the collision pairs. This means that the only deleterious effect of choosing large cells is a lack of discrimination in the macroscopic properties, but this effect will show up on scales smaller than the typical length associated with the macroscopic flow gradients. When computing the transport properties, it turns out that they are essentially independent of the size of the computing cells and only depend on the size of the subcells.[118]

The aforementioned circumstance that the DSMC method may become competitive with the numerical solution of the Navier–Stokes equations even for fairly low values of the Knudsen number has prompted a certain number of developments, that, although not strictly referring to rarefied gases, were suggested by the kinetic description; hence it may be not out of place to give a hint at these methods here. We shall give a brief description of these techniques, starting with those based on the idea of a lattice gas, which is somehow related to the discrete velocity models discussed in the previous section.

A *lattice gas* is, as the name indicates, a gas of molecules occupying the sites of a lattice. The name *cellular automata* is related to the early visions of John von Neumann and Stanislaw Ulam, who envisaged computers in which reality is simulated by exactly solvable models, stated in terms of Boolean logic. Although most of the lattice gases in use retain the feature of cellular automata, some do not, and thus the latter term (which has acquired a technical meaning, designing discretely valued dynamical systems evolving in discrete time steps on a regular lattice by nearest neighbor interactions according to simple rules) tends to disappear in the current terminology and one simple talks of lattice gases. Perhaps one should even talk of lattice fluids, since compressibility, the typical feature of gases as opposed to liquids, is the hardest to model with lattice gases (see below).

The starting point is the remark, which should be clear by now to the readers of this book, that the detailed model of the elementary constituents (molecules) has little influence on the results obtained at a macroscopic level. Thus, in 1986, Frisch et al.[121] came out with the idea of simulating fluid motion by the dynamics of molecules, which, in the plane, may possess just six possible velocities, which may be oriented as the sides issuing from a site of a hexagonal lattice. The molecules live on these sites and, at a tick of a clock, may hop and scatter, (i.e., they move along a side of a hexagon of the lattice and, if two or more molecules arrive at the same site, a scattering collision may occur). Since care is exercised to maintain the number of molecules and momentum conserved at a scattering, the main ingredients of *incompressible fluid* dynamics are there, and the Navier–Stokes equations follow as asymptotic limits, in much the same way as in the kinetic theory of gases (Chapters 2 and 5). The most surprising feature is that the equations turn out to be isotropic, in spite of the fact that the basic lattice is highly anisotropic. In fact a predecessor of the 1986 model, proposed by Jean Hardy, Olivier de Pazzis, and Yves Pomeau[122] in 1976 and based on a square lattice, lay dormant for a decade because the corresponding hydrodynamic equations were anisotropic. It is interesting to remark that, at variance with its continuous version, the Boltzmann equation for a gas on a lattice yields a solution that is global in time with moderate effort.[123]

The lattice gas model went through a stage of wild excitement, because people thought that it would provide a cheap route to the simulation of turbulence. This turned out not to be the case, as any person, simultaneously expert in the various fields involved (kinetic theory, turbulence, and numerical fluid dynamics), could easily predict. On the contrary, none less that the *Washington Post* stated in a front-page article that the potential of the lattice gas method was "one thousand to one million times faster than previous methods" and even reported that the United States Department of Defense had considered whether the method "should be classified to keep it out of Soviet hands."[124]

A more serious debate at a scientific level started at that time. The possibility of simulating incompressible fluid dynamics – with certain, well-understood corrections – emerged. The statement concerning turbulence turned out to be seriously exaggerated, though the possibility of gaining in efficiency with special-purpose machines remains open to this date. In this respect, lattice gases are just another method to solve the incompressible Navier–Stokes equations. Important, unexpected ramifications of the use of lattice gases have, however, emerged in the realm of complex fluids. Specifically, fluid mixtures that include interfaces, exhibit phase transitions, or allow for multiphase flows can be simulated with lattice gases, thanks to the unusually rich phenomenology of the latter. This favorable feature is enhanced when the geometry is also complicated as in the case of multiphase flows through disordered media, such as a porous rock. For more details on the theory and applications of the lattice gas idea we refer to a recent book.[125]

Another application of ideas motivated by kinetic theory is that of "kinetic solvers" for Euler or Navier–Stokes equations. These can be of use, especially in flows with shock waves.[126–129]

References

[1] A. Nordsiek and B. Hicks, "Monte Carlo evaluation of the Boltzmann collision integral," in *Rarefied Gas Dynamics*, C. L. Brundin, ed., Vol. I, 695–710, Academic Press, New York (1967).

[2] S. M. Yen, B. Hicks, and R. M. Osteen, "Further development of a Monte Carlo method for the evaluation of Boltzmann collision integral," in *Rarefied Gas Dynamics*, M. Becker and M. Fiebig, eds., Vol. I, A.12-1–A.12-10, DFVLR-Press, Porz-Wahn (1974).

[3] V. Aristov and F. G. Tcheremissine, "The conservative splitting method for solving the Boltzmann equation," *U.S.S.R. Computational Mathematics and Mathematical Physics* **20**, 208–225 (1980).

[4] F. G. Tcheremissine, "Numerical methods for the direct solution of the kinetic Boltzmann equation," *U.S.S.R. Computational Mathematics and Mathematical Physics* **25**, 156–166 (1985).

[5] F. G. Tcheremissine, "Advancement of the method of direct numerical solving of the Boltzmann equation," in *Rarefied Gas Dynamics: Theoretical and Computational Techniques*, E. P. Muntz, D. P. Weaver, and D. H. Campbell, eds., 343–358, AIAA, Washington (1989).

[6] C. Cercignani and A. Frezzotti, "Numerical simulation of supersonic rarefied gas flows past a flat plate: Effects of the gas–surface interaction model on the flow-field," in *Rarefied Gas Dynamics: Theoretical and Computational Techniques*, E. P. Muntz, D. P. Weaver, and D. H. Campbell, eds., 552–566, AIAA, Washington (1989).

[7] G. A. Bird, "Direct simulation of the Boltzmann equation," *Phys. Fluids* **13**, 2676–2681 (1970).

[8] G. A. Bird, "Monte-Carlo simulation in an engineering context," in *Rarefied Gas Dynamics*, S. S. Fischer, ed., Vol. I, 239–255, AIAA, New York (1981).

[9] K. Koura and H. Matsumoto, "Variable soft sphere molecular model for inverse power-law or Lennard–Jones potential," *Phys. Fluids A* **3**, 2459–2465 (1991).

[10] H. A. Hassan and D. B. Hash, "A generalized hard-sphere model for Monte Carlo simulations," *Phys. Fluids A* **5**, 738–744 (1993).

[11] A. Kersch, W. J. Morokoff, and C. Werner, "Selfconsistent simulation of sputter deposition with the Monte Carlo method," *J. Appl. Phys.* **75**, 2278–2285 (1994).

[12] K. Koura, "Null-collision technique in the direct simulation Monte Carlo technique," *Phys. Fluids* **29**, 3509–3511 (1986).

[13] O. M. Belotserkovskii and V. Yanitskii, "Statistical particle-in-cell method for solving rarefied gas dynamics problems," *Zhurnal Vychitelnik Matematiki i Matematicheskii Fiziki* **15**, 1195–1203 (1975) (in Russian).

[14] S. M. Deshpande, Dept. Aeronaut. Engng. Indian Inst. Science Rep. **78**, FM4, (1978).

[15] K. Nanbu, "Direct simulation scheme derived from the Boltzmann equation," *J. Phys. Soc. Japan* **49**, 2042–2049 (1980).

[16] K. Koura, "Comment on 'Direct simulation scheme derived from the Boltzmann equation. I. Monocomponent gases,'" *J. Phys. Soc. Japan* **50**, 3829–3830 (1981).

[17] K. Nanbu, "Reply to a comment on 'Direct simulation scheme derived from the Boltzmann equation. I. Monocomponent gases,'" *J. Phys. Soc. Japan* **50**, 3831–3832 (1981).

[18] H. Babovsky, "A convergence proof for Nanbu's Boltzmann simulation scheme," *Euro. J. Mech. B: Fluids* **8(1)**, 41–55 (1989).

[19] H. Babovsky and R. Illner: "A convergence proof for Nanbu's simulation method for the full Boltzmann equation," *SIAM J. Numerical Analysis* **26**, 45–65 (1989).
[20] W. Wagner, "A convergence proof for Bird's direct simulation Monte Carlo method for the Boltzmann equation," *J. Stat. Phys.* **66**, 1011–1044 (1992).
[21] M. Pulvirenti, W. Wagner, and M. B. Zavelani, "Convergence of particle schemes for the Boltzmann equation," *Euro. J. Mech. B* **7**, 339–351 (1994).
[22] A. L. Garcia, "Nonequilibrium fluctuations studied by a rarefaction simulation," *Phys. Rev. A* **34**, 1454–1457 (1986).
[23] F. Gropengießer, N. Neunzert, and J. Struckmeier, "Computational methods for the Boltzmann equation," in *Venice 1989: The State of Art in Applied and Industrial Mathematics*, R. Spigler, ed., 111–140, Kluwer, Dordrecht (1990).
[24] C. Cercignani and M. Lampis, "Kinetic models for gas–surface interactions," *Transp. Theory Stat. Phys.* **1**, 101–114 (1971).
[25] R. G. Lord, "Some extensions to the Cercignani–Lampis gas scattering kernel," *Phys. Fluids A* **3**, 706–710 (1991).
[26] R. S. Simmons and R. G. Lord, "Application of the C–L model to vibrational transitions of diatomic molecules during DSMC gas–surface interaction," in *Rarefied Gas Dynamics*, J. Harvey and G. Lord, eds., Vol. 2, 906–912, Oxford University Press, Oxford (1995).
[27] M. S. Woronowicz and D. F. G. Rault, "Cercignani–Lampis–Lord gas–surface interaction model: Comparison between theory and simulation," *J. Spacecraft Rockets* **31**, 532–534 (1994).
[28] J. K. Haviland, "Determination of shock-wave thicknesses by the Monte Carlo method," in *Rarefied Gas Dynamics*, J. A. Laurmann, ed., Vol. 1, 274–296, Academic Press, New York (1963).
[29] G. A. Bird, "Shock wave structure in a rigid sphere gas," in *Rarefied Gas Dynamics*, J. H. de Leeuw, ed., Vol. 1, 216–222, Academic Press, New York (1965).
[30] G. A. Bird, "Aspects of the structure of strong shock waves," *Phys. Fluids* **13**, 1172–1177 (1970).
[31] B. Schmidt, "Electron beam density measurements in shock waves in argon," *J. Fluid Mech.* **39**, 361–373 (1970).
[32] H. Alsmeyer, "Density profiles in argon and nitrogen shock waves measured by the absorption of an electron beam," *J. Fluid Mech.* **74**, 497–513 (1976).
[33] G. C. Pham-Van-Diep, D. A. Erwin, and E. P. Muntz, "Nonequilibrium molecular motion in a hypersonic shock wave," *Science* **245**, 624–626 (1989).
[34] K. O. Stewartson, "On the flow near trailing edge of a flat plate. II," *Mathematika* **16**, part 1, 106–121 (1969).
[35] A. F. Messiter, "Boundary-layer flow near the trailing edge of a flat plate," *SIAM J. Appl. Math.* **18**, 241–257 (1970).
[36] W. J. McCroskey, S. M. Bogdonoff, and J. G. McDougall, "An experimental model for sharp flat plate in rarefied hypersonic flow," *AIAA J.* **4**, 1580–1587 (1966).
[37] P. J. Harbour and J. H. Lewis, "Preliminary measurements of the hypersonic rarefied flow field on a sharp plate using an electron beam probe," in *Rarefied Gas Dynamics*, C. L. Brundin, ed., Vol. II, 1031–1046, Academic Press, New York, (1967).
[38] W. W. Joss, I. E. Vas, and S. M. Bogdonoff, "Hypersonic rarefied flow over a flat plate," AIAA Paper 68–5 (January 1968).

[39] H. Oguchi, "The sharp leading edge problem in hypersonic flow," in *Rarefied Gas Dynamics*, L. Talbot, ed., 501–524 Academic Press, New York (1961).
[40] M. Shorenstein and R. F. Probstein, "The hypersonic leading-edge problem," *AIAA J.* **6**, 1898–1902 (1968).
[41] W. L. Chow, "Hypersonic rarefied flow past the sharp leading-edge of a flat plate," *AIAA J.* **5**, 1549–1557 (1967).
[42] W. L. Chow, "Hypersonic slip flow past the leading-edge of a flat plate," *AIAA J.* **4**, 2062–2063 (1968).
[43] S. Rudman and S. G. Rubin, "Hypersonic viscous flow over slender bodies with sharp leading-edges," *AIAA J.* **6**, 1883–1890 (1968).
[44] H. K. Cheng, S. Y. Chen, R. Mobly, and C. Huber, "On the hypersonic leading-edge problem in the merged-layer regime," in *Rarefied Gas Dynamics*, L. Trilling and H. Y. Wachman, eds., Vol. I, 451–463, Academic Press, New York (1969).
[45] S. S. Kot and D. L. Turcotte, "Beam-continuum model for hypersonic flow over a flat plate," *AIAA J.* **10**, 291–295 (1972).
[46] A. B. Huang and P. F. Hwang, "Supersonic leading-edge problem according to the ellipsoidal model," *Phys. Fluids* **13**, 309–317 (1970).
[47] A. B. Huang, "Kinetic theory of the rarefied supersonic flow over a finite plate," in *Rarefied Gas Dynamics*, L. Trilling and H. Y. Wachman, eds., Vol. I, 529–544, Academic Press, New York (1969).
[48] A. B. Huang and P. F. Hwang, "Kinetic theory of the sharp leading-edge flow, II," *IAF Paper RE 63* (October 1968).
[49] F. W. Vogenitz, J. E. Broadwell, and G. A. Bird, "Leading-edge flow by the Monte Carlo direct simulation method," *AIAA J.* **8**, 504–510 (1972).
[50] S. C. Metcalf, D. C. Lillicrap, and C. J. Berry, "A study of the effect of surface temperature on the shock-layer development over sharp-edged shapes in low-Reynolds-number high speed flow," in *Rarefied Gas Dynamics*, L. Trilling and H. Y. Wachman, eds., Vol. I, 619–638, Academic Press, New York (1969).
[51] J. K. Harvey, "Direct simulation Monte Carlo method and comparison with experiment," *Progr. Astronaut. Aeronaut.* **103**, 25–42 (1986).
[52] J. N. Moss and G. A. Bird, "Direct simulation of transitional flow for hypersonic re-entry conditions," *Progr. Astronaut. Aeronaut.* **96**, 113 (1985).
[53] D. F. G. Rault, "Aerodynamics of the Shuttle Orbiter at high altitudes," *J. Spacecraft Rockets* **31**, 944–952 (1994).
[54] M. S. Ivanov, S. F. Gimelshein, and A. E. Beylich, "Hysteresis effect in stationary reflection of shock waves," *Phys. Fluids A* **7**, 685–687 (1995).
[55] M. S. Ivanov, S. F. Gimelshein, and G. N. Markelov, "Statistical simulation of the transition between regular and Mach reflection in steady flows," *Computers Math. Appl.* **35**, 113–125 (1998).
[56] H. W. Liepmann and A. Roshko, *Elements of Gasdynamics*, Wiley, New York (1957).
[57] R. Courant and K. O. Friedrichs, *Supersonic Flow and Shock Waves*, Wiley Interscience, New York (1948).
[58] H. G. Hornung and M. L. Robinson, "Transition from regular to Mach reflection of shock waves. Part 2. The steady flow criterion," *J. Fluid Mech.* **123**, 155–164 (1982).
[59] H. G. Hornung, H. Oertel, and R. J. Sandeman, "Transition to Mach reflection of shock waves in steady and pseudosteady flow and without relaxation," *J. Fluid Mech.* **90**, 541–560 (1979).

[60] M. C. Celenligil and J. N. Moss, "Hypersonic rarefied flow about a delta wing – direct simulation and comparison with experiment," *AIAA J.* **30**, 2017–2023 (1992).
[61] H. Legge, "Force and heat transfer on a delta wing in rarefied flow," in *Workshop on Hypersonic Flows for Reentry Problems*, Part II, Antibes, France (1991).
[62] J. N. Moss, C. H. Chun, and J. M. Price, "Hypersonic rarefied flow about a compression corner – DSMC simulation and experiment," *AIAA Paper 91-1313* (1991).
[63] B. Chanetz, "Study of axisymmetric shock wave–boundary layer interaction in hypersonic laminar flow," *ONERA Report RT 42/4365 AN* (1995).
[64] J. N. Moss, V. K. Dogra, and J. M. Price, "DSMC simulation of viscous interactions for a hollow cylinder-flare configuration," *AIAA Paper 94-2015* (1994).
[65] J. Allegre and D. Bisch, "Experimental study of a blunted cone at rarefied hypersonic conditions," *CNRS Report RC 94-7* (1994).
[66] H. Legge, "Experiments on a 70 degree blunted cone in rarefied hypersonic wind tunnel flow," *AIAA Paper 95-2140* (1995).
[67] J. N. Moss, J. M. Price, V. K. Dogra, and D. B. Hash, "Comparison of DSMC and experimental results for hypersonic external flow," *AIAA Paper 95-2028* (1995).
[68] K. Koura, "Null collision Monte Carlo method. Gas mixtures with internal degrees of freedom and chemical reactions," in *Rarefied Gas Dynamics: Physical Phenomena*, E. P. Muntz, D. P. Weaver, and D. H. Campbell, eds., 25–39, AIAA, Washington (1989).
[69] K. Koura, "Improved null-collision technique in the Direct Simulation Monte Carlo method: Application to vibrational relaxation of nitrogen," *Computers Math. Appl.* **35**, 139–154 (1998).
[70] S. Stefanov and C. Cercignani, "Monte Carlo simulation of Bénard's instability in a rarefied gas," *Euro. J. Mech. B* **11**, 543–553 (1992).
[71] C. Cercignani and S. Stefanov, "Bénard's instability in kinetic theory," *Transp. Theory Stat. Phys.* **21**, 371–381 (1992).
[72] A. Garcia, "Hydrodynamic fluctuations and the direct-simulation Monte Carlo method," in *Microscopic Simulations of Complex Flows*, M. Mareschal, ed., 177–188, Plenum Press, New York (1990).
[73] A. Garcia and C. Penland, "Fluctuating hydrodynamics and principal oscillation pattern analysis," *J. Stat. Phys.* **64**, 1121–1132 (1991).
[74] S. Stefanov and C. Cercignani, "Monte Carlo simulation of the Taylor–Couette flow of a rarefied gas," *J. Fluid Mech.* **256**, 199–213 (1993).
[75] S. Stefanov and C. Cercignani, "Taylor–Couette flow of a rarefied gas," in *Proceedings of the International Symposium on Aerospace and Fluid Science*, 490–500, Institute of Fluid Science, Tohoku University, Sendai (1994).
[76] S. Stefanov and C. Cercignani, "Monte Carlo simulation of a channel flow of a rarefied gas," *Euro. J. Mech. B* **13**, 93–114 (1994).
[77] E. Meiburg, "Comparison of the molecular dynamics method and the direct simulation Monte Carlo technique for flows around simple geometries," *Phys. Fluids* **29**, 3107–3113 (1986).
[78] G. A. Bird, "Direct-simulation of high-vorticity gas flows," *Phys. Fluids* **30**, 364–366 (1987).
[79] K. Nanbu, Y. Watanabe, and S. Igarashi, "Conservation of angular momentum in the Direct Simulation Monte Carlo Method," *J. Phys. Soc. Japan* **57**, 2877–2880 (1988).
[80] G. A. Bird, *Molecular Gas Dynamics and the Direct Simulation of Gas Flows*, Clarendon Press, Oxford (1994).

[81] G. A. Bird, "The initiation of centrifugal instabilities in an axially symmetric flow," in *Rarefied Gas Dynamics Symposium 20,* Ching Shen, ed., 624–629, Peking University Press, Beijing (1997).

[82] D. Reichelman and K. Nanbu, "Monte Carlo direct simulation of the Taylor instability in a rarefied gas," *Phys. Fluids A* **5**, 2585–2587 (1993).

[83] A. R. Kulthau, "Recent low-density experiments using rotating cylinder techniques," in *Rarefied Gas Dynamics*, F. M. Devienne, ed., 192–200, Pergamon, London (1960).

[84] S. Stefanov, V. Roussinov, C. Cercignani, M. C. Giurin, and J. Struckmeier, "Rayleigh–Bénard chaotic convection in a rarefied gas," *Proceedings of the 21st Symposium on Rarefied Gas Dynamics* (to appear, 1999).

[85] H. Neunzert and J. Struckmeier, "Particle methods for the Boltzmann equation," in *Acta Numerica 1995*, A. Iserles, ed., 417–457, Cambridge University Press, Cambridge (1995).

[86] S. Stefanov, P. Gospodinov, and C. Cercignani, "Monte Carlo simulation and Navier–Stokes finite difference calculation of unsteady-state rarefied gas flows," *Phys. Fluids* **10**, 289–300 (1998).

[87] Y. Sone, "Asymptotic theory of flow of rarefied gas over a smooth boundary I," in *Rarefied Gas Dynamics*, D. Dini, ed., Vol. I, 243–253, Editrice Tecnico Scientifica, Pisa (1974).

[88] Y. Sone, "Flow induced by thermal stress in rarefied gas," *Phys. Fluids* **15**, 1418–1423 (1972).

[89] T. Ohwada and Y. Sone, "Analysis of thermal stress slip flow and negative thermophoresis using the Boltzmann equation for hard-sphere molecules," *Euro. J. Mech. B* **11**, 389–414 (1992).

[90] M. N. Kogan, V. S. Galkin, and O. G. Fridlender, "Stresses produced in gases by temperature and concentration inhomogeneities. New type of free convection," *Sov. Phys. Usp.* **19**, 420–438 (1976).

[91] Y. Sone, K. Aoki, S. Tanaka, H. Sugimoto, and A. V. Bobylev, "Inappropriateness of the heat-conduction equation for the description of a temperature field of a stationary gas in the continuum limit: Examination by asymptotic analysis and numerical computation of the Boltzmann equation," *Phys. Fluids A* **8**, 628–638 (1996).

[92] T. Ohwada, Y. Sone, and K. Aoki, "Numerical analysis of the Poiseuille and thermal transpiration flows between two parallel plates on the basis of the Boltzmann equation for hard-sphere molecules," *Phys. Fluids A* **1**, 2042–2049 (1989); "Erratum," *Phys. Fluids A*, **2**, 639 (1990).

[93] S. Takata and Y. Sone, "Flow induced around a sphere with a nonuniform surface temperature in a rarefied gas, with application to the drag and thermal force problems of a spherical particle with an arbitrary thermal conductivity," *Euro. J. Mech. B* **14**, 487–518 (1995).

[94] K. Aoki, Y. Sone, and Y. Waniguchi, "A rarefied gas flow induced by a temperature field: Numerical analysis of the flow between two coaxial elliptic cylinders with different uniform temperatures," *Computers Math. Appl.* **35**, 15–28 (1998).

[95] K. Aoki, Y. Sone, and N. Masukawa, "A rarefied gas flow induced by a temperature field," in *Rarefied Gas Dynamics*, J. Harvey and G. Lord, eds., Vol. 1, 35–41, Oxford University Press, Oxford (1995).

[96] F. Golse, "Applications of the Boltzmann equation within the context of upper atmosphere vehicle aerodynamics," *Computer Meth. Eng. Appl. Mech.* **75**, 299–316 (1989).

[97] A. Klar, "Domain decomposition for kinetic problems with nonequilibrium states," *Euro. J. Mech. B* **15**, 203–216 (1996).

[98] C. Cercignani, "A nonlinear criticality problem in the kinetic theory of gases," in *Mathematical Problems in the Kinetic Theory of Gases*, D. C. Pack and H. Neunzert, eds., 129–146, P. Lang, Frankfurt (1980).

[99] M. D. Arthur and C. Cercignani, "Nonexistence of a steady rarefied supersonic flow in a half-space," *J. Appl. Math. Phys. (ZAMP)* **31**, 634–645 (1980).

[100] R. Gatignol, *Théorie Cinétique des Gaz à Répartition Discrète de Vitesses*, Lectures Notes in Physics, **36**, Springer-Verlag, Berlin (1975).

[101] H. Cabannes, *The Discrete Boltzmann Equation (Theory and Application)*, Lecture notes, University of California, Berkeley (1980).

[102] D. Goldstein, B. Sturtevant, and J. E. Broadwell, "Investigations of the motion of discrete-velocity gases," in *Rarefied Gas Dynamics: Theoretical and Computational Techniques*, E. P. Muntz, D. P. Weaver, and D. H. Campbell, eds., 110–117, AIAA, Washington (1989).

[103] T. Inamuro and B. Sturtevant, "Numerical study of discrete velocity gases," *Phys. Fluids A* **2**, 2196–2203 (1990).

[104] F. Rogier and J. Schneider, "A direct method for solving the Boltzmann equation," *Transp. Theory Stat. Phys.* **23**, 313–338 (1994).

[105] C. Buet, "A discrete-velocity scheme for the Boltzmann operator of rarefied gas-dynamics," *Transp. Theory Stat. Phys.* **25**, 33–60 (1996).

[106] A. V. Bobylev, A. Palczewski, and J. Schneider, "Discretization of the Boltzmann equation and discrete velocity models," in *Rarefied Gas Dynamics 19*, J. Harvey and G. Lord, eds., Vol. II, 857–863, Oxford University Press, Oxford (1995).

[107] A. V. Bobylev and C. Cercignani, "Discrete velocity models for mixtures," *J. Stat. Phys.* **91**, 327–342 (1998).

[108] C. Cercignani and H. Cornille, "Shock waves for a discrete velocity gas mixture," to appear (1999).

[109] D. B. Goldstein, "Discrete-velocity collision dynamics for polyatomic molecules," *Phys. Fluids A* **4**, 1831–1839 (1992).

[110] A. V. Bobylev, A. Palczewski, and J. Schneider, "On approximation of the Boltzmann equation by discrete velocity models," *C.R. Acad. Sci. Paris, Ser. I* **320**, 639–644 (1995).

[111] A. Palczewski, J. Schneider, and A. V. Bobylev, "A consistency result for a discrete-velocity model of the Boltzmann equation," *SIAM J. Numerical Analysis* **34**, 1865–1883 (1997).

[112] G. H. Hardy and E. M. Wright, *An Introduction to the Theory of Numbers*, Oxford University Press, Oxford (1938).

[113] W. Duke and R. Schulze-Pillot, "Representation of integers by positive ternary quadratic forms and equidistribution of lattice points on ellipsoids," *Invent. Math.* **99**, 49–57 (1990).

[114] C. Buet, S. Cordier, and P. Degond, "Regularized Boltzmann operators," *Computers Math. Appl.* **35**, 55–74 (1998).

[115] H. Cornille, "Exact solutions of the general two-velocity discrete Illner models," *J. Math. Phys.* **28**, 1567–79 (1987).

[116] H. Cornille, "$9v_i$ and $15v_i$ discrete Boltzmann models with and without multiple collisions," *Transp. Theory Stat. Phys.* **24**, 709–29 (1995).

[117] H. Cornille and A. d'Almeida, "Criteria for inverted temperatures in evaporation–condensation processes," *J. Math. Phys.* **37**, 5476–95 (1998).

[118] K. Koura, "Direct simulation of vortex shedding in a dilute gas past an inclined flat plate," *Phys. Fluids A* **2**, 209–213 (1990).

[119] F. J. Alexander, A. L. Garcia, and B. J. Alder, "Cell size dependence of transport coefficents in stochastic particle algorithm," *Phys. Fluids A* **10**, 1540–1541 (1998).
[120] J. Fang and I. D. Boyd, "Analysis of the accuracy and efficiency of the DSMC method," submitted to *Phys. Fluids A* (1999).
[121] U. Frisch, B. Haslacher, and Y. Pomeau, "Lattice-gas automata for the Navier-Stokes equations," *Phys. Rev. Lett.* **56**, 1505–1508 (1986).
[122] J. Hardy, O. de Pazzis, and Y. Pomeau, "Molecular dynamics of a classical lattice gas. Transport properties and time correlation functions," *Phys. Rev. A* **13**, 1949–1961 (1976).
[123] C. Cercignani, W. Greenberg, and P. F. Zweifel, "Global solution of the Boltzmann equation on a lattice," *J. Stat. Phys.* **20**, 449–462 (1979).
[124] P. J. Hilts, "Discovery in flow dynamics may aid car, plane design," *Washington Post*, November 19, 1 (1985).
[125] D. H. Rothman and S. Zaleski, *Lattice-Gas Cellular Automata: Simple Models of Complex Hydrodynamics*, Cambridge University Press, Cambridge, UK (1997).
[126] C. Kim, K. Xu, L. Martinelli, and A. Jameson, "Analysis and implementation of the gas-kinetics BGK scheme for computational gas dynamics," *Int. J. Numerical Methods* **25**, 21–49 (1997).
[127] K. Xu, C. Kim, L. Martinelli, and A. Jameson, "BGK-based schemes for the simulation of compressible flow," *Int. J. Comput. Fluid Dyn.* **7**, 213–235 (1996).
[128] K. Xu, L. Martinelli, and A. Jameson, "Gas-kinetic finite volume methods, flux-vector splitting and artificial diffusion," *J. Comp. Phys.* **120**, 48–65 (1995).
[129] C. Kim, K. Xu, L. Martinelli, and A. Jameson, "An accurate LED-BGK solver on unstructured adaptive meshes," *Paper AIAA 97-0328*, 35th AIAA Aerospace Sciences Meeting and Exhibit, Reno (January 1997).

8

Evaporation and Condensation Phenomena

8.1. Introduction

Vapor is a conventional name for the gaseous phase of a substance that is either liquid or solid at room temperature and pressure. Many substances may undergo phase changes involving their vapor. The evaporation from and the condensation on a solid or liquid surface have been the subject of several studies because of their importance in various fields of physics, chemistry, and engineering. Since the pioneering work of Hertz[1] in the nineteenth century and Knudsen[2] in the early part of the twentieth century on the evaporation of liquid mercury into a vacuum, it has been known that these processes require using the kinetic theory of gases for an accurate description. We shall use the term evaporation (rather than sublimation) in the case of a solid condensed phase as well, because the process of formation of a gaseous phase is essentially the same for both solid and liquid interfaces.

The phase change is a surface phenomenon, because the molecules separate from or attach to the surface of the condensed phase. When the phase change is spread through a volume, this is due to the fact that there is a large number of interfaces finely distributed in a volume; the treatment of these cases requires particular care because of the high level of inhomogeneity, but the basic processes are not different.

It is thus clear that, if we want to deal with evaporation and condensation phenomena, the boundary conditions for the Boltzmann equation become of paramount importance. We must modify the treatment given in Chapter 1, Section 1.11, because there we assumed strict conservation of the number of molecules at the surface. If the latter is an interface across which a phase change is taking place, a fraction σ_c of the incoming flow condenses on the surface (and is thus lost for the gas), whereas the remaining fraction $1 - \sigma_c$ satisfies the scattering boundary conditions discussed in Chapter 1. The coefficient σ_c is called the *condensation coefficient*. Frequently a second coefficient σ_e, called the *evaporation coefficient*, is introduced. It is the fraction of the maximum possible evaporating flow rate that actually occurs. One can show that in an equilibrium state $\sigma_c = \sigma_e$. Since there is no significant evidence that the two coefficients are unequal in other situations, we shall assume $\sigma_c = \sigma_e = \sigma$. Then, if $\sigma = 1$, we have the maximum evaporating flow rate at the surface. Following Hertz and Knudsen[1,2] this flow is assumed to have a Maxwellian distribution M_w having the bulk velocity ($\mathbf{v}_w$; usually zero) and the temperature (T_w) of the surface, whereas the density is a function of T_w, usually assumed to be the saturated vapor density $\rho_w(T_w)$ (related to the saturated vapor pressure p_w by $p_w = R\rho_w T_w$).

Frequently σ is given a value close to unity. However, experimental evidence,[3,4] supported by more recent molecular dynamics simulation,[5–7] leads to assigning low values, such as 0.1 for water and argon, at least at high temperatures. This is a very delicate point, upon which we shall return later in this chapter.

If we take $\sigma = 1$ the boundary condition becomes very similar to that of diffusion according to a Maxwellian distribution, with an important difference. In fact, assuming the wall to be at rest, the distribution function of the evaporating molecules is

$$f(\boldsymbol{\xi}) = \rho_w M_w(\boldsymbol{\xi}), \tag{8.1.1}$$

where ρ_w is an assigned function of T_w in the case of phase change, whereas it is determined by the impinging flow in the case of a mass-conserving boundary condition.

If we assume that the vapor is so rarefied that no molecules condense, then the mass evaporating per unit area and unit time is (Problem 8.1.1)

$$\dot{m}_w = \rho_w \left(\frac{RT_w}{2\pi} \right)^{1/2}. \tag{8.1.2}$$

If we assume that the vapor is still rarefied but there are condensing molecules arriving from far away with an equilibrium distribution function with zero bulk velocity, temperature T_∞, and density ρ_∞, Eq. (8.1.2) changes to (Problem 8.1.2)

$$\dot{m}_w = \rho_w \left(\frac{RT_w}{2\pi} \right)^{1/2} - \rho_\infty \left(\frac{RT_\infty}{2\pi} \right)^{1/2}. \tag{8.1.3}$$

This is the classical Hertz–Knudsen formula. It can be shown that, in the case of nonunity condensation coefficient σ the rate is simply multiplied by σ

(Problem 8.1.3). The formula is qualitatively accurate, but the second term is the result of a simplification that neglects the dynamics of the vapor, which must, of course, be described by the Boltzmann equation.

A major improvement of the Hertz–Knudsen formula was obtained by Schrage[8] and Kucherov and Rikenglas.[9] These authors included the effect of the bulk velocity in the second term of Eq. (8.1.3). The Schrage formula has the correct behavior for low and intermediate evaporation rates, but, because collisions are not taken into account, it overestimates the net mass flow at higher evaporation rates. In addition, all these formulas fail to relate the values of the variables describing the condensing vapor to the values of the physical parameters occurring in a given problem.

The first attempt at solving the problem in a complete way is due to Shankar and Marble,[10] who derived moment equations from the Boltzmann equation. They also solved the linearized version of these equations in a half-space situation. Their solution succeeded in coupling the external flow variables and the parameters at the surface, with the consequence that their formula for the mass flow depends on just one parameter, the pressure difference $p_w - p_\infty$. With an appropriate interpretation (which is required because of the linearization), their result leads to a value for the evaporation rate twice as high as that predicted by the Hertz–Knudsen formula. This kind of result had been already found by Patton and Springer,[11] who had considered a slab rather than a half-space.

An accurate treatment of the evaporation problem in a slab was first presented by Pao,[12,13] who also pointed out a seemingly paradoxical result, which goes under the name of "inverted temperature gradient paradox." We shall discuss this aspect of the matter in Section 8.5. Further work in the line opened by Pao is due to Loyalka,[14] Kogan and Makashev,[15] Cipolla et al.,[16] and Sone and Onishi.[17,18]

Moment approaches for the nonlinear Boltzmann equation were very useful for the purpose of clarifying the qualitative picture especially in the case of strong evaporation. Edward and Collins[19] applied Grad's thirteen moment equations matched to an external Navier–Stokes description to compute evaporation from a spherical source into vacuum. Anisimov[20] and Ytrehus et al.[21] developed methods combining the features of the Liu–Lees method for boundary value problems and the Mott-Smith method for the shock wave structure (described in Chapters 2 and 4, respectively) to compute the strong evaporation from a plane boundary. In particular, the method supplied results for strong evaporation into a vacuum[20,21,23] and the general case, including the Knudsen layer structure and the limiting behavior at $\mathrm{M}_\infty = 1$[23] (please distinguish between M and M; the former denotes the Mach number, the latter a Maxwellian distribution). This approach captured the essential features of strong condensation for an impinging subsonic flow as well as provided the first indications for the supersonic case[24–26] (in particular the two-parameter nature of the solution). The moment approach of Soga[27] differs from the papers quoted above, because it uses an entropy balance instead of a nonconserved moment equation, in a way reminiscent of the method suggested by Lampis[28] for the shock wave structure, mentioned in Chapter 4.

Kogan and Makashev[15] made the first attempt at a numerical solution of the problem by means of the BGK model in the nonlinear regime. Their work was continued by Gajewski et al.,[29] who indicated that the inversion of the temperature gradient is not due to the simplifications related to linearization. Yen[30,31] and Yen and Ytrehus[32] obtained numerical solutions for both the BGK model and the Boltzmann equation for the nonlinear evaporation–condensation problem in a slab in a variety of flow situations. Murakami and Oshima[33] made Monte Carlo simulations of the time-dependent vapor motion following a sudden change of the phase equilibrium in the half-space problem. Their results, together with the moment solution by Ytrehus,[26,27] suggested that the limiting Mach number downstream must be unity. This was stressed by Cercignani,[34] who offered an explanation, based on an asymptotic analysis, and later, with Arthur,[35] confirmed it with an exact analytical treatment of the BGK model linearized about a drifting Maxwellian distribution.

The range of existence of steady solutions for both condensation and evaporation in a half-space was clarified in a series of papers by Siewert and Thomas,[36] Sone et al.,[37,38] Abramov and Kogan,[39–41] Aoki et al.,[42] and Kryukov[43]. In the case of evaporation, the process is determined by an external-flow parameters, usually taken to be the pressure p_∞ or the Mach number M_∞. In the case of condensation, the process is determined by two external-flow parameters, usually taken to be the pressure p_∞ and the temperature T_∞, and is unique only if the external flow is subsonic. In the case of supersonic evaporation, there are infinitely many solutions for each pair of values of p_∞ and T_∞, and a third parameter is required to characterize the external flow in such a way as to make the solution unique. In the case of weak evaporation, one may linearize the Boltzmann equation, and the solutions for evaporation and condensation in the Knudsen layer just differ in the sign of the bulk velocity. A difference arises, however, when matching the inner and outer solutions. In the case of evaporation, one can match the Knudsen layer solutions with the Euler equations, whereas, as shown by Sone[44] and Sone and Onishi,[18] matching with Navier–Stokes equations is required in the case of condensation.

All the papers that have been quoted so far refer to a vapor behaving as a monatomic gas. The role of internal degrees of freedom has been investigated and will be discussed in Section 8.6, where the role of an inert gas mixed with the vapor will also be discussed.

Interest in the evaporation–condensation phenomena stems not only from the desire to understand the fundamental aspects of the problem but also from many applications, which range from condensation heat transfer to vacuum distillation,[45] from vacuum vapor deposition[46,47] to laser induced isotope separation[48] and laser-sputtering at surfaces,[20] and from outgassing from surfaces to the formation of the dusty coma around active comets.[49,50]

As one can see from this introduction, the literature on the subject is rather vast. For this reason the detailed presentation in this chapter will be mainly based on the case of evaporation or condensation in a half-space and occasionally between parallel plates. More complex geometries will be briefly surveyed in Section 8.8. Also, for brevity, calculations will be based on the moment equations, with the ansatz for the distribution function originally proposed by Anisimov[20] and systematically used by Ytrehus.[21,23,26] The degree of agreement with other methods and the relevant papers will also be discussed.

Problems

8.1.1 Prove that if the evaporating molecules have a distribution function given by Eq. (8.1.1) and no molecules condense, then the mass evaporating per unit area and unit time is given by (8.1.2).

8.1.2 Prove (8.1.2).

8.1.3 Prove (8.1.3).

8.2. The Knudsen Layer near an Evaporating Surface

A method that discloses many qualitative features about the Knudsen layer near an evaporating surface and may be considered reasonably accurate from a quantitative viewpoint is the method of moments. As we indicated in Chapters 2, 4, and 5, this method is not so good if one pretends to assign an a priori simple expression for the distribution function, valid for any problem. Rather, one must carefully select this expression on the basis of criteria dictated by a specific problem and use the moment equations to determine the parameters left unspecified in the aforementioned expression. This approach to the moment method, which is essentially due to Liu and Lees,[50,51] becomes very complicated when the geometry is not a simple one, but it is well suited for flows in the presence of plane boundaries, in particular to the study of the Knudsen layer near an evaporating or condensing plane interface. In this case one must solve the one-dimensional steady Boltzmann equation

$$\xi \partial_x f = Q(f, f)(\mathbf{x}, \boldsymbol{\xi}), \tag{8.2.1}$$

where x varies in the half-space $x > 0$ and ξ denotes the first component of $\boldsymbol{\xi}$ (which varies in the entire velocity space). If we assume that the evaporation-condensation coefficient σ is unity, the boundary conditions read as follows:

$$f = M_w^+ = \rho_w (2\pi R T_w)^{-3/2} \exp[-|\boldsymbol{\xi}|^2/(2RT_w)] \quad \text{at } x = 0 \text{ for } \xi > 0, \tag{8.2.2}$$

$$f \to M_\infty = \rho_\infty (2\pi R T_\infty)^{-3/2} \exp[-|\boldsymbol{\xi} - \mathbf{i}u_\infty|^2/(2RT_\infty)] \quad \text{for } x \to \infty. \tag{8.2.3}$$

The Maxwellian distribution at the wall M_w^+ is affected by a + superscript to underline the circumstance that it is a restriction of a Maxwellian distribution to $\xi > 0$.

The idea of Anisimov[20,22] was to use a trimodal ansatz for the molecular distribution function:

$$f(x, \boldsymbol{\xi}) = a_w^+(x) M_w^+ + a_\infty^+(x) M_\infty^+ + a_\infty^-(x) M_\infty^-, \tag{8.2.4}$$

where the ± superscripts in each Maxwellian distribution indicate that they are identically zero when the sign of ξ disagrees with the superscript and coincide with the corresponding Maxwellian distribution when the sign in question agrees with the superscript. The aforementioned authors solved[20,22] the Boltzmann equation for one set of flow conditions (sonic) downstream and estimated the thickness of the Knudsen layer with the BGK model. Ytrehus[23] carried the method to completion, solving appropriate moment equations of the Boltzmann equation with a distribution function given by (8.2.4). We essentially follow his approach here. We remark that the choice (8.2.4) is not simply an application of the Liu and Lees method, since the latter would rather lead to a two-term distribution function with $a_\infty^+ = 0$ and Maxwellians with nonconstant parameters; Eq. (8.2.4) embodies also Mott-Smith's idea for the problem of the shock wave structure.

The advantage of the choice (8.2.4) is that the boundary conditions may be satisfied exactly by imposing the following conditions on the three functions $a_w^+(x), a_\infty^\pm(x)$:

$$a_w^+(0) = 1, \quad a_\infty^+(0) = 0, \tag{8.2.5}$$

$$a_w^+(\infty) = 0, \quad a_\infty^+(\infty) = a_\infty^-(\infty) = 1. \tag{8.2.6}$$

Now we must determine the three functions by suitable equations obtained by taking suitable moments of the Boltzmann equation. The first three equations are clearly the conservation equations, which are integrated at once (Problem 8.2.1) to give

$$\rho_w u_w a_w^+ + \rho_\infty \left(\frac{RT_\infty}{2\pi}\right)^{1/2} (F^+ a_\infty^+ - F^- a_\infty^-) = \rho_\infty u_\infty, \tag{8.2.7}$$

$$\frac{1}{2}\rho_w RT_w a_w^+ + \frac{1}{2}\rho_\infty RT_\infty (G^+ a_\infty^+ + G^- a_\infty^-) = \rho_\infty u_\infty^2 + \rho_\infty RT_\infty, \tag{8.2.8}$$

$$2\rho_w u_w RT_w a_w^+ + 2\rho_\infty RT_\infty \left(\frac{RT_\infty}{2\pi}\right)^{1/2} (H^+ a_\infty^+ - H^- a_\infty^-)$$
$$= s\rho_\infty u_\infty \left(\frac{1}{2}u_\infty^2 + \frac{5}{2}\rho_\infty RT_\infty\right), \tag{8.2.9}$$

where

$$u_w = \left(\frac{RT_w}{2\pi}\right)^{1/2}, \tag{8.2.10}$$

$$F^- = -\pi^{1/2} S_\infty \operatorname{erfc}(S_\infty) + \exp(-S_\infty^2), \tag{8.2.11}$$

$$G^- = (2S_\infty^2 + 1)\operatorname{erfc}(S_\infty) - \frac{2S_\infty}{\pi^{1/2}} \exp(-S_\infty^2), \tag{8.2.12}$$

$$H^- = -\frac{\pi^{1/2} S_\infty}{2}\left(S_\infty^2 + \frac{5}{2}\right)\operatorname{erfc}(S_\infty) + \frac{1}{2}(S_\infty^2 + 2)\exp(-S_\infty^2), \tag{8.2.13}$$

$$F^+ = F^- + 2\pi^{1/2} S_\infty, \tag{8.2.14}$$

$$G^+ = -G^- + 2(2S_\infty^2 + 1), \tag{8.2.15}$$

$$H^+ = -H^- + \pi^{1/2} S_\infty \left(S_\infty^2 + \frac{5}{2}\right). \tag{8.2.16}$$

Here S_∞ is the speed ratio related to the Mach number M_∞ by

$$S_\infty = u_\infty (2RT_\infty)^{-1/2} = \left(\frac{5}{6}\right)^{1/2} \mathrm{M}_\infty \tag{8.2.17}$$

and $\text{erfc}(S_\infty)$ denotes the complementary error function:

$$\text{erfc}(S_\infty) = 1 - \text{erf}(S_\infty) = 1 - 2\pi^{-1/2}\int_0^{S_\infty} e^{-t^2}dt = 2\pi^{-1/2}\int_{S_\infty}^{\infty} e^{-t^2}dt. \tag{8.2.18}$$

We can now write Eqs. (8.2.7–8.2.9) at the wall and use the boundary conditions provided by Eq. (8.2.5) to obtain

$$zr - \beta^- F^- = 2\pi^{1/2}S_\infty, \tag{8.2.19}$$

$$z + \beta^- G^- = 4S_\infty^2 + 2, \tag{8.2.20}$$

$$z - \beta^- rH^- = r\pi^{1/2}S_\infty\left(S_\infty^2 + \frac{5}{2}\right), \tag{8.2.21}$$

where

$$z = \frac{p_w}{p_\infty}, \quad \beta^- = a_\infty^-(0), \quad r = \left(\frac{T_\infty}{T_w}\right)^{1/2}. \tag{8.2.22}$$

Here one can proceed in two different ways. The first method, originally used by Ytrehus,[23] is based on eliminating β^- and r to obtain a second-degree equation for z with rather lengthy expressions for the coefficients. The second, which seems to have been first pointed out in a paper devoted to the evaporation of polyatomic gases,[53] is based on a straightforward calculation that provides a much simpler second-degree equation for r. We shall follow the latter method, which exploits the identity (Problem 8.2.2)

$$H^- = F^- - \pi^{1/2}\frac{S_\infty}{4}G^-. \tag{8.2.23}$$

Then, if we multiply (8.2.20) by $\pi^{1/2}S_\infty/4$ and add the result to (8.2.19), we obtain

$$z\left(r + \pi^{1/2}\frac{S_\infty}{4}\right) - \beta^- H^- = \pi^{1/2}S_\infty\left(S_\infty^2 + \frac{5}{2}\right). \tag{8.2.24}$$

If we now multiply both sides of this equation by r and subtract Eq. (8.2.21) from the result, we obtain a strikingly simple relation:

$$z\left(r^2 + \pi^{1/2}\frac{S_\infty}{4}r - 1\right) = 0, \tag{8.2.25}$$

which, since $z \neq 0$, gives us a simple second-degree equation for r. The roots have opposit signs (their product is -1); the negative root must obviously be discarded and we obtain

$$r = -\pi^{1/2}\frac{S_\infty}{8} + \left[1 + \pi\left(\frac{S_\infty}{8}\right)^2\right]^{1/2}. \tag{8.2.26}$$

Then the other two parameters are easily obtained from Eqs. (8.2.19) and (8.2.20) in terms of r:

$$z = \frac{2\exp(-S_\infty^2)}{F^- + rG^-}; \quad \beta^- = \frac{2\,(2S_\infty^2 + 1)\,r - 2\pi^{1/2}S_\infty}{F^- + rG^-}. \tag{8.2.27}$$

The ratio of the mass flow rate $\dot{m}_b$ impinging upon the wall to the evaporation rate $\dot{m}_w$ is

$$\frac{\dot{m}_b}{\dot{m}_w} = \frac{\rho_\infty}{\rho_w} r\beta^- F^- \tag{8.2.28}$$

and can be computed as a function of S_∞ (Problem 8.2.3). The function is practically flat for $S_\infty \geq 0.6$ and it reaches a minimum value ($\cong$0.18) at $S_\infty \cong 0.8$; this means that for $S_\infty \geq 0.6$ about 18% of the vapor is scattered back. In more accurate theories, based on numerical simulations,[54] the minimum occurs exactly for a downstream Mach number equal to unity.

It is easy now to assign S_∞ and compute the other quantities through the previous formulas (see Table 8.1). These results, originally computed by Ytrehus,[23] turn out to be reasonably accurate when compared with numerical solutions of the Boltzmann equation.[15,30,54]

So far we used just the values of $a_w^+(x)$, $a_\infty^\pm(x)$ at the boundary and at infinity. In order to compute these three functions, we immediately realize that the

Table 8.1. *Corresponding values of different parameters in the evaporation in a half-space.*

S_∞	p_w/p_∞	ρ_∞/ρ_w	T_∞/T_w	β^-	$\dot{m}/\dot{m}_w$
0.000	1.000	1.000	1.000	1.000	0.000
0.100	1.231	0.849	0.957	1.020	0.294
0.200	1.500	0.728	0.915	1.060	0.494
0.300	1.812	0.630	0.876	1.135	0.627
0.400	2.170	0.550	0.838	1.271	0.714
0.500	2.577	0.484	0.802	1.511	0.768
0.600	3.037	0.429	0.767	1.928	0.800
0.700	3.553	0.383	0.734	2.644	0.815
0.800	4.127	0.345	0.703	3.862	0.820
0.900	4.764	0.312	0.673	5.932	0.816
0.907	4.813	0.310	0.671	6.132	0.816

three equations (8.2.7–8.2.9) are not sufficient. This is rather obvious, because, otherwise, we would obtain three constants as a solution, and this cannot be correct, because it contradicts the boundary conditions (8.2.5–8.2.6). In addition, we remark that we know already two different (constant) solutions of the system (i.e., the values at infinity, (0, 1, 1), and those at the wall, $(1, 0, \beta^-)$). This means that the determinant of the system must be zero, as one can verify by a direct computation (Problem 8.2.4). In addition any solution must be a linear combination of the two particular ones, which have just been mentioned. Hence (Problem 8.2.5)

$$a_\infty^+(x) = 1 - a_w^+(x)\,; \quad a_\infty^-(x) = 1 + (\beta^- - 1)a_w^+(x). \tag{8.2.29}$$

Since no further information can be expected from the conservation equations, we must obtain a further moment equation. Following Ytrehus,[23] we multiply the Boltzmann equation (8.2.1) by $|\boldsymbol{\xi}|^2$ and integrate over the entire velocity space with the following result (Eqs. (8.2.29) being taken into account)

$$\frac{da_\infty^-}{dx} = -\frac{K}{\lambda_w}(a_\infty^- - 1)(a_\infty^- - s), \tag{8.2.30}$$

where

$$K = \frac{\pi \rho_\infty T_w(\beta^- - 1)\Phi_1\Phi_2}{12\rho_w(T_w - T_\infty)}, \tag{8.2.31}$$

$$s = 1 - \frac{2}{\Phi_1} + \frac{4S_\infty^2}{\Phi_2}, \tag{8.2.32}$$

$$\Phi_1 = \frac{1}{\beta^- - 1}\left(\frac{\rho_w}{\rho_\infty} - 2 + \beta^- \operatorname{erfc} S_\infty\right), \tag{8.2.33}$$

$$\Phi_2 = \frac{1}{\beta^- - 1}\left(z - 2 + \beta^- \operatorname{erfc} S_\infty\right), \tag{8.2.34}$$

and λ_w is the mean free path corresponding to the saturated vapor pressure, related to the corresponding viscosity by the usual relation:

$$\lambda_w = \frac{\mu_w}{\rho_w}\left(\frac{\pi}{2RT_w}\right)^{1/2}. \tag{8.2.35}$$

All the parameters appearing in Eq. (8.2.30) can be computed, for a given vapor, if, in addition to the temperature of the wall T_w (and the related saturated vapor pressure $p_w = \rho_w R T_w$), one assigns an external parameter, which might be r, z, or S_∞. Equation (8.2.30), with the boundary condition $a_\infty^-(0) = \beta^-$, has the following solution (for $s \neq 1$):

$$\frac{a_\infty^-(x) - 1}{a_\infty^-(x) - s} = \frac{\beta^- - 1}{\beta^- - s} \exp\left(-K \frac{1-s}{l_w} x\right). \tag{8.2.36}$$

Hence $a_\infty^-(x)$ goes to the correct value $a_\infty^-(\infty) = 1$ if $s < 1$; if $s > 1$ it goes to the value $a_\infty^-(\infty) = s$ and the problem does not have an appropriate solution. Thus $s = 1$ is the critical value for a solution to exist. According to the numerical calculations based on the previous relations, $s = 1$ corresponds to $\mathrm{M}_\infty = 0.992$; this is very close to unity and thus we can expect that an exact solution would produce $\mathrm{M}_\infty = 1$ exactly. A more convincing argument will be given below. For $s = 1$, Eq. (8.2.30), with the boundary condition $a_\infty^-(0) = \beta^-$, has the following solution:

$$a_\infty^-(x) = \frac{\beta^- - 1}{1 + (\beta^- - 1)Kx/l_w}. \tag{8.2.37}$$

Thus the decay is not exponential but algebraic. Since at $s = 1$, $\beta^- \cong 6$ and $K \cong .1$, a_∞ will reach 90% (respectively, 99%) of the downstream value at about 20 (respectively, 200) mean free paths from the wall.

Once a_∞ is computed, ρ, u, and T may be computed immediately through the formulas

$$\frac{\rho}{\rho_\infty} = \frac{u_\infty}{u} = \frac{1}{2}[a_\infty(x) - 1]\Phi_1 + 1, \tag{8.2.38}$$

$$\frac{T}{T_\infty} = \frac{\rho_\infty}{3\rho}\left\{3S_\infty^2 + 3 + [a_\infty(x) - 1]\Phi_2 - 2S_\infty^2 \frac{\rho_\infty}{\rho}\right\}. \tag{8.2.39}$$

When $\mathrm{M}_\infty \cong 1$, $S_\infty \cong .91$, $\Phi_1 \cong .48$, $\Phi_2 \cong .79$, $\rho_\infty/\rho_w \cong .31$, $T_\infty/T_w \cong .67$, and $z \cong 4.8$.

A comparison of the temperature profile with the experimental data by Mager et al.[55] confirms a jump in the temperature of about 20% at the interface, an important effect upon which we shall return.

The net evaporated flux follows from the previous formulas and is given by

$$\dot{m} = \rho_w \left(\frac{RT_w}{2\pi}\right)^{1/2} - \rho_\infty \left(\frac{RT_\infty}{2\pi}\right)^{1/2} \beta^- F^-(S_\infty). \tag{8.2.40}$$

Schrage[8] found a similar formula except for the factor β^-. This factor is close to unity for low Mach numbers (see Table 8.1), but it is about 6 for a downstream sonic condition. In addition, of course, Schrage's theory does not give values for ρ_∞ and T_∞ in terms of S_∞. The Hertz–Knudsen formula (8.1.3) misses both factors, β^- and $F^-(S_\infty)$, and is subject to the same criticism.

We discuss now the circumstance that the critical Mach number is exactly unity. One way of justifying this, first pointed out by the author[35] and subsequently discussed and brought to completion by several authors,[36] is based on the study of the Boltzmann equation linearized about a drifting Maxwellian distribution with bulk velocity equal to u_∞. Here we shall follow a shortcut,[35] based on the assumptions that the critical value depends only on the behavior far away from the wall and that the Chapman–Enskog expansion, and hence the Navier–Stokes equations, should give the correct answer for this asymptotic behavior. Thus, if we denote by ρ', u', and T' the perturbations of the asymptotic values ρ_∞, u_∞, and T_∞, the linearized equations that determine the perturbations far from the wall are

$$u_\infty \frac{d\rho'}{dx} + \rho_\infty \frac{du'}{dx} = 0,$$

$$\rho_\infty u_\infty \frac{du'}{dx} + RT_\infty \frac{d\rho'}{dx} + R\rho_\infty \frac{dT'}{dx} - \frac{4}{3}\mu_\infty \frac{d^2u'}{dx^2} = 0,$$

$$\rho_\infty u_\infty^2 \frac{du'}{dx} + \frac{5}{2}\rho_\infty u_\infty \frac{dT'}{dx} - \frac{4}{3}\mu_\infty u_\infty \frac{d^2u'}{dx^2} - \kappa_\infty \frac{d^2T'}{dx^2} = 0. \quad (8.2.41)$$

Here, of course, μ_∞ and κ_∞ are the viscosity coefficient and heat conductivity evaluated at $T = T_\infty$.

This is a system of ordinary differential equations with constant coefficients. We can thus discuss its solutions by replacing differentiation by multiplication by a scalar λ (this will tell us whether the pure exponential solutions are growing, damped, or oscillating). This leads to finding the values of λ for which the determinant of the resulting algebraic system vanishes. The resulting equation for λ reads as follows (Problem 8.2.6):

$$\lambda^3 \left[\frac{4}{3}u_\infty \kappa_\infty \mu_\infty \lambda^2 - \left(2u_\infty^2 \mu_\infty \rho_\infty R + \rho_\infty R T_\infty \kappa_\infty\right)\lambda + \frac{3}{2}Ru_\infty\left(u_\infty^2 - \frac{5}{3}RT_\infty\right)\right] = 0. \quad (8.2.42)$$

The factor λ^3 can be disregarded, since we are looking for solutions that tend to zero at ∞. Then we are left with a second-degree equation. In the case of evaporation, $u_\infty > 0$ and the first coefficient is positive. Then the sign of $u_\infty^2 - \frac{5}{3}RT_\infty = \frac{5}{3}RT_\infty(\mathrm{M}_\infty^2 - 1)$ determines whether we have two positive roots ($M_\infty^2 > 1$) or one positive and one negative ($M_\infty^2 < 1$) (in the separating case ($M_\infty^2 = 1$) one root is positive and one is zero).

Hence we must have ($M_\infty^2 \leq 1$) for a solution with the correct behavior to exist. We have included the value $M_\infty^2 = 1$, about which this analysis does not say much, because our previous nonlinear computations show that we have an acceptable behavior for the critical value as well, although the decay is not exponential.

Problems

8.2.1 Prove that the system (8.2.7–8.2.9) follows from the ansatz (8.2.4) and the conservation equations.

8.2.2 Prove the identity (8.2.23).

8.2.3 Find an expression for the backscattered flow rate (8.2.28) as a function of S_∞.

8.2.4 Verify by a direct computation that the system (8.2.7–8.2.9) has a vanishing determinant.

8.2.5 Prove (8.2.29).

8.2.6 Prove (8.2.42).

8.3. The Knudsen Layer near a Condensing Surface

We shall start the discussion of the problem of condensation exactly where we ended the discussion on evaporation in the previous section. In this case, $u_\infty < 0$ and the first coefficient in the second-degree equation, which we are left with when the factor λ^3 is disregarded in Eq. (8.2.42), is negative. Then the sign of $u_\infty^2 - \frac{5}{3}RT_\infty = \frac{5}{3}RT_\infty(M_\infty^2 - 1)$ determines whether both roots are negative ($M_\infty^2 > 1$) or one positive and one negative ($M_\infty^2 < 1$) (in the separating case $M_\infty^2 = 1$ one root is negative and one is zero).

Hence a solution with the correct behavior exists anyway. If we have $M_\infty^2 > 1$, we must expect that assigning the Mach number (or any other single parameter) is not enough to determine the solution. In fact in this case the solutions of the linearized Navier–Stokes equations form a two-dimensional manifold for any given value of the Mach number at infinity.

If we could surmise that the behavior near the wall determines just one constant (as in the case of evaporation), whereas the other is determined by an extra condition at infinity, then we could parametrize the solutions with one parameter in the subsonic case and two parameters in the supersonic case. There is, however, a difference near the wall as well. This can be seen by linearizing the Navier–Stokes equations about $u = 0$ (rather than about u_∞) and ρ_w, T_w. This linearization will hold just a few mean free paths away from the wall and will not extend to infinity, but it will supply a piece of information about the number of constants to be assigned. To be consistent, the Reynolds number must be treated as finite, and, hence, we must treat the transport coefficients μ and κ as small of the same order as u. This is irrelevant in the momentum equation because both the convection and the viscosity term vanish thanks to the equation of mass balance, and just the pressure term survives. The situation is different in the energy equation. Then we have

$$\frac{du}{dx} = 0,$$

$$\frac{dp'}{dx} = 0,$$

$$\frac{5}{2}\rho_w u R \frac{dT'}{dx} = \kappa_w \frac{d^2T'}{dx^2}, \tag{8.3.1}$$

where we have written just u and not u' because the unperturbed value is zero. Thus both u and $p = p_w + p'$ are constant in the Navier–Stokes region and

$$T' = A + B\exp\left(\frac{5\rho_w R u}{2\kappa_w}\right), \tag{8.3.2}$$

where $u < 0$ in the case of condensation and $u > 0$ in the case of evaporation. Because the solution must be close to a constant and actually become closer to it when increasing x, we see that B in (8.3.2) must be zero in the case of evaporation, but not in the case of condensation. There is thus an extra parameter and thus there are two free parameters in the subsonic case and three in the supersonic one.

One might suspect an inconsistency in the supersonic case, since we linearized about $u = 0$, but this is not the case, because the state near the wall is always subsonic.

The additional parameter can be assigned arbitrarily or related to the latent heat, as we shall do below, following Ytrehus and Alvestad.[26] All these circumstances account for the differences between evaporation and condensation that we shall find below. In particular the solutions form a two-dimensional manifold in the subsonic case and a three-dimensional manifold in the supersonic case.

The general formulation of the condensation problem in a half-space is exactly the same as that discussed in the previous section, the only difference being, as we have just remarked, the sign of u_∞. The solution, however, requires more than a trimodal ansatz, because of the remarks given above. Thus, following Ytrehus and Alvestad,[26] we start with four modes

$$f(x, \boldsymbol{\xi}) = a_w^+(x)M_w^+ + a_\infty^+(x)M_\infty^+ + a_\infty^-(x)M_\infty^- + a_*^-(x)M_*^-. \quad (8.3.3)$$

The last mode, following Ytrehus and Alvestad[26] is, more or less arbitrarily, chosen as a result of first collisions among the evaporated molecules described by M_w^+, as resulting by a calculation via the BGK model. It turns out to have the following shape:

$$M_*^- = \rho_*(2\pi RT_*)^{-3/2}\exp[-|\boldsymbol{\xi} - u_*\mathbf{i}|^2/(2RT_*)] \quad \text{for } \xi < 0 \quad (8.3.4)$$

with fixed parameters

$$\rho_* = \frac{\rho_w}{2}, \quad u_* = \left(\frac{2RT_w}{\pi}\right)^{1/2}, \quad T_* = T_w\left(1 - \frac{2}{3\pi}\right). \quad (8.3.5)$$

The speed ratio of the fourth mode is therefore given by

$$S_* = u_*(2RT_*)^{-1/2} = \left(\pi - \frac{2}{3}\right)^{-1/2} \cong 0.64. \quad (8.3.6)$$

Again, the boundary conditions may be satisfied exactly by imposing the following conditions on the four functions, $a_\infty^\pm(x)$, $a_*^-(x)$:

$$a_w^+(0) = 0, \quad a_\infty^+(0) = 0, \quad (8.3.7)$$

$$a_w^+(\infty) = a_*^-(\infty) = 0, \quad a_\infty^+(\infty) = a_\infty^-(\infty) = 1. \quad (8.3.8)$$

In addition to the parameters already used in the problem of evaporation, we must introduce a further parameter:

$$\alpha^- = a_*^-(0), \quad (8.3.9)$$

which must be determined by the solution of the problem.

We now proceed as in the previous section. We still need just three conservation equations and a further moment equation. The conservation equations read as follows:

$$\rho_w u_w a_w^+ + \rho_\infty \left(\frac{RT_\infty}{2\pi}\right)^{1/2} (F^+ a_\infty^+ - F^- a_\infty^-) - \rho_* \left(\frac{RT_*}{2\pi}\right)^{1/2} F_*^- a_*^- = \rho_\infty u_\infty, \quad (8.3.10)$$

$$\frac{1}{2}\rho_w RT_w a_w^+ + \frac{1}{2}\rho_\infty RT_\infty(G^+ a_\infty^+ + G^- a_\infty^-) + \frac{1}{2}\rho_* RT_* G_*^- a_*^-$$

$$= \rho_\infty u_\infty^2 + \rho_\infty RT_\infty, \tag{8.3.11}$$

$$2\rho_w u_w RT_w a_w^+ + 2\rho_\infty RT_\infty \left(\frac{RT_\infty}{2\pi}\right)^{1/2} (H^+ a_\infty^+ - H^- a_\infty^-)$$

$$- 2\rho_* RT_* \left(\frac{RT_*}{2\pi}\right)^{1/2} H_*^- a_*^- = \rho_\infty u_\infty \left(\frac{1}{2}u_\infty^2 + \frac{5}{2}\rho_\infty RT_\infty\right). \tag{8.3.12}$$

Here, of course, F_*^-, G_*^-, and H_*^- are the same functions as F^-, G^-, and H^-, defined in (8.2.11–8.2.13), except for the fact that they are evaluated at the speed ratio S_* rather than S_∞. Since S_* is a numerical constant given by (8.3.6), F_*^-, G_*^-, and H_*^- are also numerical constants.

As in the case of evaporation, we obtain the relations between the flow parameters at $x = 0$ and $x = \infty$ by introducing the values that the coefficients of the Maxwellians take at the wall:

$$zr - \beta^- F^- - z_* r_* - \alpha^- F_*^- = 2\pi^{1/2} S_\infty, \tag{8.3.13}$$

$$z + \beta^- G^- z_* - \alpha^- G_*^- = 4S_\infty^2 + 2, \tag{8.3.14}$$

$$z - \beta^- r H^- r z_* - \alpha^- H_*^- = r\pi^{1/2} S_\infty \left(S_\infty^2 + \frac{5}{2}\right), \tag{8.3.15}$$

where $z_* = \frac{1}{2}z(1 - \frac{2}{3\pi})$ and $r_* = r(1 - \frac{2}{3\pi})^{-1/2}$ are not further unknown quantities.

Now we must assign one of the four parameters z, r, β^-, α^- (in addition to the Mach number) or some combination of them, because we have three equations for four unknowns. One can assign two parameters arbitrarily or systematize the choice of values by assuming that the vapor is at saturation in the upstream state. This assumption implies that the Clausius–Clapeyron equation

$$\frac{dp}{dT} = \Delta H \frac{p}{RT^2} \tag{8.3.16}$$

applies. Here ΔH is the latent heat per unit mass, which equals the difference between h_g and h_l, the enthalpies in the gaseous and the liquid state. In general, ΔH is a function of temperature, but if the states corresponding to T_w and T_∞ are close enough, we can assume that ΔH is constant and integrate Eq. (3.16) between these two temperatures to obtain

$$\frac{p_\infty}{p_w} = \exp\left(\Delta H \frac{p}{RT_w}\left[1 - \frac{T_w}{T_\infty}\right]\right). \tag{8.3.17}$$

Now we can assign the normalized latent heat $\Delta H/RT_w$ as a parameter in addition to the speed ratio (or Mach number).

Actual calculations[26] show that $|S_\infty| = |u_\infty|(2RT_\infty)^{1/2}$, for both large and small values of p_∞/p_w, is not very sensitive to the value given to the nondimensional latent heat. This finding is confirmed by an extensive numerical study by Aoki et al.[42] based on the BGK model. For many practical purposes, then, the speed ratio can be considered to be independent of the latent heat.

For a saturated vapor at temperature T_∞ to condense at the interface, the dense phase temperature T_w must be smaller than T_∞. Then one would intuitively expect the vapor temperature at the wall $T(0)$ to be less than T_∞. However, when the nondimensional latent heat is sufficiently high (larger than, say, 10) we have the opposite result in an appreciable range of values of the pressure ratio p_∞/p_w. In other words, there are situations in which the vapor increases its temperature as it approaches the cold wall and an inverted temperature gradient occurs, as first predicted by Pao[12,13] in the linearized case (see Section 8.5).

Even when this effect is not present there is a significant temperature jump at the interface.

The method of moments[26] can be used to obtain the detailed structure of the Knudsen layer as in the case of evaporation. The results are in agreement with those of Hatakeyama and H. Oguchi,[24] Sone et al.,[37] Aoki et al.,[42] Labuntsov and Kryukov,[25,43] and Abramov and Kogan.[40,41]

The main limitation of the moment method seems to be that it requires a fairly correct understanding of the behavior of the solution and in particular of the number of free parameters upon which the solution depends. Sone et al.[37] and Aoki et al.[42] have studied the time evolution of the half-space problem with the BGK model. They used numerical techniques based upon a finite-difference method to study the disturbance produced by the presence of the condensation process at the interface upon a chosen uniform background for a large number of cases. They were led to an accurate classification of the flow patterns and identified the parameter range within which steady-state solutions occur. For a fixed value of T_∞/T_w, if p_∞/p_w is a certain function of M_∞ and the latter is less than unity, there is a unique steady solution. For a supersonic condensing flow, all three external parameters (T_∞, p_∞, and $u_\infty < 0$) can be chosen freely, in contrast to the subsonic case for which only two parameters are free. In addition, an external tangential velocity can be assigned at will. The effect of the presence of a tangential motion has been studied in detail by Aoki et al.[56] They found that the influence of this motion upon the condensation process is marginal, except at transonic and high values of the Mach number ($M_\infty > 3$). Moment methods are only able to recover partially the behavior of supersonic condensation.

Results similar to those found by Sone et al.[37] and Aoki et al.[42] have been found by Abramov and Kogan,[39,41] who dispensed with the simplifications offered by the BGK model and used the Direct Simulation Monte Carlo method. In particular they showed that the details of the molecular interaction have a very slight influence on the solution of the problem, and then only for values of M_∞ close to unity.

Problems

8.3.1 Prove that the system (8.3.10–8.3.12) follows from the ansatz (8.3.3) and the conservation equations.

8.3.2 Construct tables similar to Table 8.1 by assigning values of one parameter, (e.g., the nondimensional latent heat $\Delta H/(RT)$).

8.4. Influence of the Evaporation–Condensation Coefficient

The results obtained so far refer to an evaporation–condensation coefficient of $\sigma = 1$. This was done for the sake of simplicity in the presentation. If we depart from this assumption a fraction $1 - \sigma$ of the molecules will not condense on the surface but will be scattered by the condensed phase as from an ordinary wall. Then the results will depend on the assumptions made on the scattering kernel. There are two simple assumptions: specular reflection and diffusion according to a Maxwellian distribution (or more generally a linear combination of the two, i.e., Maxwell's boundary conditions).

The case of specular reflection for an evaporating interface was treated by the author[57] by means of an extension of Anisimov's and Ytrehus's ansatz for the distribution function. The changes from the case of $\sigma = 1$ were not spectacular.

We also remark that a paper by Cercignani et al.[58] explored a fictitious boundary condition where the evaporating molecules were assumed to have temperature and bulk speed different from those of the interface. Calculations indicated that the temperature had to be assumed to be very different from that of the wall in order to obtain results significantly different from those of Ytrehus.

The most recent approach is due to Ytrehus[59] and assumes diffuse reflection according to a Maxwellian distribution having zero bulk velocity and the temperature of the wall. Thus the case $\sigma \neq 1$ is very similar to the case $\sigma = 1$. The distribution function of the evaporating molecules is exactly the same. The only difference is in the density, which, in addition to the contribution corresponding to the saturated vapor pressure, has a contribution with the same mass flow as the impinging flow. The advantage of this boundary condition is that the solution can be obtained from that of the case $\sigma = 1$ with the replacement

$$\rho_w^1 = \sigma\rho_w + (1-\sigma)j_r\left(\frac{2\pi}{RT_w}\right)^{1/2}, \tag{8.4.1}$$

where j_r is the flow of vapor scattered back onto the interface and ρ_w is the density of the Maxwellian distribution describing the evaporating molecules. From a mathematical viewpoint, ρ_w^1 is just what was simply denoted by ρ_w so far, although it is now a combination of the densities of evaporating and scattered molecules. To find j_r we use the mass balance to obtain (Problem 8.4.1)

$$j_r = \rho_w\left(\frac{RT_w}{2\pi}\right)^{1/2} - \rho_\infty u_\infty. \tag{8.4.2}$$

Inserting this in the previous expression and solving for $\rho_w^\sigma/\rho_\infty$, we get

$$\frac{\rho_w}{\rho_\infty} = \frac{\rho_w^1}{\rho_\infty} + \frac{1-\sigma}{\sigma}2\pi^{1/2}\left(\frac{T_\infty}{T_w}\right)^{1/2} S_\infty, \tag{8.4.3}$$

where u_∞ has been replaced by $S_\infty(2RT_\infty)^{1/2}$. Keeping in mind that T_w is given and T_∞ is determined in a way independent of σ, we can easily modify all the formulas obtained in the case $\sigma = 1$, by simply replacing ρ_w, whenever it occurs by the expression (8.4.3), where ρ_w^1 is the density of saturated vapor pressure (which was simply denoted by ρ_w in the previous sections). Thus it is easy to obtain a formula for the new pressure in the Maxwellian distribution of evaporating molecules:

$$\frac{p_w}{p_\infty} = \frac{p_w^1}{p_\infty} + \frac{1-\sigma}{\sigma}2\pi^{1/2}\left(\frac{T_w}{T_\infty}\right)^{1/2} S_\infty \tag{8.4.4}$$

and show that the flow rate for $\sigma \neq 1$ is always less than that corresponding to $\sigma = 1$. In particular the flow rate tends to zero for $\sigma \to 0$, as is physically obvious (no evaporation) (see Problems 8.4.2 and 8.4.3). It is thus clear that the value of σ has an extremely important influence on the results when the scattered molecules are assumed to undergo a diffusion according to M_w, at variance with the case of specular reflection. This can be explained by observing that, roughly speaking, specular reflection adds molecules that are in one of the high speed tails of a Maxwellian distribution centered at $-u_\infty$.

Similar considerations can be applied to the case of condensation. Here the solution is even more dependent on the value of σ, as first remarked by Kogan and Abramov[41] and detailed by Butkovsky.[60] This is easily understood from the rule (8.4.4), which we rewrite here in the case of condensation, taking into account the fact that we must change S_∞ into $-|S_\infty|$, because u_∞ is now negative:

$$\frac{p_w}{p_\infty} = \frac{p_w^1}{p_\infty} - \frac{1-\sigma}{\sigma} 2\pi^{1/2} \left(\frac{T_w}{T_\infty}\right)^{1/2} |S_\infty|. \tag{8.4.5}$$

This formula shows that a sonic state cannot occur for σ slightly below unity – $\sigma_{\text{crit}} = 0.97$ (Problem 8.4.4).

The condensing flow rate is very sensitive to the value given to σ. If we reduce the evaporation–condensation coefficient from 1 to, say, 0.4, for a pressure ratio $p_\infty/p_w = 10$, then the speed ratio $|S_\infty|$ is reduced from 0.9 to less than 0.15, which is, more or less, for this value of σ, the maximum value attained by $|S_\infty|$. For values of σ larger than 0.97, these phenomena do not occur, but much higher pressure ratios than in the ideal case $\sigma = 1$ are required to attain Mach number unity.

Although these strange results may appear to be a consequence of the approximate nature of the solution, they are similar to those found by the Monte Carlo simulations by Kogan and Abramov[41] and by the BGK model solution by Aoki et al.[42]

The results found in this and the previous section can be used to obtain a modified mass flow formula of the type first considered by Schrage[8] (Problem 8.4.5).

Problems

8.4.1 Prove Eq. (8.4.2).

8.4.2 Prove Eq. (8.4.5).

8.4.3 Find a formula for the mass flow rate at an evaporating wall for $\sigma \neq 1$.

8.4.4 Show that a sonic state cannot occur for $\sigma < 0.97$.

8.4.5 Discuss a modified Schrage formula, starting from the pure evaporation case, Eq. (8.2.40), and adding the effects of condensation as in the Hertz–Knudsen approach. Start from the case $\sigma = 1$ and then generalize the results to the case $\sigma \neq 1$.

8.5. Moderate Rates of Evaporation and Condensation

In this section we consider the case of evaporation and condensation phenomena for Mach numbers much smaller than unity. For finite Reynolds numbers this corresponds to small Knudsen numbers and hence also to the continuum limit ($\mathrm{Kn} \cong \mathrm{Ma}/\mathrm{Re}$). This means that we can use the study of the Knudsen layer to solve the Navier–Stokes equations. Mathematical justifications at various levels of rigor have been given by Sone,[44] Onishi and Sone,[18] and Caflisch.[61]

Within the accuracy of the moment solution discussed in the previous section, we can obtain the results for this moderate regime by writing

$$r = 1 + \Delta r, \quad z = 1 + \Delta z, \quad \beta^- = 1 + \Delta\beta^-, \tag{8.5.1}$$

where r, z, and β^- were defined in (8.2.22) and Δr, Δz, and $\Delta\beta^-$ are assumed to be small and of the same order as S_∞. The linearization is easily performed by looking at the expansions of the functions of S_∞, F^-, G^-, and H^-, defined in Eqs. (8.2.11)–(8.2.13), near the origin:

$$F^- = 1 - \pi^{1/2} S_\infty + O\left(S_\infty^2\right), \tag{8.5.2}$$

$$G^- = 1 - 4\pi^{-1/2} S_\infty + O\left(S_\infty^2\right), \tag{8.5.3}$$

$$F^- = 1 - \frac{5}{4}\pi^{1/2} S_\infty + O\left(S_\infty^2\right). \tag{8.5.4}$$

If we neglect higher order terms Eqs. (8.2.19)–(8.2.21) become

$$\Delta z + \Delta r - \Delta\beta^- = \pi^{1/2} S_\infty, \tag{8.5.5}$$

$$\Delta z + \Delta\beta^- = 4\pi^{-1/2} S_\infty, \tag{8.5.6}$$

$$\Delta z - \Delta r - \Delta\beta^- = \frac{5}{4}\pi^{1/2} S_\infty, \tag{8.5.7}$$

where we shall consider S_∞ as an independent variable and Δr, Δz, and $\Delta\beta^-$ as unknowns.

The system is readily solved by subtracting the third equation from the first to yield

$$\Delta r = -\frac{\pi^{1/2}}{8} S_\infty. \tag{8.5.8}$$

Then we substitute this value into the first equation and solve the system of the first two equations (Problem 8.5.1) to yield

$$\Delta z = \left(2\pi^{-1/2} + \frac{9}{16}\pi^{1/2}\right) S_\infty, \tag{8.5.9}$$

$$\Delta\beta^- = \left(\frac{2}{\pi} - \frac{9}{16}\right)\pi^{1/2} S_\infty. \tag{8.5.10}$$

It is convenient to introduce the physical quantities, recalling that, because of (8.2.22),

$$\Delta z \cong 1 - \frac{p_\infty}{p_w} = \frac{\Delta p}{p_w}, \tag{8.5.11}$$

$$\Delta r \cong -\frac{1}{2}\frac{T_w - T_\infty}{T_w} = -\frac{1}{2}\frac{\Delta T}{T_w}. \tag{8.5.12}$$

The temperature and pressure differences are obviously related by

$$\frac{\Delta T}{T_w} = \frac{\pi^{1/2}}{4} S_\infty = \frac{\pi^{1/2}}{4}\left(2\pi^{-1/2} + \frac{9}{16}\pi^{1/2}\right)^{-1}\frac{\Delta p}{p_w}. \tag{8.5.13}$$

If we look at the linearized form of Eq. (8.2.40) we find for the net mass flow:

$$\dot{m} = \rho_w \left(\frac{RT_w}{2\pi}\right)^{1/2}\left(\frac{\Delta p}{p_w} - \frac{\Delta T}{T_w} - \Delta\beta^- + \pi^{1/2} S_\infty\right). \tag{8.5.14}$$

Because of the last term, Eq. (8.5.14) provides a result that is about twice the mass flow given by the Hertz–Knudsen formula.

We can now express $\Delta T/T_w$, $\Delta\beta^-$, and S_∞ in terms of $\Delta p/p_w$ by means of the above relations to yield

$$\dot{m} = \rho_w \left(\frac{RT_w}{2\pi}\right)^{1/2}\frac{32\pi}{32 + 9\pi}\frac{\Delta p}{p_w}. \tag{8.5.15}$$

If we recall Eq. (8.4.4), we can now pass from the case of an evaporation-condensation coefficient $\sigma = 1$ to the generic case $\sigma \neq 1$. The linearized version of Eq. (8.4.4) reads

$$\begin{aligned}\frac{\Delta p}{p_w} &= \frac{\Delta p^1}{p_w} + \frac{1-\sigma}{\sigma} 2\pi^{1/2} S_\infty \\ &= \frac{\Delta p^1}{p_w}\left[1 + \frac{1-\sigma}{\sigma} 2\pi^{1/2}\left(2\pi^{-1/2} + \frac{9}{16}\pi^{1/2}\right)^{-1}\right],\end{aligned} \tag{8.5.16}$$

where Δp^1 is obviously the pressure difference in the ideal case $\sigma = 1$. Solving for $\Delta p^1/p_w$ and inserting the resulting expression in place of $\Delta p/p_w$ in Eq. (8.5.15), we finally obtain

$$\dot{m} = \rho_w \left(\frac{RT_w}{2\pi}\right)^{1/2} \frac{32\pi}{32+9\pi} \frac{\Delta p}{p_w} \frac{\sigma}{\sigma + [32\pi/(32+9\pi)](1-\sigma)}. \tag{8.5.17}$$

We can repeat the entire treatment for the case of condensation. This is, however, unnecessary, because the problem is linear and the solutions for evaporation and condensation differ just by the sign of S_∞ and hence one of them is easily obtained from the other. There is, however, one basic difference, due to the fact that in the case of condensation there is an extra parameter. We can think of assigning both the pressure and the temperature at infinity, which is impossible in the case of evaporation. In other words, the linearization does not work and the trick of using an extra term as done by Ytrehus and Alvestad[26] (see Eq. (8.3.3)) does not easily fit into a linearized description. A way out of the difficulty was proposed by Ytrehus:[59,62] We can assume that the linearized solution holds in a finite Knudsen layer near the interface. Then what we denoted by the suffix ∞ so far must be denoted by K to emphasize that we are dealing with the value at the outer edge of the Knudsen layer. Then the value of the temperature T_K is taken as a starting value at $x \cong 0$ for a solution given by the Navier–Stokes equations linearized about the state at space infinity. These equations were already written in (8.2.41), but there is a simplification owing to the small Mach number assumption; because of this we can linearize about $u = 0$ rather than about u_∞. The linearization is practically similar to that considered in the early part of Section 8.3 where we discussed the number of free parameters for condensation. Because of the finite Reynolds number assumption, we must treat the transport coefficients μ_∞ and κ_∞ as small of the same order as u_∞. This is, as in Section 8.3, irrelevant in the momentum equation because both the convection and the viscosity terms vanish thanks to the equation of mass balance, and just the pressure term survives. The situation is different in the energy equation. Then we have

$$\begin{aligned} \frac{du}{dx} &= 0, \\ \frac{dp'}{dx} &= 0, \\ \frac{5}{2}\rho_\infty u R \frac{dT'}{dx} &= \kappa_\infty \frac{d^2T'}{dx^2}, \end{aligned} \tag{8.5.18}$$

where we have written just u and not u' because the unperturbed value is zero.

Thus both $u = u_\infty$ and $p = p_\infty$ are constant in the Navier–Stokes region and

$$T' = T'_K \exp\left(\frac{5\rho_\infty R u_\infty}{2\kappa_\infty}\right), \tag{8.5.19}$$

where $u_\infty < 0$ and hence $T' \to 0$ for $x \to \infty$. We remark that this solution is not available in the case of evaporation because $u_\infty > 0$ and this offers another indication of the difference between evaporation and condensation.

We can now rewrite Eq. (8.5.13) in the following form:

$$\frac{T_K - T_w}{T_w} = \frac{\sigma \Delta p}{\beta_c + (1-\sigma)\alpha_c}, \tag{8.5.20}$$

where

$$\beta_c = \frac{32 + 9\pi}{4\pi}, \quad \alpha_c = \frac{23\pi - 32}{4\pi} \tag{8.5.21}$$

are two numerical constants.

We assume, as we did before, that the vapor is saturated at space infinity (which does not coincide now with the region outside the Knudsen layer). Then we can apply the Clausius–Clapeyron equation (8.3.16), which, in an appropriate linearized form, reads as follows:

$$\Delta p = \frac{\Delta H}{RT_w} \Delta T. \tag{8.5.22}$$

Equations (8.5.20) and (8.5.22) lead to a relation between the temperature difference in the Knudsen layer and the temperature difference between the interface and space infinity:

$$\frac{T_K - T_w}{T_w} = \frac{\sigma \Delta T}{\beta_c + (1-\sigma)\alpha_c} \frac{\Delta H}{RT_w}. \tag{8.5.23}$$

To appreciate this result, let us consider the case $\sigma = 1$, which was, in practice, the only case considered before the recent work by Ytrehus.[59,62] Then we have

$$\frac{T_K - T_w}{T_w} = \frac{\Delta T}{\beta_c} \frac{\Delta H}{RT_w}. \tag{8.5.24}$$

This means that if the normalized latent heat $\Delta H / RT_w$ is larger than $\beta_c \cong 4.7965$, the temperature drop $T_K - T_w$ exceeds the total temperature drop ΔT. This implies that a region with inverted temperature gradient (i.e., a gradient opposed to ΔT) must occur in the Navier–Stokes region. This is the effect mentioned in the introduction and first discussed by Pao[12,13] for the problem of evaporation and condensation in a slab between two interfaces of the same substance, where the effect is particularly spectacular at moderate and small Reynolds numbers. In this case, as first stressed by Aoki and Cercignani,[63] the temperature change across the Knudsen layer dominates the entire bulk flow in the slab.

The value of the parameter β_c found above is essentially correct; all methods give a value between 4.6 and 4.8. Now, it turns out that most substances have a normalized latent heat significantly higher than 5 (according to an empirical rule given by Trouton, $\Delta H/(RT_w)$ is close to 10 for most substances; it is close to 13 for water at boiling temperature).

Cercignani et al.[58] investigated the possible mechanisms that could change this paradoxical result. Most vapors are polyatomic; the presence of internal degrees of freedom gives an increase of the value of β_c[58] (which is then estimated to be 8.206 for three internal degrees of freedom), but this increase is not enough to rule out the paradoxical effect for water. Another possibility is offered by a boundary condition different from the ideal one; as mentioned above, a study of a fictitious boundary condition where the evaporating molecules were assumed to have a temperature and a bulk velocity different from those of the wall indicated that the temperature or some other parameter had to be assumed to be very different from that of the wall in order to obtain a value for β_c significantly different from 4.8.

A clarification of the matter has come with the recent work of Ytrehus,[59,62] who obtained the formula indicated above, Eq. (8.5.20). This formula indicates that the critical value for the normalized latent heat becomes

$$\beta_c^\sigma = \frac{\beta_c}{\sigma} + \frac{1-\sigma}{\sigma}\alpha_c = \beta_c + 8\frac{1-\sigma}{\sigma}, \tag{8.5.25}$$

where β_c^σ is the critical value of the normalized latent heat for an arbitrary value of σ and β_c is, as above, the critical value corresponding to $\sigma = 1$. In addition, we have taken into account that $\alpha_c + \beta_c = 8$ according to (8.5.21). Thus it is enough to have a value of σ less than 0.5 to avoid the inverted temperature gradient for water (even without taking into account the correction for internal degrees of freedom).

It should be stressed that Sone et al.[64] had considered the same problem for a gas of hard spheres between two parallel plates rather than in a half-space. They solved the problem numerically on the basis of the linearized Boltzmann equation, first for $\sigma = 1$ and then for an arbitrary σ. They had already noticed the possibility of relating the second problem with the first. In the case of a Knudsen number (based on the distance between the plates) going to infinity, their result for the critical value β_c reads as follows:

$$\beta_c^\sigma = \frac{\beta_c}{\sigma} + \frac{1-\sigma}{\sigma}\alpha_c, \tag{8.5.26}$$

where $\beta_c \cong 4.6992$ and $\alpha_c \cong 3.0611$. This should be compared with the values $\beta_c \cong 4.7965$ and $\alpha_c \cong 3.2035$, obtained by Ytrehus for Maxwell molecules, as discussed above. The difference between the values just quoted is of the order of 2% and part of it might be due to the use of a diffrent molecular model; this confirms once more that, in this case, the moment method, with a good ansatz for the distribution function, leads not only to a correct qualitative picture, but also to rather accurate numerical values.

A comment can be made on the fact that if we try to look at the matching between the Navier–Stokes region and the Knudsen layer, we find that there is necessarily a discontinuous derivative, because we match an asymptotically flat curve with a decreasing exponential with a positive amplitude. This is not contradictory, however, because the slope of the exponential is vanishingly small on the scale of the mean free path. In fact,

$$\frac{5\rho_\infty R u_\infty \lambda}{2\kappa_\infty} \cong S_\infty, \tag{8.5.27}$$

which is small by assumption.

The real question now becomes the value of σ. In a recent review, Eames et al.[65] survey several experiments on this coefficient for water performed between 1931 and 1989. Only two out of seventeen experiments surveyed give values of σ definitely higher than 0.50, whereas seven give values below 0.1. For the sake of clarity, one must remark that Eames et al.[65] conclude that the most realistic values of the evaporation–condensation coefficient are the higher ones, mainly because the other results are affected by erroneous temperature and pressure measurements or inappropriate assumptions behind most of them.

Today, another tool is available to investigate the value of σ: molecular dynamics computer simulations of phase change phenomena.[66–68] These computations reveal important information going beyond the level of the experiments performed so far. A remarkable feature of these simulations is the discovery that condensation is not a unimolecular process.[67,69] In other words, there is a rather complex interaction between the molecules of the vapor impinging upon the wall and those of the dense phase. This reduces the experimentally observable condensation coefficient. The effect seems to be more pronounced for associating liquids, such as water, and values of σ as low as 0.1, or even smaller, are found.[68,70] For spherical molecules such as argon, higher values, between 0.6 and 0.8, are obtained.[67] Another important result is that σ decreases with the dense-phase temperature and appears to depend on the degree of nonequilibrium of the system.

It should be clear that more accurate measurements are required in this area. In particular, accurate measurements of the temperature profile in a condensing vapor can produce rather stringent bounds on the upper bound for the evaporation–condensation profile, should an inverted temperature gradient be ruled out by these experiments (assuming that the latent heat is known with sufficient accuracy).

Problem

8.5.1 Obtain (8.5.9) and (8.5.10).

8.6. Effects due to the Presence of a Noncondensable Gas and to the Internal Degrees of Freedom

In addition to the effects due to a nonunity evaporation–condensation coefficient, there are two other important features to be taken into account when comparing theory and experiment. One of them is the fact that evaporation and condensation always occur, in practice, in the presence of another gas, such as air, which may be noncondensable. The other feature is the presence of internal degrees of freedom. These two effects will be briefly considered in this section.

The behavior of a gas–vapor mixture between two parallel plates of the condensed phase has been studied by many authors.[71–80] With the exception of the paper by Frezzotti,[76] who extended the method based on the moment equations to the evaporation of mixtures, and the recent paper by Aoki et al.,[80] the papers quoted above treat the case of weak evaporation and condensation and are mainly based on the use of model equations.

In the aforementioned paper Aoki et al.[80] investigated the problem of evaporation and condensation for a mixture of a vapor and a noncondensable gas by the Direct Simulation Monte Carlo method. They also gave an asymptotic analysis for small values of the Knudsen number (based on the mean free path for the collisions between vapor molecules). This analysis gives a striking result: When the average number density of the noncondensable gas is vanishingly small with respect to the average number density of the vapor, the vapor is still affected by the presence of the gas. To give an example, if the plane interfaces are located at a distance of 10 cm and we consider water vapor at a temperature of 350 K, the saturated vapor pressure of the vapor is 300 Torr, and thus the mean free path is about 10^{-5} cm (i.e., $\mathrm{Kn} = 10^{-6}$), a rather small value. It is sufficient to put a very small amount of noncondensable gas in the gap, corresponding to a partial pressure of about 3×10^{-4} Torr, to obtain an appreciable effect on the vapor flow. This is because the gas remains confined in the Knudsen layers and thus influences the temperature drop through the layer, which, as we know, rules the behavior in the part of the gap where the Navier–Stokes equations apply. This behavior is confirmed by the Monte Carlo simulations.

It should also be remarked that, in the case of condensation in a half-space, the presence of a noncondensable gas changes the range of existence of the solution, if the amount of noncondensable gas exceeds a critical value. If this value is exceeded, then there is a range of values of the Mach number, around $M_\infty = 1$, depending on the amount of noncondensable gas, for which no solution exists.[81,82]

As for the effect of internal degrees of freedom on evaporation, we follow the approach introduced by the author.[53,57] This method generalizes the approach proposed by Anisimov,[20] and brought to completion by Ytrehus,[23] to the case of internal degrees of freedom that may be described in a fully classical way. Then a logical development of Anisimov's ansatz reads as follows

$$f(x,\xi) = a_w^+(x)M_w^+ + a_\infty^+(x)M_\infty^+ + a_I(x)M_I + a_\infty^-(x)M_\infty^-, \tag{8.6.1}$$

where now

$$M_w^+ = \rho_w(2\pi RT_w)^{-3/2}\left(\Gamma\left(\frac{j}{2}\right)\right)^{-1}(k_BT_w)^{\frac{-j}{2}}E_i^{\frac{j}{2}-1}$$
$$\times \exp[-|\xi|^2/(2RT_w) - E_i/(k_BT_w)] \quad \text{for } \xi > 0, \tag{8.6.2}$$

$$M_\infty = \rho_\infty(2\pi RT_\infty)^{-3/2}\left(\Gamma\left(\frac{j}{2}\right)\right)^{-1}(k_BT_\infty)^{\frac{-j}{2}}E_i^{\frac{j}{2}-1}$$
$$\times \exp[-|\xi - \mathbf{i}u_\infty|^2/(2RT_\infty) - E_i/(k_BT_\infty)]. \tag{8.6.3}$$

Here E_i is the internal energy, j is the number of the internal degrees of freedom (assumed to be indistinguishable from each other), and $\Gamma(j/2)$ denotes the Gamma function, which is introduced for normalization purposes. In fact,

$$\int \frac{f_w}{\rho_w}\, d\xi dE_i = \int \frac{f_\infty}{\rho_\infty}\, d\xi dE_i = 1. \tag{8.6.4}$$

The distribution function M_I must be chosen to represent those molecules which have the internal degrees of freedom in equilibrium with the wall, whereas their translational degrees of freedom have almost reached equilibrium with the gas streaming far away from the wall. The related part of the distribution function only influences the structure of the Knudsen layer but has no influence on the discussion of the general features of the solution. We assume

$$M_l = \rho_\infty (2\pi R T_\infty)^{-3/2} \left(\Gamma\left(\frac{j}{2}\right)\right)^{-1} (k_B T_w)^{-\frac{j}{2}} E_i^{\frac{j}{2}-1} \times \exp[-|\boldsymbol{\xi} - \mathbf{i}u_l|^2/(2RT_\infty) - E_i/(k_B T_w)]. \tag{8.6.5}$$

The streaming velocity $\mathbf{i}u_l$ must be chosen so as to comply with the conservation laws. The latter now reads

$$\rho_w u_w a_w^+ + \rho_\infty \left(\frac{RT_\infty}{2\pi}\right)^{1/2} (F^+ a_\infty^+ - F^- a_\infty^-) + a_l \rho_\infty u_\infty \frac{S_l}{S_\infty} = \rho_\infty u_\infty, \tag{8.6.6}$$

$$\frac{1}{2}\rho_w R T_w a_w^+ + \frac{1}{2}\rho_\infty R T_\infty (G^+ a_\infty^+ + G^- a_\infty^-) + a_l (\rho_\infty u_l^2 + \rho_\infty R T_\infty) = \rho_\infty u_\infty^2 + \rho_\infty R T_\infty, \tag{8.6.7}$$

$$2\rho_w u_w R T_w a_w^+ + 2\rho_\infty R T_\infty \left(\frac{RT_\infty}{2\pi}\right)^{1/2} (H^+ a_\infty^+ - H^- a_\infty^-) + \frac{j}{2}\rho_\infty R T_\infty \left(\frac{RT_\infty}{2\pi}\right)^{1/2} (F^+ a_\infty^+ - F^- a_\infty^-) + \frac{j}{2}\rho_w u_w R T_w a_w^+ + a_l \left[\rho_\infty u_l \left(\frac{1}{2}u_l^2 + \frac{5+j}{2}\rho_\infty R T_\infty\right)\right] = \rho_\infty u_\infty \left(\frac{1}{2}u_\infty^2 + \frac{5+j}{2}\rho_\infty R T_\infty\right). \tag{8.6.8}$$

The advantage of the choices we have made for the various terms of the distribution function is that, as in the case of a monatomic gas, the boundary conditions may be satisfied exactly by imposing the following conditions on the four functions $a_w^+(x)$, $a_\infty^\pm(x)$, $a_l(x)$:

$$a_w^+(0) = 1, \quad a_\infty^+(0) = 0, \quad a_l(0) = 0, \tag{8.6.9}$$

$$a_w^+(\infty) = 0, \quad a_\infty^+(\infty) = a_\infty^-(\infty) = 1, \quad a_l(\infty) = 0. \tag{8.6.10}$$

As in Section 8.2, we can now write Eqs. (8.6.6)–(8.6.8) at the wall and use the boundary conditions provided by Eq. (8.6.9) to obtain

$$zr - \beta^- F^- = 2\pi^{1/2} S_\infty, \tag{8.6.11}$$

$$z + \beta^- G^- = 4S_\infty^2 + 2, \tag{8.6.12}$$

$$z\frac{j+4}{4} - \beta^- r H_j^- = r\pi^{1/2} S_\infty \left(S_\infty^2 + \frac{5+j}{2} \right), \tag{8.6.13}$$

where we have used the same notation as in Section 8.2 and let

$$H_j^- = H^- + \frac{j}{4} F^-. \tag{8.6.14}$$

We can easily generalize the solution found in Section 8.2 in the case $j = 0$. Thus (8.2.26) is replaced by

$$r = -\pi^{1/2} \frac{S_\infty}{(2(j+4))} + \left[1 + \pi \left(\frac{S_\infty}{2(j+4)} \right)^2 \right]^{1/2} \tag{8.6.15}$$

and the relations (8.2.27) remain true (of course now r is given by (8.6.15)).

If we follow Anisimov's approach, then we must compute the above quantities for

$$S_M = \left(\frac{j+5}{2(j+3)} \right)^{1/2}. \tag{8.6.16}$$

We thus obtain Table 8.2.

To study the structure of the evaporating layer it is necessary to determine the constant u_I as well as the x dependence of the coefficients a_w^+, $a_\infty^\pm$, and a_I. The

Table 8.2. *Values of different parameters in the evaporation in a half-space of polyatomic molecules. The values are calculated for sonic conditions downstream for different values of j, the number of internal degrees of freedom. The number in the last column, γ_b, gives the fraction of backscattered molecules.*

j	S_∞	z	β^-	T_∞/T_w	γ_b
0	0.913	4.850	6.286	0.669	0.184
1	0.866	4.479	5.789	0.737	0.201
2	0.836	4.256	5.475	0.781	0.212
3	0.816	4.108	5.260	0.814	0.219
4	0.801	4.003	5.105	0.837	0.224

constant u_I is determined as follows: We multiply Eq. (8.6.6) by $-RT_\infty(j+4)/2$ and Eq. (8.6.7) by $-u_\infty/2$ and add the results to (8.6.8). If Eq. (8.6.15) is taken into account we obtain

$$0 = a_I \left[\rho_\infty RT_\infty u_I \frac{j+4}{2}\frac{u_\infty}{2}\left(\rho_\infty u_I^2 + \rho_\infty RT_\infty\right) - \rho_\infty u_I \left(\frac{1}{2}u_I^2 + \frac{5}{2}RT_\infty\right) - \rho_\infty u_I \frac{j}{2} RT_\infty\right]. \quad (8.6.17)$$

If we want to have $a_I \neq 0$ we must let the expression in square brackets vanish. In nondimensional form we have

$$2S_I^3 - 2S_\infty S_I^2 + [1 + j(1-r)]S_I - S_\infty = 0. \quad (8.6.18)$$

This cubic equation has at least one real root. To show that there is no other real root in the range [0,1] we compute the first derivative of the polynomial P in the left-hand side of (8.6.18). We obtain

$$P'(S_I) = 6S_I^2 - 4S_\infty S_I^2 + [1 + j(1-r)] \quad (8.6.19)$$

Since r is clearly less than unity thanks to (8.6.15), the last polynomial has no real roots for $S_\infty^2 \leq 3/2$ (Problem 8.6.1), which is more than needed. Thus the real root of (8.6.18) is unique and positive because its product with the two other (complex conjugate) roots is positive (Problem 8.6.2).

To determine $a_w^+(x)$, $a_\infty^\pm(x)$, and $a_I(x)$, we must first solve the linear system (8.6.6–8.6.8). It is not hard to express $a_\infty^\pm(x)$ in terms of $a_w^+(x)$ and $a_I(x)$ (Problem 8.6.3). The we must obtain two moment equations to determine the latter quantities. We need to choose a collision model. Following two papers by the author[53,57] we can choose the model proposed by Morse.[83] This model has two relaxation times. A simple but lengthy discussion[53,57] of the relevant moment equations shows that the condition for the existence of an acceptable solution is formally the same as discussed in Section 8.2 for a monatomic gas (Problem 8.6.3). The numerical results show that for two internal degrees of freedom and sonic conditions downstream, the maximum deviation between the internal and translational temperatures occurs at the surface and is of the order of 14%, but it reduces to 1/10 of this amount at a distance of 10 mean free paths from the wall.

Problems

8.6.1 Prove that the polynomial (8.6.15) has no real roots for $S_\infty^2 \leq 3/2$, using the fact that $r < 1$ because of (8.6.15).

8.6.2 Prove that the cubic equation (8.6.18) has a unique real root, which is positive, for $S_\infty^2 \leq 3/2$.

8.6.3 Complete the solution of the evaporation problem for a polyatomic gas (see Refs. 53 and 57).

8.7. Evaporation from a Finite Area

The result that evaporation from a plane into a half-space cannot produce a supersonic flow at infinity may look surprising but is absolutely correct if one insists that the evaporation surface is an infinite plane. A short discussion of the case of evaporation from a finite area (say, a disk) is of both theoretical and practical interest and leads to a different result as we shall presently explain, following again two papers mentioned before.[53,57]

To explain this aspect of the matter, let us consider an experimental situation, where evaporation is restricted to a circular area on a plane and the vapor is surrounded by vacuum. The diameter D of the evaporating disk is assumed to be much larger than the mean free path. Then we can argue, as we have done several times in this book, that the gas can be treated as a continuum a few mean free paths away from the boundary and the treatment considered in the previous sections applies only to a short cylinder a few mean free paths (say, 100) high, having the disk as basis. We must then match this Knudsen layer with a continuum flow described, at the lowest order, by the Euler equations. This expanding flow must start exactly at Mach number unity, because this is required when imposing that not only the functions describing the state of the gas but also the boundary of the region filled by vapor be smoothly matched at the boundary between the inner and the outer expansion. Of course, there are a few molecules escaping from the boundary of the aforementioned cylinder, but their number is negligible, being a fraction of the order of the ratio between the mean free path and the diameter D of the disk.

We thus have to compute a sonic jet of the kind mentioned in Chapter 5, (Section 5.9). What is said henceforth is valid for any jet and not only for a jet produced by strong evaporation.

Since in the exit plane the solution corresponds to Mach number unity, there is a singularity because the two families of characteristic lines coincide in that plane. Thus one cannot use the simple method of characteristics to perform the calculations in an immediate way. Hill and Pack[84] were the first to analyze a method for carrying the solution away from the sonic plane in a two-dimensional case. Smith[85] extended the analysis to the axially symmetrical case.

The computation of the solution for a sonic jet of a diatomic gas with an axial symmetry was first carried out by Owen and Thornhill.[86] They did not start from an analysis such as Smith's[85] but avoided the singular characteristics in the sonic plane by the assumption of an initial Mach angle of 85° rather than 90°.

Important properties of a jet expanding into the vacuum are the following: The Mach number of the jet increases monotonically along the axis, tending to infinity downstream.

The speed of the gas has an upper bound, achieved when all the thermal energy of the gas has been converted into kinetic energy.

The computed flow exhibits a rather simple and self-similar development in the inertia-dominated region of high Mach number flow, where the streamlines appear to radiate from a "source" at a distance r_0 downstream from the sonic disk and the density decreases along each streamline in proportion to the inverse square of the distance from the source.

The variation of the density with the polar angle θ at a constant distance $r - r_0$ from the source is approximately independent of r_0.

The analogy to a purely radial source flow was noticed by Sherman, who, in a joint paper with Ashkenas,[87] suggested the following rather accurate fitting formula for the centerline Mach number of a free jet:

$$\mathrm{M} = A\left(\frac{r-r_0}{D}\right)^{\gamma-1} - \frac{1}{2}\left(\frac{\gamma+1}{\gamma-1}\right)A^{-1}\left(\frac{r-r_0}{D}\right)^{-(\gamma-1)}. \qquad (8.7.1)$$

Here A is a constant and γ, as usual, is the ratio of specific heats ($\gamma = 5/3$ for a monatomic gas). Ashkenas and Sherman[87] were able to find values of A and r_0 that cause Eq. (8.7.1) to reproduce the computer data within a maximum deviation of about 0.5% for $r > \bar{r}$, with $\bar{r}$ of the order of a few sonic diameters ($A = 3.26$, $r_0/D = 0.075$, $\bar{r}/D = 2.5$ for monatomic gases). For monatomic and diatomic gases, Eq. (8.7.1) can even be improved by adding a third term of the form $C[(r - r_0)/D]^{-3(\gamma-1)}$, which gives an accurate data fit, extending upstream to surprisingly small values of r/D ($r > r_*$; for monatomic gases $C = 0.31$, $r_* = D$).

The approximately self-similar behavior of the density field in the inertia-dominated region consists in ρr^2 being a function of θ only; this result is actually exact (by mass conservation) at the lowest order in an expansion in inverse powers of r. According to Ashkenas and Sherman[87] one can fit the numerical results by the simple formula

$$\frac{\rho(r,\theta)}{\rho(r,0)} = \cos^2\left(\frac{\pi\theta}{2\theta_0}\right) \qquad (8.7.2)$$

with an accuracy of about 3% of $\rho(r, 0)$. The constant θ_0 depends on the ratio of specific heats and equals 1.365 for monatomic gases. If Eq. (8.7.2) is extrapolated to the jet boundary, one finds that ρ should vanish for $\theta = \theta_0$ ($\cong 78.2°$ for a monatomic gas, as opposed to the theoretical value of 90°; but the density near a jet boundary is so small that it is not clearly defined by a formula possessing an accuracy of about 3% of $\rho(r, 0)$).

The results of early experimental work on free jet expansions into a vacuum confirmed the usefulness of the computations of Owen and Thornhill,[86] but different experimenters starting with Anderson and Fenn[88] showed that the increase in the Mach number does not continue indefinitely, as predicted by continuum mechanics. M tends to a constant value. Thus the flow, which is a collision-dominated expansion just after the sonic disk, develops into an almost collision-free hypersonic jet downstream. Since the speed of the gas is practically constant, the constancy of the Mach number suggests that the temperature also tends to a constant value. The jet is said to "freeze"; the situation is similar to that discussed in Section 5.9 of Chapter 5 for the case of an expansion from a spherical source and a similar analysis can be repeated.[89,57] The main difference is that the constant ρ_1 of Chapter 5, is now replaced by a function of the polar angle θ (Problem 8.7.1).

Problem

8.7.1 Extend the treatment of Chapter 5, Section 5.9, from the case of a spherical source to the case of an axial jet (see Refs. 85 and 57).

8.8. Concluding Remarks

The study of evaporation and condensation is a very active area of rarefied gas dynamics. This circumstance is due to its importance in technology and also to the fact that many aspects are not completely clarified, especially because of the lack of reliable experimental data.

In order to keep this chapter within acceptable bounds, the detailed presentation of phenomena and methods was mainly based on the case of evaporation or condensation in a half-space and occasionally between parallel plates.

Important work has also been devoted, however, to other geometries, in particular spheres and cylinders.[90–92] These problems are interesting because of their relation with the problem of evaporation from a finite area, discussed in the previous section.

We finally remark that important problems in the general area of evaporation and condensation arise in free molecular situations. They are related to the formation of fogs and other phenomena associated with coagulation processes. We quote the paper by Kogan et al.[93] as a typical example of these problems.

References

[1] H. Hertz, "Ueber die Verdunstung der Flüssigkeiten, insbesondere des Quecksilbers, im luftleeren Raume," *Ann. Phys. und Chemie* **17**, 177–200 (1882).

[2] M. Knudsen, "Die Maximale Verdampfunggeschwindigkeit des Quecksilbers," *Ann. Phys. und Chemie* **47**, 697–708 (1915).

[3] A. F. Mills and R. A. Seban, "The condensation coefficient of water," *Int. J. Heat Mass Transfer* **10**, 1815–1827 (1967).

[4] B. Paul, "Compilation of evaporation coefficients," *ARS J.* **32**, 1321–1328 (1962).

[5] M. Matsumoto, K. Yasuoka, and Y. Kataoka, "Microscopic features of evaporation and condensation at liquid surfaces: molecular simulation," *Ther. Sci. & Eng.* 64–69 (1994).

[6] S. Fujikawa, M. Kotani, and H. Sato, "Molecular study of evaporation and condensation of an associating liquid (shock-tube experiment and molecular dynamics simulation)," *Heat Transfer – Japanese Research* **23**, 595–610 (1994).

[7] B. Hafskjold and T. Ikeshoji, "Nonequilibrium molecular dynamics simulations of coupled heat and mass transport across liquid vapor interphases," *Molecular Simulations* **16**, 139–150 (1996).

[8] R. W. Schrage, *A Theoretical Study of Interphase Mass Transfer*, Columbia Univ. Press, New York (1953).

[9] R. Y. Kucherov and L. E. Rikenglas, "On hydrodynamic boundary conditions with evaporation of a solid absorbing radiant energy," *Soviet Phys. JETP* **37**, 10–11 (1960).

[10] P. N. Shankar and F. E. Marble, "Kinetic theory of transient condensation and evaporation at a plane surface," *Phys. Fluids* **14**, 510–516 (1971).

[11] A. J. Patton and G. S. Springer, "A kinetic theory description of liquid vapor phase change," in *Rarefied Gas Dynamics*, L. Trilling and H. Y. Wachman, 1497–1504, Academic Press, New York (1969).

[12] Y.-P. Pao, "Application of kinetic theory to the problem of evaporation and condensation," *Phys. Fluids* **14**, 306–311 (1971).

[13] Y.-P. Pao, "Temperature and density jumps in the kinetic theory of gases and vapors," *Phys. Fluids* **14**, 1340–1346 (1971).

[14] S. K. Loyalka, "Approximate method in the kinetic theory," *Phys. Fluids* **14**, 2291–2294 (1971).

[15] M. N. Kogan and N. K. Makashev, "Knudsen layer in the theory of heterogeneous reactions," *Akad. Nauk SSR, Mech. Zhidk i Gasa* **6**, 3–11 (1971) (in Russian; English translation in *Fluid Dyna.* **6**, 913–921 (1971)).

[16] J. W. Cipolla, J. H. Lang, and S. K. Loyalka, "Kinetic theory of evaporation and condensation," in *Rarefied Gas Dynamics*, K. Karamcheti, ed., 179–185, Academic Press, New York (1974).

[17] Y. Sone and Y. Onishi, "Kinetic theory of evaporation and condensation – Hydrodynamic equations and slip boundary conditions," *J. Phys. Soc. Japan.* **44**, 1981–1994 (1978).

[18] Y. Onishi and Y. Sone, "Kinetic theory of slightly strong evaporation and condensation – Hydrodynamic equations and slip boundary conditions for finite Reynolds numbers," *J. Phys. Soc. Japan.* **47**, 1676–1685 (1979).

[19] R. H. Edwards and R. L. Collins, "Evaporation for a spherical source into a vacuum," in *Rarefied Gas Dynamics*, L. Trilling and H. Y. Wachman, eds., Vol. II, 1489–1496, Academic Press, New York (1969).

[20] S. I. Anisimov, "Metal evaporation by laser radiation," *J. Eksperim. Teor. Fisika* **54**, 339–342 (1968).

[21] T. Ytrehus, J. J. Smolderen, and J. Wendt, "Mixing of thermal molecular jets produced from Knudsen effusion," *Entropie* **42**, 33–39 (1971).
[22] A. V. Luikov, T. L. Perleman, and S. I. Anisimov, "Evaporation of a solid into vacuum," *Int. J. Heat Mass Transfer* **14**, 177–183 (1971).
[23] T. Ytrehus, "Theory and experiments on gas kinetics in evaporation," in *Rarefied Gas Dynamics*, J. L. Potter, ed., Part II, 1197–1212, AIAA, New York (1977).
[24] M. Hatakeyama and H. Oguchi, "Kinetic approach to nonlinear condensation in flowing vapor," in *Rarefied Gas Dynamics*, R. Campargue, ed., Part II, 1293–1303, CEA, Paris (1979).
[25] D. A. Labuntsov and A. P. Kryukov, "Analysis of intensive evaporation and condensation," *Int. J. Heat Mass Transfer* **22**, 989–1002 (1979).
[26] T. Ytrehus and J. A. Alvestad, "A Mott–Smith solution for nonlinear condensation," in *Rarefied Gas Dynamics*, S. Fischer, ed., Part I, 330–345, AIAA, New York (1981).
[27] T. Soga, "Quasi-steady one-dimensional evaporation problem using entropy-balance relation," in *Rarefied Gas Dynamics*, J. L. Potter, ed., Part II, 1184–1196, AIAA, New York (1977).
[28] M. Lampis, "New approach to the Mott–Smith method for shock waves," *Meccanica* **12**, 171–173 (1977).
[29] P. Gajewski, A. Kulicki, A. Wisniewski, and M. Zgorelski, "Kinetic theory approach to the vapor-phase phenomena in a non-equilibrium condensation process," *Phys. Fluids* **17**, 321–327 (1974).
[30] S. M. Yen, "Numerical solutions of non-linear kinetic equations for one-dimensional evaporation-condensation problems," *Computers and Fluids* **1**, 357–377 (1973).
[31] S. M. Yen, "Numerical solution of the non-linear Boltzmann equation for non-equilibrium gas flow problems," *Annu. Rev. Fluid Mech.* **16**, 67–97 (1984).
[32] S. M. Yen and T. Ytrehus, "Treatment of the non-equilibrium vapor motion near an evaporating interface boundary," *Chem. Eng. Commun.* **10**, 357–367 (1984).
[33] M. Murakami and K. Oshima, "Kinetic approach to the transient evaporation and condensation problem, "in *Rarefied Gas Dynamics*, M. Becker and M. Fiebig, eds., Vol. II, F.3-1-12, DFLVR Press, Porz-Wahn (1974).
[34] C. Cercignani, "A nonlinear criticality problem in the kinetic theory of gases," in *Mathematical Problems in the Kinetic Theory of Gases*, D. C. Pack and H. Neunzert, eds., 129–146, P. Lang, Frankfurt (1980).
[35] M. D. Arthur and C. Cercignani, "Nonexistence of a steady rarefied supersonic flow in a half-space," *J. Appl. Math. Phys. (ZAMP)* **31**, 634–645 (1980).
[36] C. E. Siewert and R. J. Thomas, Jr., "Strong evaporation into a half space," J. *Appl. Math. Phys. (ZAMP)* **32**, 421–433 (1981).
[37] Y. Sone, K. Aoki, and I. Yamashita, "A study of strong condensation on a plane condensed phase with special interest in formation of steady profile," in *Rarefied Gas Dynamics*, V. Boffi and C. Cercignani, eds., Vol. II, 323–333, B. G. Teubner, Stuttgart (1986).
[38] Y. Sone, K. Aoki, and T. Yamada, "Steady evaporation and condensation on the plane condensed phase," *Theoretical and Applied Mechanics* **14**, 89–93, Bulgarian Academy of Sciences, Sofia (1988).
[39] A. A. Abramov and M. N. Kogan, "Strong subsonic condensation of a monatomic gas," *Izv. Akad. Nauk SSR, Mech. Zhidk. i Gasa* **N1**, 165–169 (1989) (in Russian; English translation in *Fluid Dyna.* **24**, 139–143 (1989)).
[40] A. A. Abramov and M. N. Kogan, "Strong evaporation and condensation," in *Proc. 9th Sov. Symp. Rarefied Gas Dynamics* **2**, 104–110, Ural University Press, Sverdlovsk (1987) (in Russian).

[41] M. N. Kogan and A. A. Abramov, "Direct simulation of the strong evaporation and condensation problem," in *Rarefied Gas Dynamics*, A. E. Beylich, ed., 1251–1257, VCH, Weinheim (1991).
[42] K. Aoki, Y. Sone, and T. Yamada, "Numerical analysis of steady flows of gas condensing on its plane condensed phase on the basis of kinetic theory," *Phys. Fluids A* **2**, 1867–1878 (1990).
[43] A. P. Kryukov, "Strong subsonic and supersonic condensation on a plane surface," in *Rarefied Gas Dynamics*, A. E. Beylich, ed., 1278–1284, VCH, Weinheim (1991).
[44] Y. Sone, "Kinetic theory of evaporation and condensation – linear and nonlinear problems," *J. Phys. Soc. Japan.* **45**, 315–320 (1978).
[45] J. Ferron, "Four-moment model of one-dimensional multicomponent distillation at high vacuum," in *Rarefied Gas Dynamics*, V. Boffi and C. Cercignani, eds., Vol. II, 301–322, B. G. Teubner, Stuttgart (1986).
[46] K. Nanbu, "Rarefied gas dynamics problems on fabrication process of semiconductor film," in *Rarefied Gas Dynamics*, V. Boffi and C. Cercignani, eds., Vol. I, 410–419, B. G. Teubner, Stuttgart (1986).
[47] A. Kersch, W. J. Morokoff, and C. Werner, "Selfconsistent simulation of sputter deposition with the Monte Carlo method," *J. Appl. Phys.* **75**, 2278–2285 (1994).
[48] Soubbaramayer, "Interface liquide-vapeur et couches limites adjacentes dans un evaporateur d'uranium," *Annales de Physique* **13**, 75–80 (1988).
[49] J. F. Crifo, "Comets as large dirty snowballs subliming in interplanetary space," in *Rarefied Gas Dynamics*, V. Boffi and C. Cercignani, eds., Vol. II, 229–250, B. G. Teubner, Stuttgart (1986).
[50] T. Ytrehus, "Moment solutions in the kinetic theory of strong evaporation and condensation: Application to dusty gas-dynamics," in *Rarefied Gas Dynamics: Theory and Simulations*, B. D. Shizgal and D. P. Weaver, eds., 503–513, AIAA, Washington (1994).
[51] L. Lees, "A kinetic theory description of rarefied gases," GALCIT Hypersonic Research Project Memo. No. 51, California Institute of Technology (1959).
[52] C. Liu and L. Lees, "Kinetic theory description of plane compressible Couette flow," in *Rarefied Gas Dynamics*, L. Talbot, ed., 391–428, Academic Press, New York (1961).
[53] C. Cercignani, "Strong evaporation of a polyatomic gas," in *Rarefied Gas Dynamics*, S. S. Fisher, ed., Part I, 305–320, AIAA, New York (1981).
[54] Y. Sone and H. Sugimoto, "Strong evaporation from a plane condensed phase," in *Adiabatic Waves in Liquid-Vapor System*, G. E. A. Meier and P. A. Thompson, eds., 293–304, Springer-Verlag, Berlin (1990).
[55] R. Mager, G. Adomeit, and G. Wortberg, "Theoretical and experimental investigation of strong evaporation of solids," in *Rarefied Gas Dynamics: Physical Phenomena*, E. P. Muntz, D. P. Weaver, and D. H. Campbell, eds., 460–469, AIAA, Washington (1989).
[56] K. Aoki, K. Nishino, Y. Sone, and H. Sugimoto, "Numerical analysis of steady flows of gas condensing on or evaporating from its plane condensed phase on the basis of kinetic theory," *Phys. Fluids A* **3**, 2260–2275 (1991).
[57] C. Cercignani, "Present status of rarefied gas dynamics approach to the structure of a laser-induced evaporating jet," Report EUR 6843 EN, Commission of the European Communities, Brussels (1980).
[58] C. Cercignani, W. Fiszdon, and A. Frezzotti, "The paradox of the inverted temperature profiles between an evaporating and a condensing surface," Phys. Fluids **28**, 3237–3240 (1985).

[59] T. Ytrehus, "Molecular-flow effects in evaporation and condensation at interfaces," *Multiphase Science and Technology* **9**, 205–327 (1997).
[60] A. V. Butkovsky, "The blockage effect at strong condensation," Preprint, Central Aerohydrodynamics Institute (TsAGI), Moscow (1990) (in Russian).
[61] R. E. Caflisch, "Asymptotic expansions of solutions of the Boltzmann equation," *Transp. Theory Stat. Phys.* **16**, 701–725 (1987).
[62] T. Ytrehus, "The inverted temperature gradient solutions in condensation – revisited," in *Proceedings of the 21st Symposium on Rarefied Gas Dynamics*, to appear (1999).
[63] K. Aoki and C. Cercignani, "Evaporation and condensation on two parallel plates at finite Reynolds numbers," *Phys. Fluids* **26**, 1163–1168 (1983).
[64] Y. Sone, T. Ohwada, and K. Aoki, "Evaporation and condensation of a rarefied gas between its two parallel plane condensed phases with different temperatures and negative temperature-gradient phenomenon – Numerical analysis of the Boltzmann equation for hard-sphere molecules," in *Mathematical Aspects of Fluid and Plasma Dynamics*, G. Toscani, V. Boffi, and S. Rionero, eds., LNM 1460, 186–202, Springer-Verlag, Berlin (1991).
[65] I. W. Eames, N. J. Marr, and H. Sabir, "The evaporation coefficient of water: a review," *Int. J. Heat Mass Transfer*, **40**, 2963–2973 (1997).
[66] M. Matsumoto and S. Fujikawa, "Non-equilibrium vapor condensation. Molecular simulation and shock-tube experiment," *Microscale Thermophys. Eng.* **1**, 119–126 (1997).
[67] K. Yasuoka and M. Matsumoto, "Evaporation and condensation at a liquid surface. I. Argon," *J. Chem. Phys.* **101**, 7904–7911 (1994).
[68] M. Matsumoto, in *Int. Symp. on Molecular Thermodynamics and Molecular Simulation. MTMS* **97**, 34–41 (1997).
[69] A.F. Mills, *Heat and Mass Transfer*, Irvin Inc. (1995).
[70] M. Matsumoto and K. Yasuoka, "Evaporation and condensation at a liquid surface. II. Methanol," *J. Chem. Phys.* **101**, 7912–7917 (1994).
[71] Y.-P. Pao, "Evaporation in a vapor–gas mixture," *J. Chem. Phys.* **59**, 6688–6689 (1973).
[72] J. C. Hass and G. S. Springer, "Mass transfer through binary gas mixtures," *Trans. ASME Ser. C* **95**, 263 (1973).
[73] T. Matsushita, "Evaporation and condensation in a vapor–gas mixture," in *Rarefied Gas Dynamics*, J. L. Potter, ed., Part II, 1213–1225, AIAA, New York (1977).
[74] T. Soga, "Kinetic theory analysis of evaporation and condensation in a vapor–gas mixture," *Phys. Fluids* **25**, 1978–1986 (1982).
[75] Y. Onishi, "A two-surface problem of evaporation and condensation in a vapor–gas mixture," in *Rarefied Gas Dynamics*, H. Oguchi, ed., 875–884, Univ. Tokyo Press, Tokyo (1984).
[76] A. Frezzotti, "Kinetic theory study of the strong evaporation of a binary mixture," in *Rarefied Gas Dynamics*, V. Boffi and C. Cercignani, eds., Vol. II, 313–322, B. G. Teubner, Stuttgart (1986).
[77] Y. Onishi, "Nonlinear analysis for evaporation and condensation of a vapor–gas mixture between the two plane condensed phases. Part I: Concentration of inert gas $\cong O(1)$," in *Rarefied Gas Dynamics: Physical Phenomena*, E. P. Muntz, D. P. Weaver, and D. H. Campbell, eds., 470–491, AIAA, Washington (1989).
[78] Y. Onishi, "Nonlinear analysis for evaporation and condensation of a vapor–gas mixture between the two plane condensed phases. Part I: Concentration of inert gas $\cong O(\mathrm{Kn})$," in *Rarefied Gas Dynamics: Physical Phenomena*, E. P. Muntz, D. P. Weaver, and D. H. Campbell, eds., 492–513, AIAA, Washington (1989).

[79] D. Bedeaux, J. A. M. Smit, L. J. F. Hermans, and T. Ytrehus, "Slow evaporation and condensation II. A dilute mixture," *Physica A* **182**, 388–418 (1992).
[80] K. Aoki, S. Takata, and S. Kosuge, "Vapor flows caused by evaporation and condensation on two parallel plane surfaces: Effect of the presence of a noncondensable gas," *Phys. Fluids A* **10**, 1519–1533 (1998).
[81] Y. Sone, K. Aoki, and T. Doi, "Kinetic theory analysis of gas flows condensing on a plane condensed phase," *Transp. Theory Stat. Phys.* **21**, 297–328 (1992).
[82] K. Aoki, and T. Doi, "High speed vapor flows condensing on a plane condensed phase: Case of a mixture of a vapor and a noncondensable gas," in *Rarefied Gas Dynamics: Theory and Simulations*, B. D. Shizgal and D. P. Weaver, eds., 521–536, AIAA, Washington (1994).
[83] T. F. Morse, "Kinetic model equations in a fluid," *Phys. Fluids* **7**, 2012–2013 (1964).
[84] R. Hill and D. C. Pack, "An investigation by the method of characteristics of the lateral expansion of the gases behind a detonating slab of explosive," *Proc. Roy. Soc. A* **191**, 524–541 (1947).
[85] M. G. Smith, "The behavior of an axially symmetric sonic jet near to the sonic plane," *Quart. J. Mech. Appl. Math.* **12**, 287–297 (1959).
[86] P. L. Owen and C. K. Thornhill, *Aero. Res. Council Reports and Memoranda* 2616 (1968).
[87] H. Ashkenas and F. S. Sherman,"The structure and utilization of supersonic free jets in low density wind tunnels," in *Rarefied Gas Dynamics*, J. H. de Leeuw, ed., Vol. **2**, 84–105, Academic Press, New York (1965).
[88] J. B. Anderson and J. B. Fenn, "Velocity distributions in molecular beams from nozzle sources," *Phys. Fluids* **8**, 780–787 (1965).
[89] R. E. Grundy, "Axially symmetric expansion of a monatomic gas from an orifice into vacuum," *Phys. Fluids* **12**, 2011–2018 (1969).
[90] H. Sugimoto and Y. Sone, "Numerical analysis of steady flows of a gas evaporating from its cylindrical condensed phase on the basis of the kinetic theory," *Phys. Fluids A* **4**, 419–440 (1992).
[91] Y. Sone and H. Sugimoto, "Kinetic theory analysis of steady evaporating flows from a spherical condensed phase into a vacuum," *Phys. Fluids A* **5**, 1491–1511 (1993).
[92] Y. Sone and H. Sugimoto, "Evaporation of a rarefied gas from a cylindrical condensed phase into a vacuum," *Phys. Fluids A* **7**, 2072–2085 (1995).
[93] M. N. Kogan, I. N. Bobrov, C. Cercignani, and A. Frezzotti, "Interaction of evaporating and condensing particles in the free-molecular regime," *Phys. Fluids A* **7**, 1775–1781 (1995).

Index

Printed in the United States
By Bookmasters